AF782766

Logistics Management and Industrial Engineering

Logistics Management and Industrial Engineering

Editor: Courtney Hoover

NY RESEARCH PRESS
New York

Published by NY Research Press
118-35 Queens Blvd., Suite 400,
Forest Hills, NY 11375, USA
www.nyresearchpress.com

Logistics Management and Industrial Engineering
Edited by Courtney Hoover

© 2017 NY Research Press

International Standard Book Number: 978-1-63238-530-7 (Hardback)

This book contains information obtained from authentic and highly regarded sources. Copyright for all individual chapters remain with the respective authors as indicated. All chapters are published with permission under the Creative Commons Attribution License or equivalent. A wide variety of references are listed. Permission and sources are indicated; for detailed attributions, please refer to the permissions page and list of contributors. Reasonable efforts have been made to publish reliable data and information, but the authors, editors and publisher cannot assume any responsibility for the validity of all materials or the consequences of their use.

The publisher's policy is to use permanent paper from mills that operate a sustainable forestry policy. Furthermore, the publisher ensures that the text paper and cover boards used have met acceptable environmental accreditation standards.

Trademark Notice: Registered trademark of products or corporate names are used only for explanation and identification without intent to infringe.

Cataloging-in-publication Data

Logistics management and industrial engineering / edited by Courtney Hoover.
p. cm.
Includes bibliographical references and index.
ISBN 978-1-63238-530-7
1. Business logistics--Management. 2. Industrial engineering. 3. Logistics--Data processing. I. Hoover, Courtney.
HD38.5 .L64 2017
658.7--dc23

Printed in China.

Contents

Preface

The planning, creation and implementation of supply chains, manual and automated, that seek to implement and control goods and services can be called as logistics management. Supply chains are an integral part of industrial engineering. This book on logistics management and industrial engineering discusses topics related to the distribution of products, the control of inflow of raw materials and the safe storage of finished products. This book aims to emphasize the role played by logistics in the growth of the economy. Some of the diverse topics covered herein address the varied branches that fall under this category. This book includes contributions of experts and scientists which will provide innovative insights into this field. It is a vital tool for all researching and studying in this field.

This book aims to highlight the current researches and provides a platform to further the scope of innovations in this area. This book is a product of the combined efforts of many researchers and scientists from different parts of the world. The objective of this book is to provide the readers with the latest information in the field.

I would like to express my sincere thanks to the authors for their dedicated efforts in the completion of this book. I acknowledge the efforts of the publisher for providing constant support. Lastly, I would like to thank my family for their support in all academic endeavors.

Editor

1

A mobile application for knowledge-enriched short-term scheduling of complex products

D. Mourtzis[1] · M. Doukas[1] · E. Vlachou[1]

Abstract The ever-increasing product complexity, especially for the case of engineer-to-order products, highly affects the performance of manufacturing systems. Therefore, a high degree of flexibility is needed during daily decision-making activities, such as production scheduling. For addressing this challenge, this research work proposes a knowledge-enriched short-term job-shop scheduling mechanism, which is implemented into a mobile application. More precisely, it focuses on the short-term scheduling of the resources of the machine shop, through an intelligent algorithm that generates and evaluates alternative assignments of resources to tasks. Based on the requirements of a new order, a similarity mechanism retrieves successfully executed past orders together with a dataset that includes the processing times, the job and task sequence, and the suitable resources. In addition to that, the similarity mechanism is used to calculate the due-date assignments of the orders based on the knowledge stored in past cases. Afterwards, it adapts these parameters to the requirements of the new order so as to evaluate the alternative schedules and identify a good alternative in a timely manner. The deriving schedule can be presented on mobile devices, and it can be manipulated by the planner on-the-fly respecting tasks precedence constraints and machine availability. A case study from the mould-making industry is used for validating the proposed method and application.

Keywords Manufacturing systems · Scheduling · Mobile applications

1 Introduction

Modern manufacturing relies on the reuse of previous knowledge that is either contained in data repositories and IT systems or exists in the form of tacit human knowledge. Knowledge constitutes a key factor for improving manufacturing performance, during design, planning, and operational phases [1]. Most contemporary manufacturing industries acknowledge that the exploitation of the existing knowledge is necessary to enhance the performance of manufacturing [2]. Indicatively, the importance of knowledge reuse for a system's design and planning phase is evident, as rough estimations indicate that more than 20 % of an engineer's time is spent on searching and absorbing information for a new project [3].

A particular type of manufacturing system, which essentially relies on the knowledge and expertise of human assets to improve its performance, is a job-shop that produces engineer-to-order (ETO) products. Usually, the incorporation of new orders in the schedules of such systems is performed empirically and using rules of thumb, even when the system operates near its maximum capacity. However, with the rising complexity of production requirements and the increased penetration of IT systems in manufacturing, knowledge reuse is an enabler to reduce the product development cycle and increase manufacturing performance. On the contrary, in current practice, this valuable knowledge generated and associated with

This article is part of a focus collection on "Robust Manufacturing Control: Robustness and Resilience in Global Manufacturing Networks".

✉ D. Mourtzis
mourtzis@lms.mech.upatras.gr

1 Laboratory for Manufacturing Systems and Automation, Department of Mechanical Engineering and Aeronautics, University of Patras, 26500 Rio, Greece

products and processes in a daily basis remains tacit and its reusability is confined to a specific operator or planner [4].

Further to that, another requirement today is pervasive access to information and decision-making. One enabling technology to achieve ubiquitous access to knowledge and assist decision-making is mobile apps, i.e. applications developed for mobile devices. The general market of apps is expected to reach revenues of $70B by 2017 [5]. More and more companies are starting to base their business on mobility; however, the adoption of apps in the manufacturing domain is yet at a primary stage [6].

Motivated by the above, in this research work, a scheduling method that is enhanced with an integrated knowledge reuse mechanism is proposed. The knowledge reuse mechanism retrieves historical scheduling cases and through the case-based reasoning (CBR) methodology extracts information related to the modelling of the scheduling workload. The deriving workload model includes necessary input for a scheduler, such as the job structure and the task breakdown, the precedence constraints, and the processing and set-up times. Alternative schedules are generated and are evaluated using multiple conflicting criteria, such as flowtime and tardiness. The scheduling is performed using an intelligent search algorithm (ISA) with three tuneable parameters, which are adjusted through a parametric investigation, using a statistical design of experiments method. All functionalities are exposed through a developed mobile app.

The remainder of the paper is structured as follows. Section 2 includes a literature survey on knowledge-enriched scheduling (KES) applications. Section 3 analyses the proposed methodology. Section 4 describes the design of the scheduling app. Section 5 demonstrates a real-life case study in a mould-making industry. Section 6 draws the conclusions and describes the future work directions.

2 State of the art

Following the main topics of the research work, this section discusses knowledge reuse in manufacturing, knowledge-enriched scheduling methods, mobile apps in manufacturing, and lead time estimation approaches.

Throughout the years, several methods have been proposed for knowledge reuse in the manufacturing domain with the aim to support designers and engineers in decisions related to modelling, design, prediction, monitoring, and optimisation. Knowledge reuse is considered to have a major impact on several manufacturing domains, offering productivity gains [7]. There are two main ways to reuse past knowledge: reuse the past case solution and reuse the past method that constructed the solution [8]. A widely used artificial intelligence (AI) method that can effectively enable reuse of past solutions is CBR, which retrieves past experience to reuse for a target problem; of course, the solutions of past cases may need to be revised for applying in the new case. The successful problem-solving experience is then retained for further reuse [9]. CBR utilises similarity mechanisms in order to compare past cases stored in the repository and the target case, i.e. a new product, based on case's features or attributes [10]. The most similar cases to the new one are recalled in order to provide recommendations [11]. Li et al. 2011 presented CBR as a tool for knowledge management during product development and reported that reuse of past product knowledge can be used to improve the problem-solving capabilities [12]. The CBR method is utilised in this research work due to its suitability for complex ill-defined concepts, with unstructured knowledge, and because case generalisation is necessary [9].

The second area of interest in this research work is job-shop scheduling. Scheduling is one of the most critical issues for a manufacturing enterprise. In most SMEs, who cannot afford costly investments in IT software solutions, scheduling is carried out empirically. However, the definition of a performing solution is quite difficult, depending on the job-shop environment, process constrains, and performance indicators [13]. Numerous approaches have been reported for the modelling and solving of the job-shop scheduling problem [14]. Wang et al. [15] proposed the development of an application using a genetic algorithm including a chromosome representation in seven different machines of a job floor that enables a dynamic job-shop scheduling within complex production systems. Chryssolouris et al. [39], considering the limitations of static scheduling, proposed a dynamic scheduling problem to accurately reflect a real job-shop scheduling environment. This dynamic nature of these scheduling problems [16] constitutes KES approaches essential, as knowledge reuse could assist in incorporating and scheduling new jobs in an ever-changing environment. However, literature findings that focus on knowledge reuse as an enabler for improving scheduling performance are scarce. Motivated by empirical knowledge, [17] proposes an efficient search method for the multi-objective flexible job-shop scheduling in order to reach high automation levels towards generating optimal or near-optimal production schedules. Another study exploiting previous knowledge proposed a data mining technique for discovering dispatching rules that improve scheduling performance [18]. The job-shop scheduling problem has been addressed using a knowledge-enriched genetic algorithm in [19]. The idea was to imbue production system knowledge during the formulation of the initial population of the algorithm with the potential of faster and better convergence. The authors in [20] utilise data mining for optimising a basic aspect of production scheduling, i.e.

the assignment of due dates to orders dispatched in a dynamic job-shop. Moreover, a knowledge-based algorithm for flexible job-shop scheduling is presented in [21]. The authors in this approach combine the variable neighbourhood search with a knowledge management module in order to reach the optimum solution more efficiently. In addition to these techniques, group technology could be also utilised in order to minimise the makespan and the sum of the completion times of a generated schedule based on position-dependent learning effects [22]. Concluding, according to [23], research should be shifted more towards knowledge reuse for decision support tools, within safety, reliability, and maintainability. Several of the above-mentioned studies utilise knowledge reuse techniques to improve scheduling performance. However, most of them focus on reusing knowledge to define the tuneable parameters of the algorithm that perform and determine scheduling attributes including dispatching rules or ordering details without offering actual reuse of meaningful scheduling data [22]. There is little existing work on knowledge reuse for determining the processing times of new tasks or the due dates of new jobs. Nevertheless, this can lead to the realisation of accurate and quick dynamic scheduling.

The third area of interest is mobile technology. Mobile technology evolves rapidly; in the last decade, the use of mobile apps has outpaced traditional PC-based web-browsing [24]. The usage of apps doubled on average over the last 2 years, with utility and productivity apps ranking second in growth [25]. The necessary components of apps in order for them to be fully leveraged in manufacturing are presented in [26], where architecture, development, infrastructure, security, portfolio, and privacy issues are investigated. Estimations speak of apps boosting productivity by 5–10 % [27]. The growth of mobility and mobile apps is highly influenced by the growth and the adoption of the cloud technology in manufacturing [28]. Cloud technology acts as an enabler to adopt mobile devices in manufacturing not only for the provision of applications but also for production data management purposes [29]. However, the adoption of apps focused on core manufacturing processes was up to now limited [25]. Cloud and mobile technologies are not sufficiently adopted in manufacturing systems yet, despite the productivity boost that they can offer [28, 30]. Nevertheless, apps are finding their way into activities such as manufacturing network design [31] and other scientific domains [32, 33].

The last area of interest is the manufacturing lead time estimation. Based on the literature, the most robust methods for lead time estimation are AI methods [34]. Ozturk et al. [34] used data mining as an AI method and attribute tables in order to calculate manufacturing lead time. Among AI methods, CBR, which focuses on solving problems by adapting acceptable solutions and comparing differences and similarities between previous and current products, has been utilised for lead time estimation. An approach based on CBR was applied during product development in [35], and it effectively reduced lead time and improved the problem-solving capabilities. The literature review makes apparent that the CBR and the data mining techniques should be further considered for supporting decision supports tools [22], especially in determining machining sequence and processing times.

Building upon the literature on the field, the combination of knowledge reuse techniques together with intelligent scheduling algorithms under the umbrella of mobile and cloud technology is considered as a necessary step towards the next generation of decision support tools. Addressing these challenges, the proposed research work provides a methodology that utilises a knowledge reuse mechanism for extracting manufacturing information related to machining sequence and orders due dates in order to support a short-term scheduling application. Moreover, the scheduling mechanism is developed into an app, motivated by the fact that the adoption of the mobile devices in manufacturing, and specifically in decision-making activities, can lead to easier access to information, as well as quick and accurate visualisation and interaction with the generated scheduling and planning information [6, 30, 31]. Finally, the proposed methodology is applied in a real manufacturing environment utilising data from a mould-making SME and the scheduling algorithm is compared with others dispatching rules in order to benchmark its performance. This work extends the research presented in [36–38] by enhancing the scheduling algorithm with knowledge reuse capabilities and by verifying the method in a case coming from the domain of ETO products.

3 The knowledge-enriched scheduling method

The knowledge-enriched scheduling (KES) engine consists of two mechanisms, namely: (1) the knowledge extraction and reuse mechanism and (2) the short-term scheduling mechanism (Fig. 1). Regarding the workflow of the first mechanism, once a new order enters the system, a breakdown of the product components into a bill of materials (BoM) structure is performed. The product is characterised by a number of attributes (product features) that are used by the similarity mechanism of CBR for a pairwise attribute comparison. The result of the similarity comparison is an ordered list that contains the past cases ranked from the most to the least similar. By reusing the knowledge stored in these past cases, the expert planner is allowed to extract valuable information that helps introduce the new order into the production system with the needed adaptations.

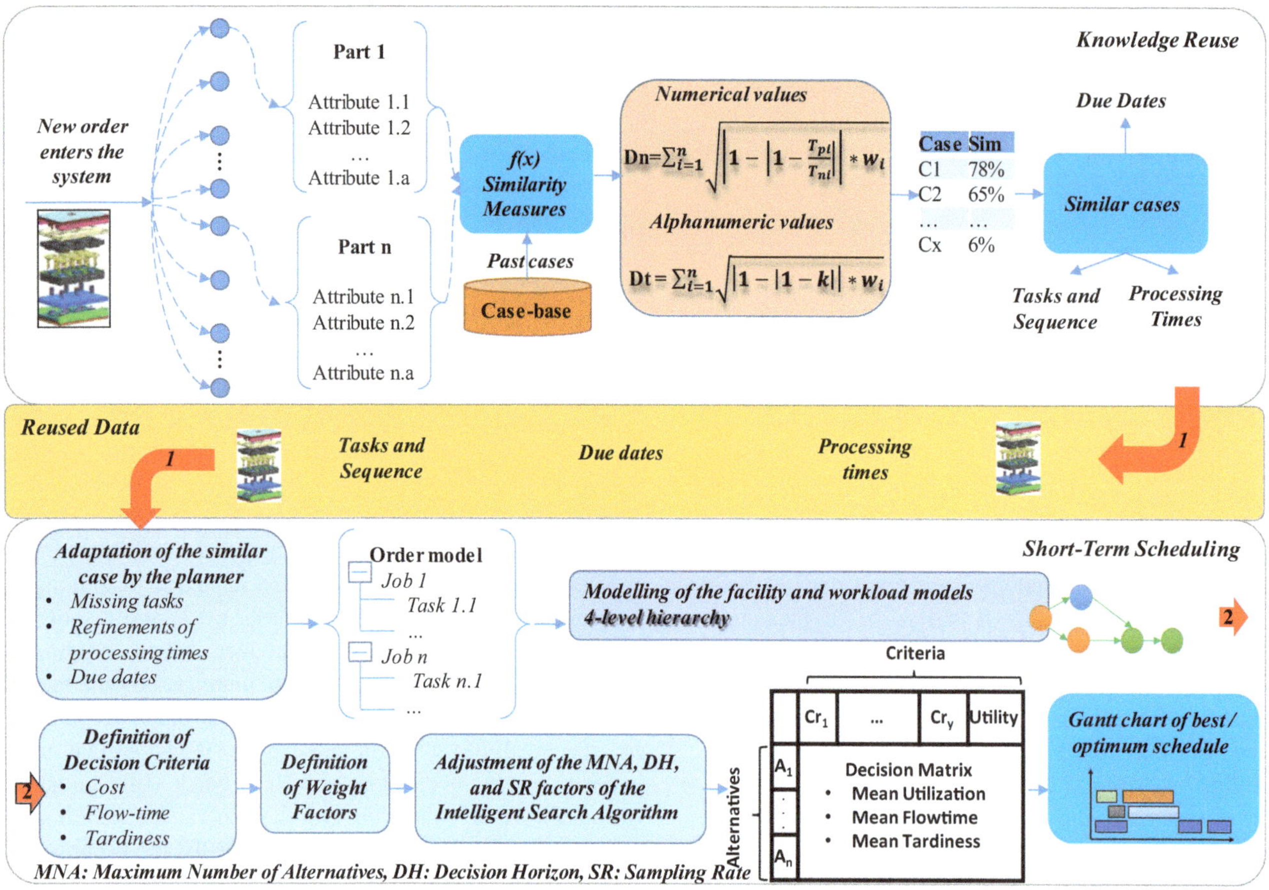

Fig. 1 Workflow of the knowledge-enriched short-term scheduling (KES) method. The method consists of two mechanisms, namely: (1) the knowledge extraction and reuse mechanism and (2) the short-term scheduling mechanism

The reusable information includes the required number and type of jobs, the number of tasks for each job and their precedence constraints, the processing times for each task in specific machines, and finally the due dates of the tasks. The output of this process comprises the necessary input for a scheduling engine. It is noted that specific process planning information, such as cutter selection, process parameters, and fixture specification, is beyond the direct scope of the proposed work.

The latter component of the KES is the short-term scheduling mechanism. After the identification of the most similar cases, the expert planner aggregates information that can be reused in the new case and adapts it. The adaptation is required in order to compensate for missing tasks that were not identified during the similarity measurement, or in order to imbue to the dataset the actual situation of the shop-floor (machine breakdowns and availability). The result of the adaptation is the workload and the facility models. These models are imported into the intelligent scheduling engine. The planner defines the decision-making criteria and their weight factors, which reflect the design and planning objectives of the company. Following on that, the definition of the tuneable parameters of the scheduling algorithm is defined. The tuneable parameters are the maximum number of alternatives (MNA), the decision horizon (DH), and the sampling rate (SR). The description of the function of these parameters is provided in Sect. 3.2. The scheduling algorithm generates alternative schedules, selects a good alternative in a timely manner, and displays it in the form of a Gantt chart.

The two mechanisms are designed in a modular way. The scheduling engine is capable to function without input provided by the knowledge mechanism, if the latter is not available. Similarly, the knowledge mechanism is decoupled from the scheduling engine and can be used for extracting manufacturing information for different purposes, such as for the estimation of the delivery time of an injection mould [38].

3.1 Modelling of the facility and the workload

The production facility is hierarchically divided into job-shops that contain work-centres, which in turn contain a number of resources. The latter are individual processors

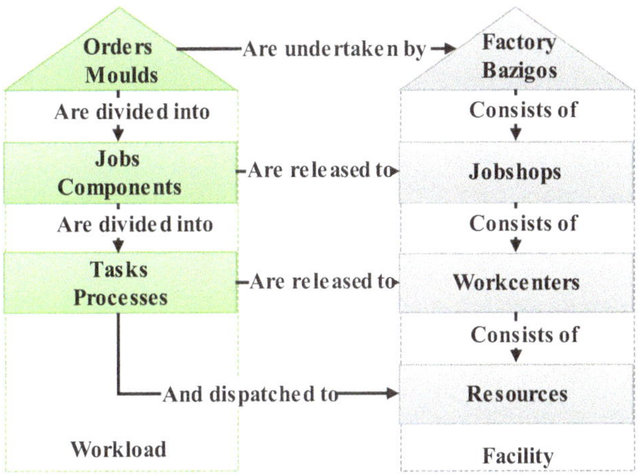

Fig. 2 Four-level hierarchical workload and facility model

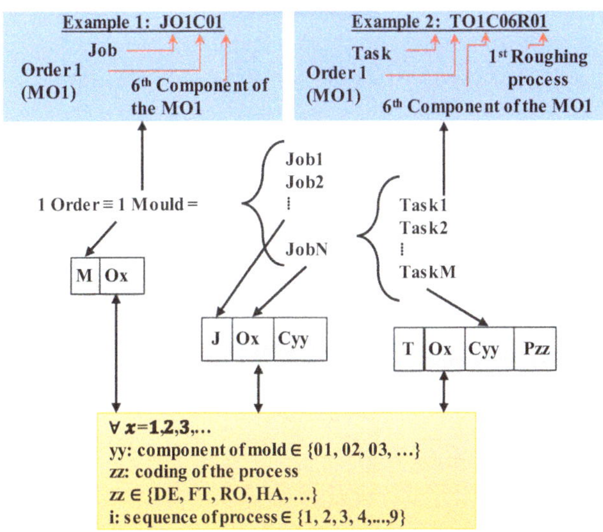

Fig. 3 Modelling of the mould order, jobs, and tasks

(resources with diverse processing capabilities (machining technology, cycle times, investment costs, fixed operating costs, etc.)). Similarly, the workload model includes orders that are broken down into jobs, each containing a number of tasks that are processed by the resources. As Fig. 2 shows, orders are dispatched to the facility, jobs to jobshops, and tasks to work-centres' resources. Resources are not parallel processors, and their availability is subject to the system workload. The release of tasks considers finite capacity, precedence relations, and availability constraints.

The job and task modelling is shown in Fig. 3. Specifically, each mould consists of a number of components. Each mould is represented as an order, and each component is represented as a job. Each job is composed of a number of tasks that need to be performed in order to manufacture each component of an order. Finally, the generation, evaluation, and selection of task to resource assignment are performed by an intelligent multi-criteria search algorithm (ISA) as described in [39]. ISA evaluates the alternatives in a decision matrix based on set-up cost and processing time criteria. A utility function is used for ranking the alternatives and for selecting the highest performing one.

3.2 Description of the knowledge reuse mechanism

The first step in the workflow of the knowledge reuse mechanism is the comparison of the new order against past cases in order to identify similar cases and reuse their data. This similarity measurement emphasises on the differences exhibited between the basic attributes that characterise old and new orders alike. CBR is functioning on the premise that a comparison between cases is feasible. To accomplish that, a set of attributes of the product that enters the system as a new order is selected, to characterise the case. For every new ETO product, the engineers together with the customer complete a web-based form with predefined fields that comprise the product's specifications. The attributes used to characterise the ETO product cases include: type of product, geometry stacks, slides, type of hardening, core cap, stacks material, profile rings, data provided, polishing, side of injection, tamper evident, gating type, way of ejection, ejector rings type, no of cavities, and wall thickness, depth, width, diameter, and length.

The past cases are retrieved using the CBR methodology and are compared with similarity mechanisms. The type of attributes considered takes both numeric and alphanumeric values. The alphanumeric attributes are mapped into discrete values represented by numbers in [0–1] for normalisation reasons. Moreover, both attribute types are multiplied with weight factors, considering their influence on the actual similarity between cases. Equations (1) and (2) are used for measuring the Euclidean distance through a pairwise comparison between the attributes of past and new cases. Equation (3) aggregates the results of the two distance metrics.

$$D_n = \sum_{i=1}^{n} \sqrt{\left|1 - \left|1 - \frac{T_{pi}}{T_{ni}}\right| \times w_i\right|}, \quad \text{numerical values} \tag{1}$$

$$D_t = \sum_{i=1}^{n} \sqrt{|1 - |1 - k|| \times w_i}, \quad \text{alphanumeric values} \tag{2}$$

$$S = (D_n + D_t)^2 \tag{3}$$

where D_n numerical distance, D_t text distance, n number of attributes, T_{ni} ith attribute of the new case n, T_{pi} ith attribute of the past case p, k mapping for alphanumeric attributes, and w_i the weight of attributes.

The past case with the highest similarity index is analysed first (Fig. 4). The planner may retrieve the process sequence, precedence constraints, the components, and the resources used in the past case, as well as processing and set-up times per task and resource. Moreover, based on their experience, expert planners have the capability to recognise whether the retrieved data are adequate to describe the new order. In case they are insufficient, the planner adapts the dataset to the requirements of the new case. Furthermore, in cases when the new product requires a different amount of components/processes than the retrieved most similar past case, then the second most similar case can be consulted, then the third, and so on. Either way, the similarity index between two cases must always remain above the threshold of 60 %, which is calculated based on historical observations; otherwise, the retrieved information would be misleading. Indicatively, if the best match in terms of similarity index is fairly aged in comparison with the new case, it is probable that adaptations would be required to compensate for changes in the shop-floor, such as the addition of new manufacturing resources and technologies. In this case, engineers are aware of the current state of the shop-floor and can replace the old resources with the new similar ones in the new process plan. Having decided the matching past similar cases, the task sequences are retrieved, the availability of the machines is confirmed, and then, the final combination of the new sequence of processes and components is settled.

3.3 Description of the short-term scheduling mechanism

Schedules are constructed on the basis of events occurring sequentially through time. Thus, the next scheduling decision is identified by moving along the time horizon until an event (release of a new order in the system or the completion of a task) is scheduled to occur that will initiate a change in the status of the system [40]. The set of pending tasks becomes eligible for release at the time a resource becomes available. The operational policy behind the assignment of a task to a specific resource can be either

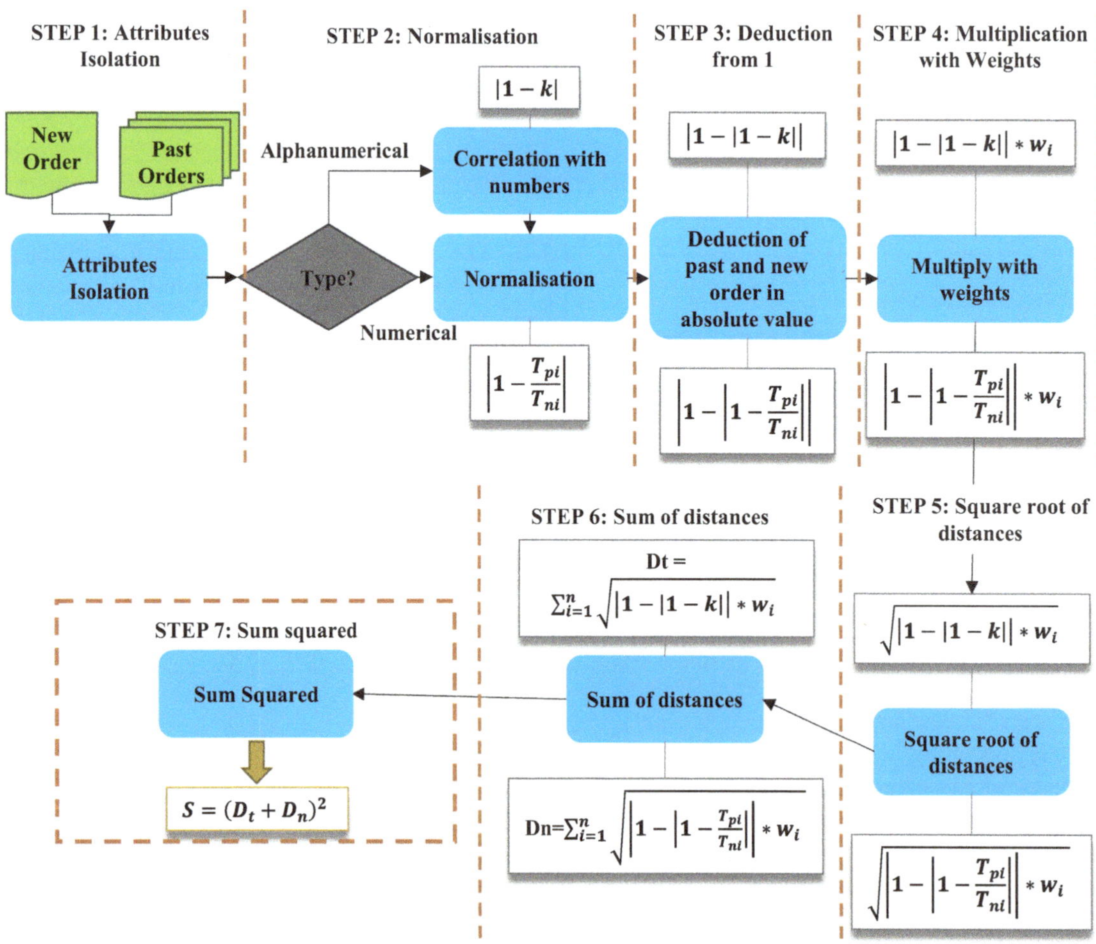

Fig. 4 Main steps of the case-based reasoning (CBR) mechanism

a simple dispatching rule or a multiple-criteria decision-making technique as described below. The advantages of dispatching rules derive from their simplicity, since they do not attempt to predict the future, but rather make decisions based on the present. Thus, these rules are very useful in factories that are extremely unpredictable, such as job-shops. Also, they are spatially local, requiring only the information available at the location where the decision will be implemented. Finally, they are easily understood by human operators and are easy to implement [41]. On the other hand, the multiple-criteria decision-making technique involves the formation of several alternatives and their evaluation before assigning the available resources to pending production tasks.

Since the method considers a finite capacity problem, in case multiple jobs are competing for a resource, ISA, and the decision matrix with the criteria and their weighting factors are used to determine which task will be dispatched to which resource, considering the planning objectives. The weighting factors that are used in the scheduling algorithm are determined based on the planner's knowledge of the business in order to satisfy business's production requirements and goals. Moreover, in case of a tie between two identical resources that are both suitable and available for a task assignment, the intelligent search algorithm selects randomly one of them and evaluates the generated path. Nevertheless, other tie breaking policies can be used such as the one proposed in [42].

In the ISA algorithm, the search of the solution space is guided by three adjustable control parameters, namely the maximum number of alternatives (MNA), the decision horizon (DH), and the sampling rate (SR). MNA controls the breadth of the search, DH controls the depth, and SR directs the search towards branches of high-quality solutions [43]. The proper selection of MNA, DH, and SR allows the identification of a good solution by examining a limited portion of the search space, thus effectively reducing computational time. For that purpose, a statistical design of experiments [44] has been carried out to reduce the number of experiments and to identify the preferable values of these factors in order to obtain the results of the highest possible quality [37, 45]. The workflow of the algorithm follows (Fig. 5):

Step 1: Start at the root and generate alternatives by random assignments for DH layers until MNA.
Step 2: For each branch (Step 1), create SR random samples until all the branch nodes are searched.
Step 3: Calculate the criteria scores for all the samples belonging to the same alternative of Step 1.
Step 4: Calculate the score of the branch as the average of the scores achieved by its samples.
Step 5: Calculate the utility values of each alternative/branch.
Step 6: Select the alternative with the highest utility value.
Step 7: Repeat Steps 1–6 until an assignment has been done for all the nodes of the selected branch.

The nodes mentioned in the workflow steps above represent decision points where a task is assigned to a resource. Once a task to resource assignment is made in each one of the nodes, an alternative production scheme is generated as shown in Fig. 5. More specifically, the ISA follows consecutive steps during the decision-making phase. The first step is the determination of the alternatives. An alternative is defined as a set of possible assignments of tasks to resources. The second step is the determination of the attributes, which are the criteria used to evaluate the alternatives. The multi-criteria ISA take into consideration a number of conflicting criteria including flowtime (4), cost (5), quality (6), and tardiness (7):

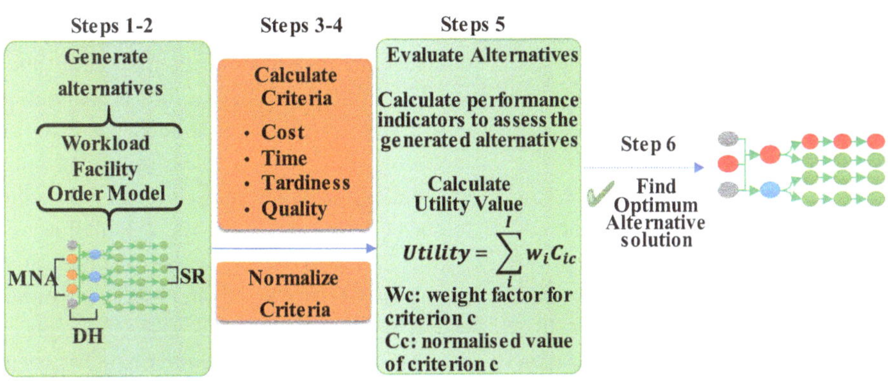

Fig. 5 Six main steps of the intelligent search algorithm (ISA). *Step 1* generate alternatives for DH until MNA, *Step 2* for each branch (*Step 1*) create SR random samples, *Step 3* calculate the criteria scores of each alternative, *Step 4* calculate the score of the branch based on its samples scores, *Step 5* calculate the utility value of each alternative, and *Step 6* select the alternative with the highest utility value

$$\text{FLOW}\,(\text{alt}_q) = \frac{\sum_{i=1}^{L} T_i^{\text{comp}}(alt_q) - T_i^{\text{arr}}}{L} \tag{4}$$

$$\text{TARD}\,(\text{alt}_q) = \frac{\sum_{i=1}^{L} T_i^{\text{comp}}(alt_q) - T_i^{\text{dd}}}{L} \tag{5}$$

$$\text{COST}\,(\text{alt}_q) = \frac{\sum_{i=1}^{L} t_i^{\text{proc}}(alt_q) \times R_i^{\text{proc}}}{L} \tag{6}$$

$$\text{QUALITY}\,(\text{alt}_q) = \frac{\sum_{i=1}^{L}(QL_i)}{L} \tag{7}$$

where R_i *i*th resource, alt_q *q*th alternative formed at the decision point, L number of completed tasks in the work-centre/job-shop at a decision point, T_i^{comp} (alt_q) completion time of the *i*th pending task if alt_q is implemented, T_i^{arr} time at which the *i*th pending task arrived at the work-centre, $t_i^{\text{proc}}(\text{alt}_q)$ estimated time required to process *i*th pending task if alt_q is implemented, R_i^{proc} cost of resource i to process the pending task if alt_q is implemented, QL_i quality index of the resource R_i to perform a *i*th pending task if alt_q is implemented, and T_i^{dd} due date of the *i*th pending task.

Once the criteria are determined, the consequences need to be defined. Consequences are the values of the attributes at the time the decisions and are performed in order to evaluate the selected alternatives. The set of alternatives selected during the decision-making phase is assessed using a set of performance indicators including production flowtime, resource utilisation, and mean tardiness.

A main benefit offered by the proposed scheduling mechanism is its ability to adapt to new order arrivals and quickly reschedule the job-shop. More specifically, when a new schedule needs to be generated due to the arrival of a new order, the running tasks of the existing jobs are fixed in their current positions and the rest of the tasks that are planned ahead of them are rescheduled together with the new ones. Through this functionality, the job-shop can adapt to unforeseen demand, rush orders, and other disruption in production such as machine breakdowns.

3.4 Order due-date assignment

Scheduling that incorporates due dates is of permanent interest [46]. The problem has been investigated since 1965 in the work of Jackson [47]. In academic research, the incorporation of due dates in scheduling problem acts in a twofold manner: it increases the constraints of the problem and thus it increases computational complexity, and it serves as a milestone against which important indicators are calculated, such as tardiness, earliness, and slack time. In a similar manner, in an industrial context, where the satisfaction of customer needs is the primary concern, the assignment and adherence of due dates determine the efficiency of a factory. Therefore, the assignment of realistic due dates is utterly important. In the literature, few works treat scheduling problems with the due-date assignment decision being of primary focus. Among the most significant contributions, Cheng and Gupta [48] review due-date assignment approaches up to 1990, whereas Gordon et al. [46] review more recent publications (up to 2002). In most works, the due-date assignment problem is treated using benchmarking instances [49], single-resource problems [50, 51] or is focused on assembly shops [52]. Moreover, the calculation of due dates in ETO environments is relatively mistreated. Most studies focus on static problem instances where jobs do not arrive continuously in the system [46], or require significant modelling efforts [53] and simulation experiments [54], which are rarely feasible in actual daily practice for due-date calculation.

In ETO industries, and particularly in mould-making, the actual practice implies for delivery dates to be negotiated with the customer during quotation. This agreement, later on during the job dispatching phase, will dictate how the due dates for the order will be set. The customer of a mould-making SME cannot tolerate delays in mould deliveries from the latter since this may delay the entire production of the former. Deviations in the expected delivery date of a mould can cause perturbations across the value chain and lead to supplying bottlenecks in the subsequent value-added phases of the customer, in the interlinked economy [55]. Therefore, it is utterly important for an SME to provide a customer with a solid estimation of when the mould will be available. However, the planner must have an estimation of the manufacturing lead time in the first place to use it in customer negotiations. The due date that will be defined by the lead time estimation of an order will be considered together with the customer's due date in order to satisfy the customer's demand according to the estimated due date calculated by the mould-making industry.

Therefore, the mechanism described in Sect. 3.2 is also utilised for the accurate calculation of the due date of a new order. It is reasonable that the manufacturing of two similar ETO products in a similarly configured manufacturing system will require the use of similar resources and will be completed in approximately the same time. Utilising the similarity mechanism, accurate estimations about the delivery dates can be produced for ETO products manufacturing. Once the complete set of attributes is submitted as described above, a sales agreement is achieved, the order is considered active, and the calculation of lead time initiates. The proposed due date estimation method reuses previous knowledge from executed orders, which are stored as cases in a case base. First, previous cases are retrieved and a pairwise comparison between the new order and each of the stored ones is performed. Based on the procedure described previously, the most similar case is obtained together with the calculated similarity index. The obtained case is revised in order to fit the new case requirements. Then, the manufacturing lead time of the past case is multiplied with the similarity index to derive the estimated lead time for the new case. The estimated manufacturing lead time is essentially the due date of the order. The estimation is finally checked by the planner and if accepted is inserted in the scheduling module as the due date of the order. If rejected, the procedure described in Sect. 3.2 is followed. Finally, after production is completed, multiple scheduling adherence methods (integration with an MES system, machine monitoring techniques, manual reporting, etc.) can be utilised in order to validate the accuracy of this estimation. If the accuracy is acceptable, the case is retained and stored in the case base. The mechanism which is shown to yield high-quality results is a similar industrial case as reported by Mourtzis et al. [38].

4 Development of the knowledge-enriched scheduling app

The two sections below present the KES app and describe its architecture and the software tools used for its development.

4.1 Description of the knowledge-enriched scheduling app

The scheduling and the similarity mechanisms have been implemented in Java for validation purposes. The integrated KES engine has been designed for implementation into a mobile app for the Android OS. The designed app allows data entry, selection of decision-making criteria, definition of weight factors and tuneable parameters of the ISA, and visualisation of results, as shown in the screenshot of Fig. 6.

The planner, through the data entry menu, is capable to insert information related to the facility, jobs, and workload and model their interrelationships in the form of precedence constraints (pre- and post-conditions). The planner can also provide information related to the working calendar not only of the resources but also of the entire factory. Moreover, the data related to the workload are defined according to the order due-date assignment. The due date of each order is specified by the knowledge reuse mechanism and is taken into consideration during the generation of the alternative schedules.

The app also allows the operator to interact with the proposed schedule. This rescheduling is necessary in cases

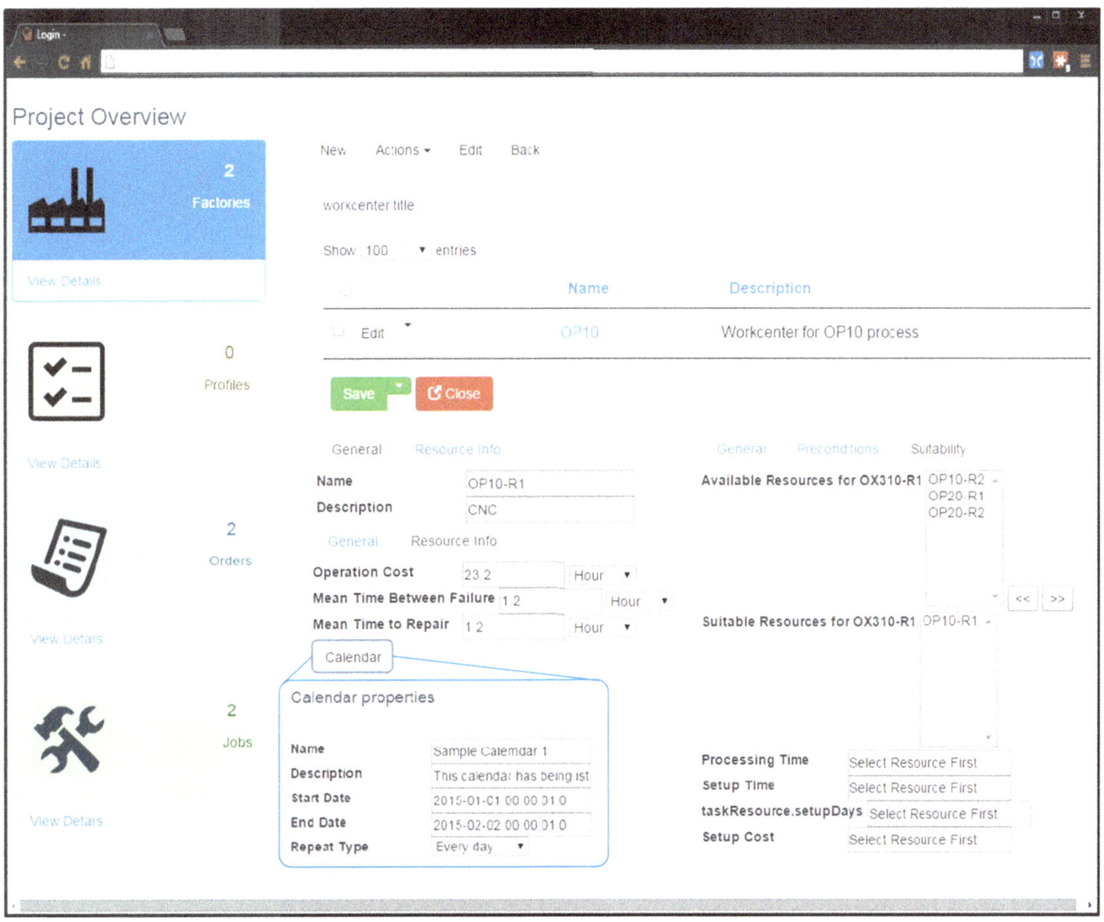

Fig. 6 Screenshot of the data entry fields of the scheduling app

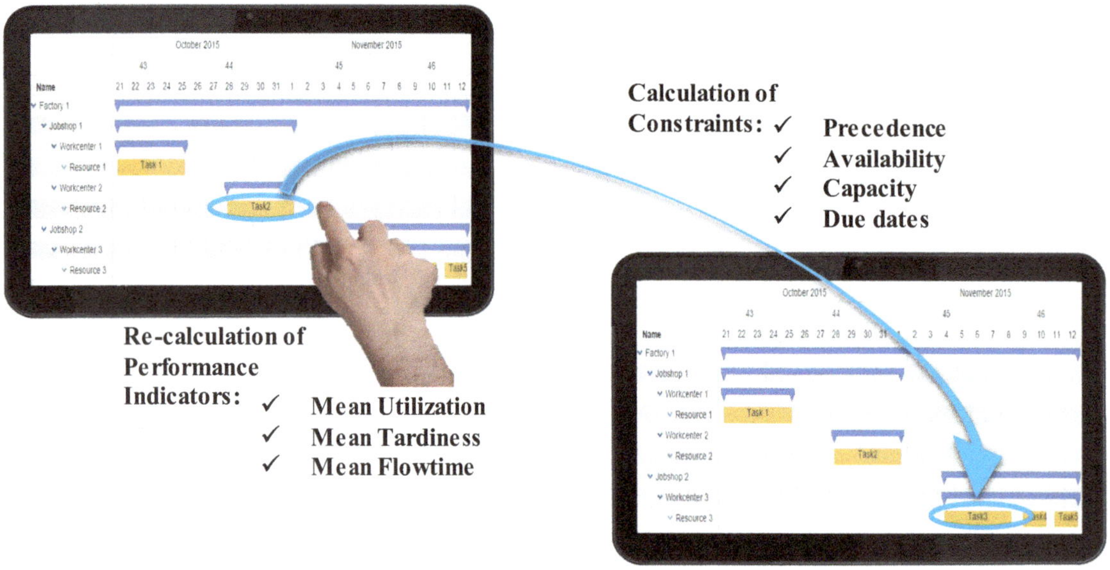

Fig. 7 Rescheduling performed by the operator through a tablet using drag and drop commands

that the derived schedule needs refinements due to order prioritisation and machine breakdowns among other reasons (Fig. 7). The precedence constrains, machine availability and capacity, and due dates are respected during this human-triggered rescheduling action. Moreover, performance indicators are recalculated each time a rescheduling occurs. Finally, the alternative with the highest utility value is displayed together with the scheduling Gantt chart and the mean values of the performance indicators (utilisation, flowtime, and tardiness).

4.2 Software architecture and development tools

Mobile apps deployed on the Android OS are based on a three-tier architecture that consists of three layers (data, business, and presentation) following the rules of the Model–view–controller architectural pattern. The presentation layer includes the graphical user interfaces of the app, and the data layer retrieves data from the back end. Finally, the business layer handles the data exchange between these two layers. For the programming of the platform, the Android software development kit (SDK) was used, which provides the developers the API libraries and tools necessary to build, test, and debug apps for Android. The back end is implemented with the Apache Tomcat version 7.0.19, since it is fully compliant with the latest advances in web programming and servlet specifications. The supporting data model of the app is based on requirements' collection from a mould manufacturer. The application runs on devices with ARM-based processors, 512 MB minimum memory, 300 MB free minimum storage space, and OS Android 4.0™ or later.

5 Industrial case study: experiments and results

The case study uses real data from a high-precision mould-making machine shop. The mould-shop best fits to the engineer-to-order (ETO) business model, where custom moulds and dies are designed and manufactured based on particular customer needs. Injection moulds produced by this mould-shop are one of a kind, first-time-right products that vary greatly in terms of quality needs, tolerances, and mainly functionality. Evidently, the mould-making industry is highly specialised and knowledge dependant. Once a new production order is released, a scheduling of its tasks must follow. Work is delegated among engineers, based on their expertise, who are usually in charge of a project from start to end. The resources needed are determined by the project's particularities. The manufacturing lead time is identified as a major competitive factor of mould-making industries [56]. Machinists together with designers usually perform a first estimation of manufacturing lead time, but the accuracy is empirical [38]. Similarly, in the current business model of the company, unofficial oral meetings take place in order to schedule resources, and, if the situation demands it, the management department is involved in the decision-making and work prioritisation. However, no software tools are used to support short-term scheduling or to document the decisions made. Therefore, by using the developed knowledge-enriched scheduling mechanism, not only will the scheduling of the tasks be more accurate, flexible, and easily reconfigurable to handle unpredictable events, but also the employees will be able to receive notifications related to schedule changes and other job-shop-related information through mobiles devices.

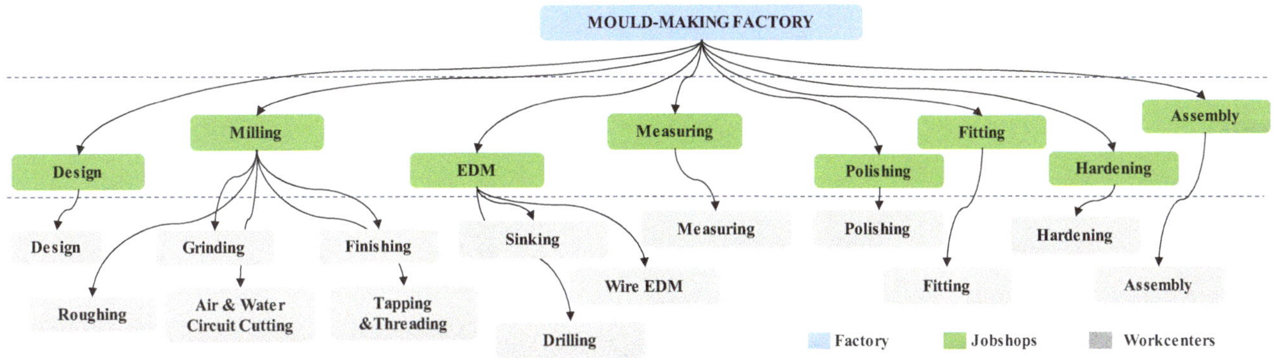

Fig. 8 Facility model of the mould-making job-shop

Table 1 Hierarchical model of the mould-making job-shop

Job-shop	Work-centre	No. of resources
Design	Design	2
Milling	Roughing	14
	Grinding	5
	Air and water circuit cutting	3
	Tapping and threading	4
	Finishing	6
EDM	Sinking	3
	Wire EDM	5
	Drilling	1
Measuring	Measuring	1
Polishing	Polishing	1
Fitting	Fitting	1
Hardening	Hardening	2
Assembly	Assembly	1

The shop-floor of the case study consists of eight job-shops, which include 14 work-centres that are formed by 40 individual resources in total (Fig. 8). The resources include on the one hand high-precision CNC machines that are capable to perform milling, drilling, turning, electro-discharge wire cutting, sinking, grinding, tapping, roughing, polishing, and hardening operations and on the other human operators that perform manually operations of design, fitting, assembly, measuring, and polishing (Table 1). The hierarchical model of the production facility is shown in Fig. 8. The utilised historical dataset includes the processing times, tasks, sequences, and resources used for the manufacturing of thirty (30) moulds. The dataset was collected within a time span of approximately 3 years.

In actual production terms, five orders (moulds) are simultaneously executed in the shop-floor on average, as the analysis of the historical data depicted. Therefore, in the experiments below, it is considered that four orders are already under processing and a new order enters the system eight calendar days later. The schedules for these orders have already been generated previously and are currently being executed. The system must then be rescheduled in order to accommodate the new order. The new mould order carries the identification number "13.23". The basic attributes of "13.23" are given in Table 2, and the different components and tasks required for this mould are shown in Fig. 9.

As described in Sect. 3.2, the new order that triggers the scheduling mechanism is first compared against documented past cases for the reuse of knowledge related to processes and product structure. In the presented work, the

Table 2 Attributes of the compared mould cases

Attributes	Mould 13.23	Mould 12.20	Mould 11.38
Number of cavities	6	2	4
Type of hardening	Very good	Very good	Very good
Side of injection	Moving side	Moving side	Moving side
Mould size	Medium	Large	Large
Core cap	No	Yes	Yes
Ejector rings	6	2	4
Temper evident	No	No	No
Type of data	Idea	Idea	Idea
Surface's quality	Mirrors	Mirrors	Mirrors
Number of basic components	9	12	11

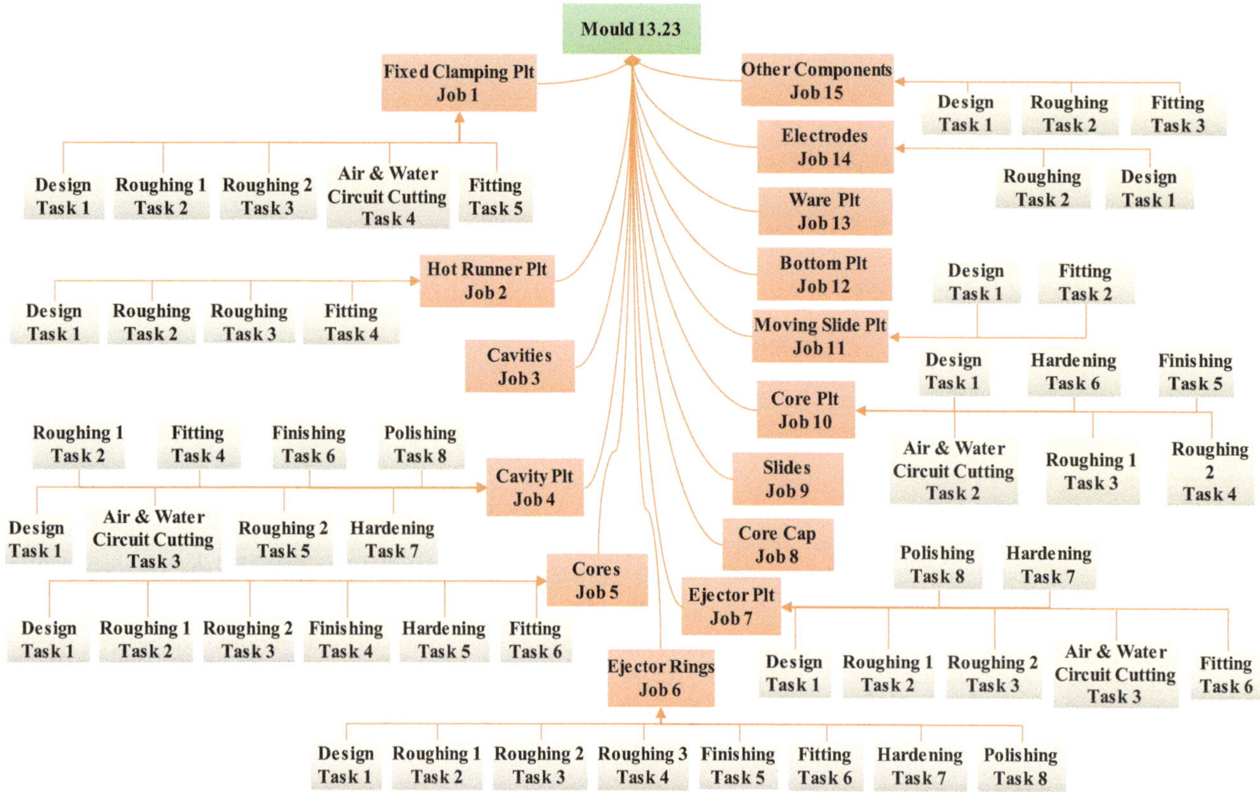

Fig. 9 Components, jobs, and tasks of the "13.23" order

past case is the complete dataset of a previously executed mould-making order, including the mould specifications/attributes, scheduling parameters used, the schedule followed (policy, sequencing, etc.), and the documented processing times and sequencing of tasks. In the case study, the new order is compared against all 30 documented past cases. Ten attributes are considered by CBR for the pairwise comparison of cases, namely the: number of cavities, type of hardening, side of injection, mould size, core cap, ejector rings, tamper evident, type of data, surface's quality, and number of basic components, with the following weight factors: 0.15, 0.05, 0.1, 0.1, 0.1, 0.1, 0.1, 0.05, 0.1, and 0.15. The weight factors were defined and fine-tuned together with the experts of the case study. The resulting values denote the potential of an attribute to reveal actual production-related similarities between any two cases. For instance, the number of basic components that need to be manufactured directly affects the required number of manufacturing processes, and therefore, the attribute "number of basic components" is attributed with a large weight value, since the more the number of required process, the longer the flowtime of production. Another fundamental parameter that is taken into consideration is the shape of the stacks. There are two options, namely moulds with cylindrical or rectangular stacks. Since the mould "13.23" has cylindrical stacks, the attribute "length" is not considered during similarity. After a similarity calculation, the results indicate a similarity index of 83 % between moulds "13.23" and "12.20". The planner then adapts the process plan of the latter in order to prepare the dataset for scheduling the former. The next most similar mould is "11.38", which is 75 % similar to "13.23". As given in Table 2, mould "13.23" differs from "12.20" in the attributes: "type of hardening", "number of basic components", and "width". The components that are needed for manufacturing the "13.23" mould are less than the components of "12.20", so the planner should reuse the sequence of processes of "11.38" mould and observe that components, such as the bottom plates, are missing. Based on his expertise, the extra components are removed and the process sequence is successfully customised for the new mould.

In a second step, the similarity mechanism is utilised in order to estimate the due date of the new mould case. The pairwise comparison of the past cases with the new one results in a ranked list of the most similar cases according to the defined attribute weights. Then, the manufacturing lead time of the past case is multiplied with the similarity index to derive the estimated lead time for the new case.

Table 3 presents the actual data of the processing times of each mould based on the company's database. Moreover, based on the data retrieved from the mould-making

Table 3 Lead time estimation produced by the CBR mechanism

Process time (h)	Mould 11.38	Mould 12.20	Mould 13.23
Roughing	406.5	333.5	212.5
Finishing	87.0	296.5	221.8
Air and water circuit	81.0	80.0	89.5
Fitting	124.5	46.5	43.5
Polishing	69.5	33.0	41.5
Hardening	504.0	504.0	504.0
EDM	34.0	20.0	30.0
Electrodes	51.0	7.5	11.5
Other processes	61.5	23.5	10.5
Assembly	72.0	86.5	72.0
Design	40.0	12.25	12.3
Lead time in hours (days)	1531 (64)	1443.25 (60)	1249 (52)

Table 4 Similarity measurements between the mould cases "13.23", "12.20", and "11.38"

Compared mould attributes	Distance 11.38 → 13.23	Distance 12.20 → 13.23
Number of cavities	0.273861279	0.273861279
Type of hardening	0.223606798	0.223606798
Side of injection	0.316227766	0.316227766
Mould size	0.223606798	0.316227766
Core cap	0	0.316227766
Ejector rings	0.223606798	0.223606798
Temper evident	0.316227766	0.316227766
Type of data	0.223606798	0.223606798
Surface's quality	0.316227766	0.316227766
Number of basic components	0.350324525	0.369274473
Sum	2.467296293	2.895094975
Similarity measure	6.087550996	8.381574914

industry and after their analysis, the average set-up time required to perform a task is 30 min. This set-up time is incorporated in the processing times in order to decrease the complexity of the model.

The results of the calculated similarity index are included in Table 4. As previously mentioned, the two stored cases "12.20" and "11.38" are the most similar to the new case "13.23". The similarity measure between "12.20" and "13.23" is calculated at $S_1 = 8.381574914$. The measure between mould "11.38" and "13.23" is $S_2 = 6.087550996$.

Afterwards, the adaptation of the case is performed in order to estimate its manufacturing lead time. The lead time of case "13.23" mould is multiplied with the similarity measure between "12.20" and "13.23", and the result is divided by 10. The resulting value for the estimation of the lead time is Lead $\text{time}_{13.23} = 1209.6708$ h (Table 5). Translating this lead time into calendar days obeying at all times the working shifts of each resource per day, the due date of the order is calculated at 51 days. It is highlighted that this estimation deviated only 3.15 % compared to the actual historical values, which depicts the accuracy of the lead time estimation method. It is also noted that by utilising a larger pool of past cases, the

Table 5 Lead time and processing times' estimation according to similarity comparison results

Processes	Processing times of 13.23
Roughing	279.52
Finishing	248.51
Air and water circuits	67.05
Fitting	38.97
Polishing	27.66
Hardening	422.43
EDM	16.76
Electrodes	6.29
Assembly	72.50
Design	10.27
Other processes	19.70
Sum (lead time)	1209.67

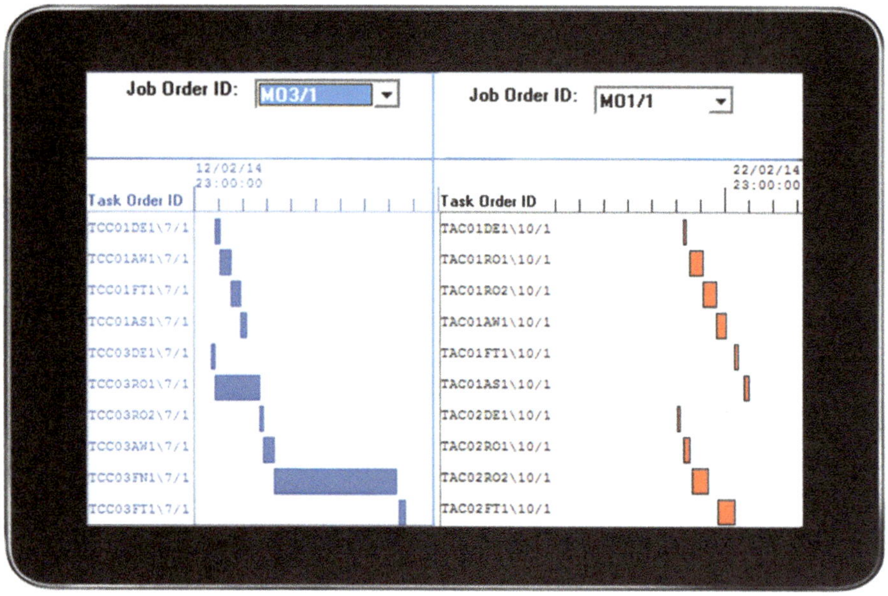

Fig. 10 Visualisation of schedule for two orders that are executed in the job-shop

method yields results of even higher accuracy. Once the adaptation of the new case is complete, the scheduling algorithm generates and evaluates scheduling alternatives and their respective performance indicators.

The tuneable parameters of the ISA are defined using a statistical design of experiments [44], which reduced the required number of experiments for determining the impact of tuneable parameters on the cardinal preference of the decision-making process. The number of experiments was 25, and each tuneable factor had five levels. The analysis of means (ANOM) diagrams are created and depicted the impact of the values of the tuneable factors to the utility value. According to ANOM diagrams, the preferable values to be used in the particular scheduling experiment are MNA = 100, DH = 15, and SR = 20. Each schedule is assessed with the mean values of the performance indicators of utilisation, flowtime, and tardiness, which are given by the following formulas (8, 9, and 10):

$$\text{Tardiness}\quad MT(t_n) = \frac{1}{N^{\text{comp}}} \sum_{i=1}^{N^{\text{comp}}} \max\left(0, t_i^{\text{comp}} - t_i^{\text{dd}}\right) \qquad (8)$$

$$\text{Flowtime}\quad MF(t_n) = \frac{1}{N^{\text{comp}}} \sum_{i=1}^{N^{\text{comp}}} \left(t_i^{\text{comp}} - t_i^{\text{arr}}\right) \qquad (9)$$

$$\text{Utilisation}\quad MU(t_n) = \frac{1}{N^{\text{comp}}} \sum_{i=1}^{N^{\text{comp}}} \left(\frac{t_i^{\text{comp}} - t_i^{\text{start}}}{t^{\text{tot}}}\right) \qquad (10)$$

where N^{comp} the number of completed jobs up to time t_n, t_i^{comp} the completion time of job i, t_i^{dd} the due date of job i, t_i^{arr} the arrival time of job i, t_i^{start} the start time of job i, t^{tot} the total operating time of the facility, and t_n the time point at which all performance measures are calculated.

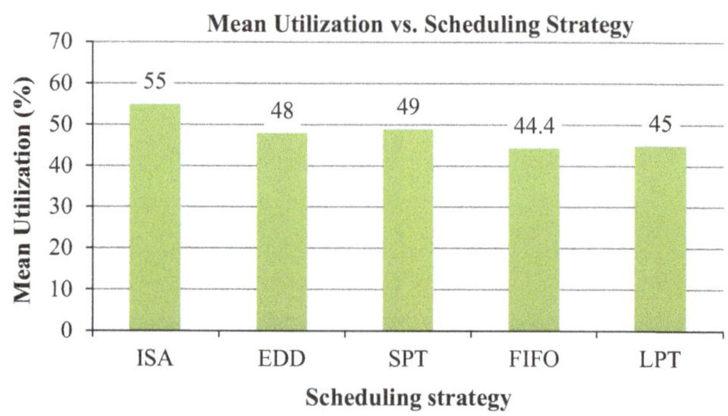

Fig. 11 Mean utilisation versus scheduling strategy

Figure 10 visualises in slip-view the schedules of an old order that was executed in the job-shop (left-hand side) (right) and the new "13.23" order that has a start date 8 days earlier (right-hand side). The figure displays the tasks for specific jobs, their duration, and the start and end time in calendar form.

In order to benchmark the performance of the ISA, a comparison against widely used dispatch rules is also performed. The rules are: first in first out (FIFO), shortest processing time (SPT), earliest due date (EDD), and least process time (LPT) [45]. The diagrams of Figs. 11, 12 and 13 reveal the superiority of the ISA in terms of the calculated performance indicator values. Still, in cases when a specific production target must be achieved, dispatch rules yielded high-quality results. For instance, EDD identified schedules with lowest flowtime and near zero tardiness compared to the other dispatch rules and ISA.

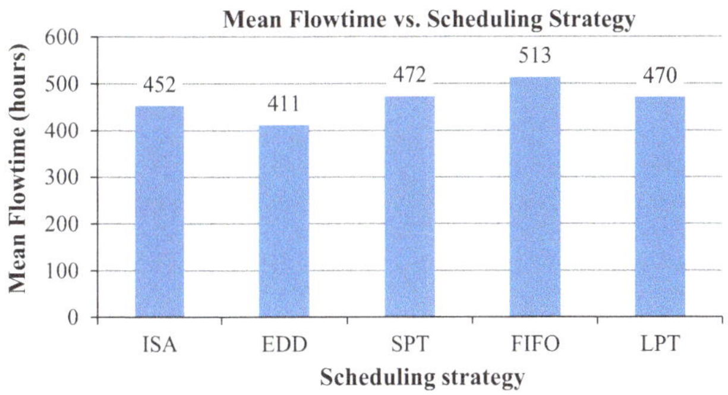

Fig. 12 Mean flowtime versus scheduling strategy

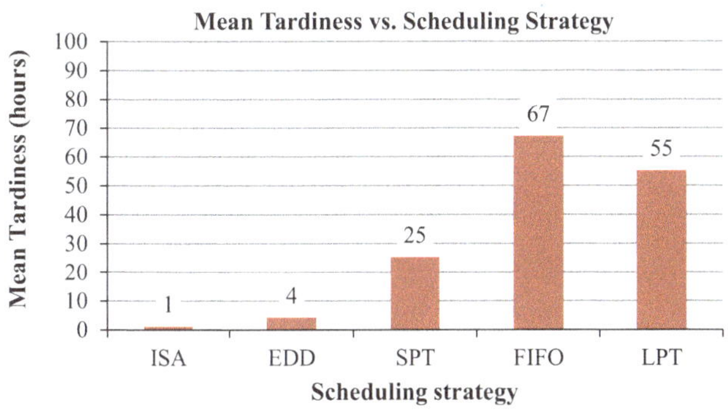

Fig. 13 Mean tardiness versus scheduling strategy

6 Conclusions and future work

The presented work focused on the enhancement of the short-term scheduling of manufacturing resources through the exploitation of historical design and planning knowledge. The scheduling of tasks for the realisation of ETO products is supported by a knowledge retrieval mechanism that is based on the CBR method and similarity measurements. The similarity index between past and new cases is measured using the Euclidean distance, and both numerical and alphanumeric attributes are considered. The scheduling is performed by the ISA that uses adjustable parameters, which configure the depth and breadth of the search, while guiding it through the solution space to identify high performing alternatives in a timely manner. The results of the application of the methodology into a real-life pilot case with data obtained from a mould-making industry verified that the short-term scheduling algorithm provides solutions of high quality in comparison to the historical values. Moreover, the deployment of the scheduling engine on mobile devices offers a certain degree of mobility, which is desired for compensating for the dynamic nature of today's turbulent manufacturing environment.

A limitation of the proposed knowledge reuse approach is the necessity for pre-existing and sufficiently documented cases. The repository of past cases in the examined case study included 30 cases with ten attributes each and provided good results. The performance of the method, in case fewer cases with partial documentation were stored in the case base, is expected to be decreased. Yet, the gathering of this amount of information about previous cases is relatively easy, since these ten attributes comprise basic characteristics of a mould, well known to the planner, and a repository with 30 products can be built in a fairly short amount of time.

Future work will focus on the quantitative evaluation of the knowledge reuse and scheduling mechanisms. The company of the case study is currently testing the developments in real-life situations. Moreover, a series of interviews with the engineers will be organised to assess the quality of the produced schedules and the accuracy of the similarity measurement results. A long-term vision is the total integration of these mechanisms in the everyday practice of the company and their utilisation through the developed app.

Acknowledgments The work reported in this paper has been partially supported by the EC FP7 Project "Applications for Advanced Manufacturing Engineering—Apps4aME" (GA No. 314156).

References

1. Chryssolouris G, Mavrikios D, Papakostas N, Mourtzis D, Michalos G, Georgoulias K (2009) Digital manufacturing: history, perspectives, and outlook. Proc Inst Mech Eng Part B J Eng Manuf 223:451–462
2. Karmarkar US (1993) Manufacturing lead times, order release and capacity loading. Handb Oper Res Manag Sci 4:287–329
3. Lowe A, McMahon C, Culley S (2004) Information access, storage and use by engineering designers. J Eng Des 30(2):30–32
4. Efthymiou K, Sipsas K, Mourtzis D, Chryssolouris G (2013) On an integrated knowledge based framework for manufacturing systems early design phase. Proc CIRP 9:121–126
5. Takahashi D (2014) Mobile apps could hit $70B in revenues by 2017. http://venturebeat.com/2014/04/29/mobile-apps-could-hit-70b-in-revenues-by-2017-as-non-game-categories-take-off/
6. Kubach U (2012) Manufacturing business web: enabling flexible and mobile business processes for the manufacturing industry. Industrial technologies 2012. http://www.industrialtechnologies2012.eu/sites/default/files/presentations_session/4_Uwe_Kubach.pdf
7. Akgun AE, Byrne JC, Keskin H, Lynn GS, Imamoglu SZ (2005) Knowledge networks in new product development projects: a transactive memory perspective. Inf Manag 42:1105–1120
8. Aamodt A, Plaza E (1994) Case-based reasoning: foundational issues, methodological variations, and system approaches. AI Commun 7:39–59

9. Kolodner JL (1992) An introduction to CBR. AI Rev 6:3–34
10. Li BM, Xie SQ, Xu X (2011) Recent development of knowledge-based systems, methods and tools for one-of-a-kind production. Knowl Based Syst 24:1108–1119
11. Lee LH, Chew EP, Frazier PI, Jia QS, Chen CH (2013) Advances in simulation optimisation and its applications. IEE Trans 45:683–684
12. Watson GI (1997) Repulsine particles on two-dimensional lattice. Phys Stat Mech Appl 246:253–274
13. Michalos G, Makris S, Mourtzis D (2011) A web based tool for dynamic job rotation scheduling using multiple criteria. CIRP Ann Manuf Tech 60(1):453–456
14. Jain AS, Meeran S (1999) Deterministic job-shop scheduling: past, present and future. Eur J Oper Res 113(2):390–434
15. Wang JF, Du BQ, Ding HM (2011) A genetic algorithm for the flexible job-shop scheduling problem. Adv Res Comput Sci Inf Eng 152:332–339
16. Ouelhadj D, Petrovic S (2009) Asurvey of dynamic scheduling in manufacturing systems. J Sched 12(4):417–431
17. Xing LN, Chen YW, Yang KW (2009) An efficient search method for multi-objective flexible job shop scheduling problems. J Int Manuf 20:283–293
18. Li X, Olafsson S (2005) Discovering dispatching rules using data mining. J Sched 8:515–527
19. Varela R, Vela CR, Puente J, Gomez A (2003) A knowledge-based evolutionary strategy for scheduling problems with bottlenecks. Eur J Oper Res 145:57–71
20. Sha DH, Liu CH (2005) Using data mining for due date assignment in a dynamic job shop environment. Int J Adv Manuf Technol 25:1164–1174
21. Karimia H, Seyed H, Rahmatia A, Zandiehb M (2012) An efficient knowledge-based algorithm for the flexible job shop scheduling problem. Knowl Based Syst 36:236–244
22. Yong H, Li S (2015) One-machine scheduling problems with deteriorating jobs and position-dependent learning effects under group technology considerations. Int J Syst Sci 26(7):1319–1326
23. Reed N, Scanlan J, Wills G, Halliday ST (2011) Knowledge use in an advanced manufacturing environment. Des Stud 32(3): 1345–1351
24. Accenture. Mobile Web Watch Survey; 2012
25. Enterprise mobile application Report, Kony; 2015. http://forms.kony.com/rs/656-WNA-414/images/Kony-Enterprise-Mobile-App-Report.pdf
26. Gröger C, Silcher S, Westkämper E, Mitschang B (2013) Leveraging apps in manufacturing. A framework for app technology in the enterprise. Proc CIRP 7:664–669
27. Lunani M Role-based mobile applications can help manufacturing companies dramatically boost productivity and growth. http://www.cognizant.com/insights/perspectives/role-based-mobile-apps-transform-manufacturing. Accessed 08 Aug 2013
28. Xu X (2012) From cloud computing to cloud manufacturing. Robot Comput Integr Manuf 28(1):75–86. doi:10.1016/j.rcim.2011.07.002
29. Hu SJ, Ko J, Weyand L, Elmaraghy HA, Kien TK, Koren Y, Bley H, Chryssolouris G, Nasr N, Shpitalni M (2011) Assembly system design and operations for product variety. CIRP Ann Manuf Technol 60(2):715–733
30. Chung S, Lee KY, Choi J (2015) Exploring digital creativity in the workspace: the role of enterprise mobile applications on perceived job performance and creativity. Comput Hum Behav 42:93–109
31. Mourtzis D, Doukas M, Vandera C (2014) Mobile applications for product customisation and design of manufacturing networks. Manuf Lett. doi:10.1016/j.mfglet.2014.01.002
32. Young HA (2012) Scientific apps are here (and more will be coming). Cytokine 59:1–2
33. (2012) Mobile ''apps'' provide parts data and calculate O-ring dimensions. Seal Technol 2012(2):3. doi:10.1016/S1350-4789(12)70037-5
34. Ozturk A, Kaya S, Ozdemirel NE (2006) Manufacturing lead time estimation using data mining. Eur J Oper Res 173(2): 683–700
35. Li BM, Xie SQ, Xu X (2011) Recent development of knowledge-based systems, methods and tools for one-of-a-kind production. Knowl Based Syst 24(7):1108–1119
36. Chryssolouris G, Subramaniam V (2001) Dynamic scheduling of manufacturing job shops using genetic algorithms. J Int Manuf 12:281–293
37. Mourtzis D, Doukas M, Psarommatis F (2014) Design of manufacturing networks for mass customisation using an intelligent search method. Int J Comput Int Manuf. doi:10.1080/0951192X.2014.900867
38. Mourtzis D, Doukas M, Fragou K, Efthymiou K, Matzorou V (2014) Knowledge-based estimation of manufacturing lead time for complex engineered-to-order products. Proc CIRP 17:499–504
39. Chryssolouris G, Dicke K, Lee M (1992) On the resources allocation problem. Int J Prod Res 30(12):2773–2795
40. Lalas C, Mourtzis D, Papakostas N, Chryssolouris G (2006) A simulation-based hybrid backwards scheduling framework for manufacturing systems. Int J Comput Int Manuf 19(8): 762–774
41. Baker A (1998) A survey of factory control algorithms which can be implemented in a multi-agent hierarchy: dispatching, scheduling, and pull. J Manuf Syst 17(4):297–320
42. Viagas F-V, Framinan JM (2014) On insertion tie-breaking rules in heuristics for the permutation flowshop scheduling problem. Comput Oper Res 45:60–67. doi:10.1016/j.cor.2013.12.012
43. Mourtzis D, Doukas M, Psarommatis F (2012) A multi-criteria evaluation of centralised and decentralised production networks in a highly customer-driven environment. CIRP Ann Manuf Technol 61(1):427–430
44. Phadke MS (1989) Quality engineering using robust design, 1st edn. Prentice Hall, Englewood Cliffs
45. Chryssolouris G (2006) Manufacturing systems: theory and practice, 2nd edn. Springer, New York
46. Gordon V, Proth JM, Chu C (2002) A survey of the state-of-the-art of common due date assignment and scheduling research. Eur J Oper Res 139:1–25
47. Jackson JR (1955) Scheduling a production line to minimize maximum tardiness. Management Science Research Project, UCLA
48. Cheng TCE, Gupta MC (1989) Survey of scheduling research involving due date determination decisions. Eur J Oper Res 38:156–166
49. Vig MM, Dooley KJ (1993) Mixing static and dynamic flowtime estimates for due-date assignment. J Oper Manag 11(1):67–79
50. Seidmann A, Panwalkar SS, Smith ML (1981) Optimal assignment of due-dates for a single processor scheduling problem. Int J Prod Res 19(4):393–399
51. Panwalkar SS, Smith ML, Seidmann A (1982) Common due date assignment to minimize total penalty for the one machine scheduling problem. Oper Res 30(2):391–399
52. Bock DB, Patterson JH (1990) A comparison of due date setting, resource assignment, and job pre-emption heuristics for the multiproject scheduling problem. Decis Sci 21(2):387–402
53. Gordon V, Strusevich V, Dolgui A (2012) Scheduling with due date assignment under special conditions on job processing. J Sched 15(4):447–456
54. Vinod V, Sridharan R (2011) Simulation modeling and analysis of due-date assignment methods and scheduling decision rules in

a dynamic job shop production system. Int J Prod Econ 129:127–146
55. Denkena B, Bluemel P, Kroening S, Roebbing J (2012) Condition based maintenance planning of highly productive machine tools. Prod Eng Res Dev 6:277–285
56. Pecas P, Henriques E (2003) The need for agile manufacturing implementation in mould making business. In: Proceedings of the business excellence I—performance measures, benchmarking and best practices in new economy. pp 321–332

2

Extended MULTIMOORA method based on Shannon entropy weight for materials selection

Arian Hafezalkotob[1] · Ashkan Hafezalkotob[2]

Abstract Selection of appropriate material is a crucial step in engineering design and manufacturing process. Without a systematic technique, many useful engineering materials may be ignored for selection. The category of multiple attribute decision-making (MADM) methods is an effective set of structured techniques. Having uncomplicated assumptions and mathematics, the MULTIMOORA method as an MADM approach can be effectively utilized for materials selection. In this paper, we developed an extension of MULTIMOORA method based on Shannon entropy concept to tackle materials selection process. The entropy concept was considered to assign relative importance to decision-making attributes. The proposed model consists of two scenarios named the weighted and entropy-weighted MULTIMOORA methods. In the first scenario, subjective weight was considered in the formulation of the approach like most of conventional MADM methods. The general form of entropy weight that is a combination of subjective and objective weighting factors was employed for the second scenario. We examined two popular practical examples concerning materials selection to show the application of the suggested approach and to reveal the effect of entropy weights. Our results were compared with the earlier studies.

✉ Ashkan Hafezalkotob
Hafezalkotob@iust.ac.ir; a_hafez@azad.ac.ir

Arian Hafezalkotob
ar_hafezalkotob@azad.ac.ir

[1] Department of Mechanical Engineering, Islamic Azad University, South Tehran Branch, Ahang Ave, Mahalati Blv, Tehran, Iran

[2] Department of Industrial Engineering, Islamic Azad University, South Tehran Branch, Entezari Aly, Oskoui St, Choobi Brg, Tehran, 1151863411, Iran

Keywords Multiple attribute decision making · MULTIMOORA · Shannon entropy · Materials selection

Introduction

More than 40,000 practical metallic alloys and a same number of nonmetallic materials like polymers, ceramics, and composites are utilized in various industries (Farag 2002). Because of the considerable number, dissimilar production techniques, and different properties of engineering materials, the selection process of materials can be regarded as a complex undertaking for an engineer or designer. If the process takes place unsystematically, many significant materials may be neglected. Therefore, a structured mathematical approach is needed for materials selection.

MADM methods can be used as effective systematic tools for materials selection. Each MADM technique has specific assumptions and principles. A number of MADM methods have been utilized in the materials selection process by earlier researchers, like the technique for order preference by similarity to ideal solution (TOPSIS) (Bakhoum and Brown 2013; Das 2012; Huang et al. 2011; Jee and Kang 2000), analytic hierarchy approach (AHP) (Chauhan and Vaish 2013; Dweiri and Al-Oqla 2006), compromise ranking also known as vlse kriterijumska optimizacija kompromisno resenje (VIKOR) (Jahan and Edwards 2013b; Liu et al. 2013), diverse versions of elimination and choice expressing the reality (ELECTRE) also recognized as outranking method (Anojkumar et al. 2014; Chatterjee et al. 2009; Shanian and Savadogo 2009), preference ranking organization method for enrichment evaluation (PROMETHEE)

(Jiao et al. 2011; Peng and Xiao 2013), graph theory and matrix approach (Rao 2006), gray relational analysis (Chan and Tong 2007; Zhao et al. 2012), various preference ranking-based techniques (Chatterjee and Chakraborty 2012; Maity et al. 2012), preference selection index (Maniya and Bhatt 2010), utility additive (UTA) (Athawale et al. 2011), weighted property index (Findik and Turan 2012), linear assignment (Jahan et al. 2010a), modified digital logic (Manshadi et al. 2007; Torrez et al. 2012), Z-transformation (Fayazbakhsh and Abedian 2010; Fayazbakhsh et al. 2009), and quality function deployment (Mayyas et al. 2011; Prasad 2013; Prasad and Chakraborty 2013). Two groups of researchers have reviewed the applications of MADM methods in materials selection (Jahan and Edwards 2013a; Jahan et al. 2010b).

Almost all the aforementioned methods have a key feature that is moderate to the extreme complexity of their mathematical models. Utilization of these techniques seems to be difficult, requiring advanced mathematical knowledge (Karande and Chakraborty 2012a). Accordingly, an undemanding MADM method can be a real blessing for decision makers. The multi-objective optimization on the basis of ratio analysis (MOORA) method proposed by Brauers and Zavadskas (2006) has uncomplicated mathematics. Therefore, it can be employed effortlessly and effectually for selection of materials. The MULTIMOORA method is a comprehensive form of the MOORA technique. As discussed by Brauers and Ginevičius (2010), because the final rank is generated by the integration of three subordinate ranks in the MULTIMOORA technique, its results can be more robust than traditional MADM methods in which a single rank is obtained. The MOORA and MULTIMOORA techniques have been used in different applications like decision making in manufacturing environment (Chakraborty 2011), robot selection (Datta et al. 2013), supplier selection (Farzamnia and Babolghani 2014; Karande and Chakraborty 2012b; Mishra et al. 2015), evaluating the risk of failure modes (Liu et al. 2014a), project selection (Rached-Paoli and Baunda 2014), selection of health-care waste treatment (Liu et al. 2014b), ranking of banks (Brauers et al. 2014), and student selection (Deliktas and Ustun 2015).

In the present paper, we extended the MULTIMOORA method using entropy weight based on Shannon information theory for application in materials selection. Our study is closely related to Karande and Chakraborty (2012a). They used the MOORA technique in the materials selection process of four practical cases. The novelties of our paper comparing the study of Karande and Chakraborty (2012a) are as follows: First, they did not calculate the final ranking of the MULTIMOORA method and only reported the three subordinate ranks. The third subordinate rank of the MULTIMOORA method, i.e., the full multiplicative form rank, was incorrectly called the MULTIMOORA ranking in their study. In this paper, we employed the dominance theory to integrate the three subordinate ranks into the final ranking, named the MULTIMOORA ranking. This aggregate final ranking is more robust than each of the subordinate ranks as stated by Brauers and Ginevičius (2010). Second, Karande and Chakraborty (2012a) did not utilize any relative significance for attributes. However, we used two forms of attributes weighting, i.e., subjective and the general Shannon entropy weights, to generate two solution modes named the weighted and entropy-weighted MULTIMOORA rankings, respectively. Third, Karande and Chakraborty (2012a) employed Voogd ratio (Voogd 1983) for normalization, whereas we utilized the original MULTIMOORA normalization equation that is the most robust option among various ratios as shown by Brauers and Zavadskas (2006). A few studies on assigning weights for the MOORA and MULTIMOORA techniques exist. Brauers and Zavadskas (2006) mentioned that giving importance to each attribute is possible, but they did not discuss on the specifications of these significance factors. Özçelik et al. (2014) assigned weight for the reference point approach of the MOORA method. In their study, the fuzzy analytic hierarchy process was utilized for the determination of significance coefficients of attributes. El-Santawy (2014) used a new form of entropy weight to develop the MOORA method. Derivation of their significance factors differs from Shannon entropy weight. In addition, they did not develop the MULTIMOORA method with their suggested weights. To the best of the authors' knowledge, no study has been conducted on combination of Shannon entropy weight with MULTIMOORA technique. In our proposed approach, the general form of entropy weight was utilized that includes subjective and objective parts. The subjective significance coefficient is obtained directly from decision makers opinions. The objective part is calculated based on the entropy concept through analyzing the data regardless of decision makers' comments. The general form of entropy weight improves the initial values of decision matrix and reliability of the ranking of alternatives obtained by the MULTIMOORA approach. We evaluated two practical examples in the field of materials selection. The results were compared with other studies that have considered these two problems. Eventually, concluding remarks were cited to make a summary of our work and to present an overview of the developed MULTIMOORA method and its application in materials selection.

The MULTIMOORA method

The MOORA method proposed by Brauers and Zavadskas (2006) is formed from two parts: the ratio system and the reference point approach. Brauers and Zavadskas (2010) developed the concept by utilizing the full multiplicative form. The updated method, called MULTIMOORA, is composed of MOORA parts and the full multiplicative form. The MULTIMOORA method begins with a decision matrix **X** in which x_{ij} presents the performance index of ith alternative respecting jth attribute, $i = 1, 2, \ldots, m$ and $i = 1, 2, \ldots, n$:

$$\mathbf{X} = [x_{ij}]_{m \times n}. \quad (1)$$

To make the performance indices dimensionless and comparable, the decision matrix is normalized. This normalization ratio is a comparison between each response of an alternative to an attribute, as a numerator, and a denominator that is a representative for all alternative performances with respect to that attribute. In the MULTIMOORA method, the dominator is selected as the square root of the sum of squares of performance indices per attribute as shown in the following:

$$x_{ij}^* = \frac{x_{ij}}{\left[\sum_{i=1}^{m} x_{ij}^2\right]^{1/2}}, \quad (2)$$

in which x_{ij}^* denotes the normalized performance index of ith alternative respecting jth attribute. Brauers and Zavadskas (2006) proved that this ratio is the most robust selection among different normalization equations for the MULTIMOORA method.

The ratio system

Equation (2) justifies the appellation of this technique as the ratio system. For this method, the normalized performance indices are added for beneficial attributes (in case of maximization) or deducted for non-beneficial attributes (in case of minimization) as follows (Brauers and Zavadskas 2006):

$$y_i^* = \sum_{j=1}^{g} x_{ij}^* - \sum_{j=g+1}^{n} x_{ij}^*, \quad (3)$$

in which g indicates the number of beneficial attributes and $(n - g)$ is the number of non-beneficial attributes. y_i^* denotes the assessment value of ith alternative regarding all attributes for the ratio system. The optimal alternative based on the ratio system has the highest assessment value (Datta et al. 2013):

$$A_{\mathrm{RS}}^* = \left\{ A_i \middle| \max_i y_i^* \right\}. \quad (4)$$

The reference point approach

As the second part of the MOORA method, the reference point approach is also based on the ratio system, i.e., Eq. (2). A maximal objective reference point is utilized in the method. The ith co-ordinate of the maximal objective reference point vector is defined as follows (Brauers and Zavadskas 2006):

$$r_j = \begin{cases} \max_i x_{ij}^* & \text{in case of maximization} \\ \max_i x_{ij}^* & \text{in case of minization} \end{cases}. \quad (5)$$

Deviation of a performance index from the reference point r_j can be obtained as $(r_i - x_{ij}^*)$. Afterwards, maximum value of the deviation for each alternative respecting all attributes can be calculated as:

$$z_i^* = \max_j \left| r_j - x_{ij}^* \right|. \quad (6)$$

To reach the optimal alternative based on the reference point approach, the minimum value of Eq. (6) among all alternatives should be found. The optimal alternative of the reference point approach can be calculated as (Datta et al. 2013):

$$A_{\mathrm{RP}}^* = \left\{ A_i \middle| \min_i z_i^* \right\}. \quad (7)$$

The full multiplicative form

Brauers and Zavadskas (2010) developed the full multiplicative form as the third part of the MULTIMOORA method. The formula of the method can be determined as follows:

$$U_i' = \frac{\prod_{j=1}^{g} x_{ij}}{\prod_{j=g+1}^{n} x_{ij}}, \quad (8)$$

in which g is defined similarly as aforementioned for the ratio system. The numerator of Eq. (8) indicates the product of performance indices of ith alternative relating to beneficial attributes. The denominator of Eq. (8) represents the product of performance indices of ith alternative relating to non-beneficial attributes.

Using the normalized decision matrix, an equivalent form of U_i' can be established as:

$$U_i^* = \frac{\prod_{j=1}^{g} x_{ij}^*}{\prod_{j=g+1}^{n} x_{ij}^*}. \quad (9)$$

The assessment values of U_i^* differ from U_i'; however, the ranking calculated by both equations is analogous. Accordingly, to preserve a harmony between all parts of the MULTIMOORA method, we use Eq. (9) as the full multiplicative form representation.

Similar to the ratio system, an optimal alternative can be distinguished by searching for maximum among all assessment values of U_i^* as:

$$A_{\mathrm{MF}}^* = \left\{ A_i \middle| \max_i U_i^* \right\}. \quad (10)$$

The final ranking of the MULTIMOORA method based on the dominance theory

The dominance theory was employed as a tool for consolidation of subordinate rankings of the MULTIMOORA method (Brauers et al. 2011; Brauers and Zavadskas 2011, 2012). After the calculation of the subordinate ranks as above, they can be integrated into a final ranking, named the MULTIMOORA rank, based on the dominance theory. For a detailed explanation of the dominance theory, readers can refer to the study of Brauers and Zavadskas (2012).

Shannon entropy weight

Entropy concept has been widely employed in social and physical sciences. Economics, spectral analysis, and language modeling are a few typical practical applications of entropy. A mathematical theory of communication was proposed by Shannon (1948). Entropy evaluates the expected information content of a certain message. Entropy concept in information theory can be considered as a criterion for the degree of uncertainty represented by a discrete probability distribution.

Entropy idea can be effectively employed in the process of decision making, because it measures existent contrasts between sets of data and clarifies the average intrinsic information transferred to decision maker.

To determine objective weight through Shannon entropy, the following procedure should be adopted (Hwang and Yoon 1981):

Step 1 Normalization of the arrays of decision matrix (performance indices) to obtain the project outcomes p_{ij}:

$$p_{ij} = \frac{x_{ij}}{\sum_{i=1}^{m} x_{ij}} \tag{11}$$

Step 2 Computation of the entropy measure of project outcomes using the following equation:

$$E_j = -k \sum_{i=1}^{m} p_{ij} \ln p_{ij}, \tag{12}$$

in which $k = 1/\ln(m)$.

Step 3 Defining the objective weight based on the entropy concept:

$$w_j = \frac{1 - E_j}{\sum_{j=1}^{n} (1 - E_j)} \tag{13}$$

Step 4 Calculating the general form of the entropy weight, if the decision maker assigns subjective weight s_j. By considering s_j, Eq. (13) transforms into the following:

$$w_j^* = \frac{s_j w_j}{\sum_{j=1}^{n} s_j w_j}, \tag{14}$$

in which subjective and objective weights (s_j and w_j) are combined to produce the general form of Shannon entropy weight w_j^*.

The extended MULTIMOORA method based on Shannon entropy weight

In the initial paper on the MOORA method, Brauers and Zavadskas (2006) allocated a section for the importance given to an attribute. They mentioned that a significance coefficient can be considered to affix more importance to a specific attribute. Their weighted form of the MOORA method confines to general representation of the main formulas and no details concerning characteristics of significance coefficient have been cited. This concept was later updated to encompass all subsections of MULTIMOORA method (Brauers and Zavadskas 2010, 2011).

Significance coefficient can be subjective weight gained directly from the decision makers similar to the routine procedure of the majority of MADM methods. The coefficient can also be regarded as an objective factor like Shannon entropy weight. The inclusive significance coefficient is the combination of subjective and objective factors like the general Shannon entropy weight.

In the present paper, we designate two forms of weight as significance coefficient of attributes. If significance coefficient only consists of subjective weight s_j earned from the decision makers, the resultant approach is named the weighted MULTIMOORA method. Application of the general Shannon entropy weight that is a combined subjective and objective significance coefficient leads to the so-called entropy-weighted MULTIMOORA technique. Based on Shannon entropy weight and the original MULTIMOORA approach, the following methodology is attained.

The extended ratio system

Significance coefficient or importance weight of attributes can be added to the ratio system. As mentioned above, the two forms of weighting were considered in this paper. By considering Eq. (3), the extended ratio system can have two parts, as follows:

$$y_i^w = \sum_{j=1}^{g} s_j x_{ij}^* - \sum_{j=g+1}^{n} s_j x_{ij}^*, \tag{15}$$

$$y_i^{ew} = \sum_{j=1}^{g} w_j^* x_{ij}^* - \sum_{j=g+1}^{n} w_j^* x_{ij}^*. \tag{16}$$

y_i^w and y_i^{ew} represent the assessment values of ith alternative regarding all attributes for the weighted and entropy-weighted ratio systems, respectively. The resultant optimal alternatives based on these techniques can be identified as follows:

$$A_{\mathrm{WRS}}^* = \left\{ A_i \middle| \max_i y_i^w \right\}. \tag{17}$$

$$A_{\mathrm{EWRS}}^* = \left\{ A_i \middle| \max_i y_i^{ew} \right\}. \tag{18}$$

The extended reference point approach

The reference point approach can also be developed using subjective and the general Shannon entropy weights:

$$z_i^w = \max_j \left| s_j r_j - s_j x_{ij}^* \right|, \tag{19}$$

$$z_i^{ew} = \max_j \left| w_j^* r_j - w_j^* x_{ij}^* \right|. \tag{20}$$

Then, alternatives can be listed in ascending order based on the assessment values of Eqs. (19) and (20) to find the optimal alternatives of the weighted and entropy-weighted reference point approaches, respectively, as:

$$A_{\mathrm{WRP}}^* = \left\{ A_i \middle| \min_i z_i^w \right\}. \tag{21}$$

$$A_{\mathrm{EWRP}}^* = \left\{ A_i \middle| \min_i z_i^{ew} \right\}. \tag{22}$$

The extended full multiplicative form

Brauers and Zavadskas (2012) showed that considering weights as coefficients is meaningless for the full multiplicative form. Instead, the weights should be employed as exponents. The weighted and entropy-weighted full multiplicative forms can be formulated, respectively, as:

$$U_i^w = \frac{\prod_{j=1}^{g} (x_{ij}^*)^{s_j}}{\prod_{j=g+1}^{n} (x_{ij}^*)^{s_j}}, \tag{23}$$

$$U_i^{ew} = \frac{\prod_{j=1}^{g} (x_{ij}^*)^{w_j^*}}{\prod_{j=g+1}^{n} (x_{ij}^*)^{w_j^*}}, \tag{24}$$

The optimal alternatives based on the two techniques have the greatest assessment value:

$$A_{\mathrm{WMF}}^* = \left\{ A_i \middle| \max_i U_i^w \right\}, \tag{25}$$

$$A_{\mathrm{EWMF}}^* = \left\{ A_i \middle| \max_i U_i^{ew} \right\}. \tag{26}$$

The final ranking of the extended MULTIMOORA method based on the dominance theory

By utilizing the dominance theory, we integrated the subordinate rankings into a final ranking.

Application of the extended MULTIMOORA method in materials selection

Karande and Chakraborty (2012a) utilized the MOORA technique to choose materials for different applications. However, they altered the original normalization ratio of the method, i.e., Eq. (2), into another form. They used Voogd ratio (Voogd 1983) as normalization formula that is $x_{ij}^* = x_{ij} / \sum_{i=1}^{m} x_{ij}$.

Brauers and Zavadskas (2006) established that among different choices for the denominator of the normalization ratio, the square root of the sum of each alternative performance index, $[\sum_{i=1}^{m} x_{ij}^2]^{1/2}$, is the most robust option. Therefore, the results of the study of Karande and Chakraborty (2012a) may not be as robust as the original MULTIMOORA method. Thus, we do not verify our results with their outcomes.

In the following subsections, we calculated the weighted and entropy-weighted MULTIMOORA rankings for two material selection problems cited in the study of Karande and Chakraborty (2012a). Besides, we compared our results with the related studies on the field.

Example 1: Material selection for flywheel

The problem addresses materials selection for a flywheel. Other studies have solved this practical case using various methods (Chatterjee et al. 2009; Jahan et al. 2010a; Jee and Kang 2000). The main requirements in the design of a flywheel are to save the maximum amount of kinetic energy as well as to prevent fatigue and fracture. Stored kinetic energy per unit mass of a thin flywheel is as follows (Lewis 1990):

$$\frac{u}{m} = \frac{s k_s}{(1 - v)\rho}, \tag{27}$$

in which u, m, s, v, and ρ are kinetic energy, mass, failure strength, Poisson ratio, and density, respectively. k_s is a factor related to the extent of material anisotropy. Fatigue strength σ_f can be considered as failure strength s for a flywheel. By ignoring effects of the values of v and k_s, a general relation obtains from Eq. (27). That is, if σ_f/ρ increases, u/m will be greater. Thus, the first attribute is specific strength σ_f/ρ. Waterman and Ashby (1991) showed that the criterion for minimization of the disc weight is s/ρ. Therefore, σ_f/ρ can concurrently be a measure for fatigue strength, kinetic energy maximization, and weight minimization. The fracture strength can be represented by fracture toughness K_{IC}. Thus, to minimize the probability of brittle fracture, K_{IC}/ρ is taken as the second attribute. The third important index can be price per unit mass. Fragmentability is an essential feature of a given flywheel that ensures safety. Hence, fragmentability is regarded as the last attribute. Only price per unit mass attribute is non-beneficial and the rest of the attributes are beneficial. Ten candidate materials for the engineering materials selection problem and their properties are gathered in Table 1. Table 2 shows the decision matrix for the problem. The performance indices can be normalized using Eq. (2) as displayed in Table 3.

E_j and w_j were calculated using Eqs. (12) and (13), respectively, as shown in Table 4. Four different sets of subjective weight s_j exist for the flywheel problem in the study of Jee and Kang (2000). In the present paper, we considered case one for the subjective weight. By applying the subjective and objective weighting factors, the general Shannon entropy weight w_j^* was obtained according to Eq. (14) as listed in Table 4.

The assessment values of the weighted and entropy-weighted MULTIMOORA methods and their resultant rankings are shown in Tables 5 and 6, respectively. Assessment values presented in Tables 5 and 6 are related to the three parts of the weighted and entropy-weighted MULTIMOORA approaches that can be obtained using Eqs. (15), (19), and (23) besides Eqs. (16), (20), and (24), respectively. In Tables 5 and 6, the rankings for the first and third parts were calculated based on descending order. In contrast, the assessment values for the second part of the proposed method that is the reference point approach were arranged in ascending order. The last columns were allocated to the final ranks determined based on the dominance theory (Brauers and Zavadskas 2012). The optimal material can be found using the related A^* equations. From the assessment values of Tables 5 and 6, $A^*_{\mathrm{WRS}} = A^*_{\mathrm{EWRS}} = A^*_{\mathrm{WRP}} = A^*_{\mathrm{EWRP}} =$ Kevlar 49–epoxy FRP and $A^*_{\mathrm{WMF}} = A^*_{\mathrm{EWMF}} =$ S glass–epoxy FRP. Final ranking has more importance because it is the integrated form of subordinate ranks. Kevlar 49–epoxy FRP, Carbon–epoxy FRP, Kevlar 29–epoxy FRP, and S glass–epoxy FRP obtain the first to third positions, respectively, in the final rankings of the both weighted and entropy-weighted MULTIMOORA methods.

Table 7 shows the final ranks of the proposed model and other approaches for the flywheel materials selection problem. The optimal material in all the methods is similar that is Kevlar 49–epoxy FRP. However, similarity or contrast may exist between our materials ranks and the others.

To show an association between the materials ranks of our methods and other approaches listed in Table 7, we utilized Spearman rank correlation coefficient. Figure 1 illustrates Spearman coefficients for Example 1. By considering the coefficients related to the weighted MULTIMOORA, because of considering entropy concept for weight calculation, the TOPSIS method (Jee and Kang 2000) has the lowest value 0.76. Because other techniques exploited subjective weights, they show more concordance with the proposed weighted MULTIMOORA results. The ELECTRE approach (Chatterjee et al. 2009) outranks with

Table 1 Candidate materials and their properties for Example 1 (Jee and Kang 2000)

Materials	Properties				
	Fatigue strength (Mpa)	Fracture toughness (Mpa·m$^{1/2}$)	Density (g/cm^3)	Price/mass (10^3 US$/t)	Fragmentability
300 M	800	68.9	8	4.2	3 (poor)
2024-T3	140	38	2.82	2.1	3 (poor)
7050-T73651	220	35.4	2.82	2.1	3 (poor)
Ti–6Al–4V	515	123	5	10.5	3 (poor)
E glass–epoxy FRP	140	20	2	2.735	9 (excellent)
S glass–epoxy FRP	330	50	2	4.095	9 (excellent)
Carbon–epoxy FRP	700	35	2	35.47	7 (fairly good)
Kevlar 29–epoxy FRP	340	40	1	11	7 (fairly good)
Kevlar 49–epoxy FRP	900	50	1	25	7 (fairly good)
Boron–epoxy FRP	1000	46	2	315	5 (good)

Table 2 Decision matrix for Example 1 (Jee and Kang 2000)

Materials	Beneficial and non-beneficial attributes			
	MAX Specific strength (kN·m/kg)	MAX Specific toughness (kPa·$m^{1/2}$/kg/m^3)	MIN Price/mass (10^3 US$/t)	MAX Fragmentability
300 M	100	8.613	4.2	3
2024-T3	49.645	13.475	2.1	3
7050-T73651	78.014	12.553	2.1	3
Ti–6Al–4V	108.879	26.004	10.5	3
E glass–epoxy FRP	70	10	2.735	9
S glass–epoxy FRP	165	25	4.095	9
Carbon–epoxy FRP	440.252	22.013	35.47	7
Kevlar 29–epoxy FRP	242.857	28.571	11	7
Kevlar 49–epoxy FRP	616.438	34.247	25	7
Boron–epoxy FRP	500	23	315	5

Table 3 Normalized decision matrix for Example 1

Materials	Beneficial and non-beneficial attributes			
	MAX Specific strength	MAX Specific toughness	MIN Price/mass	MAX Fragmentability
300 M	0.103	0.124	0.013	0.156
2024-T3	0.051	0.194	0.007	0.156
7050-T73651	0.080	0.181	0.007	0.156
Ti–6Al–4V	0.112	0.375	0.033	0.156
E glass–epoxy FRP	0.072	0.144	0.009	0.468
S glass–epoxy FRP	0.170	0.360	0.013	0.468
Carbon–epoxy FRP	0.453	0.317	0.111	0.364
Kevlar 29–epoxy FRP	0.250	0.412	0.035	0.364
Kevlar 49–epoxy FRP	0.634	0.493	0.079	0.364
Boron–epoxy FRP	0.514	0.331	0.989	0.260

Table 4 Entropy measure and weighting factors for Example 1 (Jee and Kang 2000)

Entropy and weights	Beneficial and non-beneficial attributes			
	MAX Specific strength	MAX Specific toughness	MIN Price/mass	MAX Fragmentability
s_j	0.4	0.3	0.2	0.1
E_j	0.861	0.963	0.415	0.960
w_j	0.174	0.047	0.730	0.050
w_j^*	0.296	0.060	0.623	0.021

the value 0.95. The ranking of the ELECTRE approach (Chatterjee et al. 2009) is almost similar to our outcome that can be observed in Table 7, as well. In the entropy-weighted MULTIMOORA method category, the highest 0.89 is for the TOPSIS method (Jee and Kang 2000). The reason is that among the four studies, only Jee and Kang (2000) considered entropy weight in the formulation of their method. In this category, the linear assignment method (Jahan et al. 2010a), by 0.53, has the lowest agreement with the results of the present paper.

Example 2: Material selection for cryogenic storage tank

We considered materials selection problem of a cryogenic pressure vessel for storing liquid nitrogen as the second example. The material of a cryogenic storage tank should be adequately strong and stiff. Moreover, weldability and processability of the vessel must be high. The other important properties for a pressure vessel or storage tank are density, specific heat, thermal expansion coefficient,

Table 5 Assessment values and rankings of the weighted MULTIMOORA method for Example 1

Materials	Assessment values			Rankings			
	y_i^w	z_i^w	U_i^w	y_i^w Rank	z_i^w Rank	U_i^w Rank	Final Rank
300 M	0.091	0.212	0.425	10	7	9	9
2024-T3	0.093	0.233	0.422	9	10	10	10
7050-T73651	0.101	0.222	0.495	8	8	6	7
Ti–6Al–4V	0.166	0.209	0.510	5	6	5	5
E glass–epoxy FRP	0.117	0.225	0.469	7	9	8	8
S glass–epoxy FRP	0.220	0.186	0.802	4	4	2	4
Carbon–epoxy FRP	0.290	0.072	0.724	2	2	4	2
Kevlar 29–epoxy FRP	0.253	0.154	0.779	3	3	3	3
Kevlar 49–epoxy FRP	0.422	0.014	1.014	1	1	1	1
Boron–epoxy FRP	0.133	0.197	0.482	6	5	7	6

Table 6 Assessment values and rankings of the entropy-weighted MULTIMOORA method for Example 1

Materials	Assessment values			Rankings			
	y_i^{ew}	z_i^{ew}	U_i^{ew}	y_i^{ew} Rank	z_i^{ew} Rank	U_i^{ew} Rank	Final Rank
300 M	0.033	0.157	6.404	8	6	5	8
2024-T3	0.026	0.173	8.230	9	9	3	9
7050-T73651	0.034	0.164	9.370	7	7	1	6
Ti–6Al–4V	0.038	0.155	3.965	5	5	8	5
E glass–epoxy FRP	0.035	0.167	7.774	6	8	4	7
S glass–epoxy FRP	0.074	0.138	8.233	4	4	2	4
Carbon–epoxy FRP	0.092	0.065	2.834	2	2	9	2
Kevlar 29–epoxy FRP	0.085	0.114	5.003	3	3	6	3
Kevlar 49–epoxy FRP	0.176	0.045	3.997	1	1	7	1
Boron–epoxy FRP	−0.438	0.612	0.752	10	10	10	10

Table 7 Comparison between the materials ranks of the proposed model and other methods for Example 1

Materials	Methods					
	Weighted MULTIMOORA	Entropy-weighted MULTIMOORA	TOPSIS (Jee and Kang 2000)	ELECTRE (Chatterjee et al. 2009)	VIKOR (Chatterjee et al. 2009)	Linear assignment (Jahan et al. 2010a)
300 M	9	8	5	10	9	7
2024-T3	10	9	9	9	10	10
7050-T73651	7	6	7	8	8	8
Ti–6Al–4V	5	5	6	6	6	6
E glass–epoxy FRP	8	7	8	7	7	9
S glass–epoxy FRP	4	4	3	3	5	5
Carbon–epoxy FRP	2	2	4	2	2	3
Kevlar 29–epoxy FRP	3	3	2	4	4	4
Kevlar 49–epoxy FRP	1	1	1	1	1	1
Boron–epoxy FRP	6	10	10	5	3	2

thermal conductivity, and sufficient toughness at the operating temperature (Manshadi et al. 2007). The decision matrix of Example 2 consists of seven engineering materials and their properties as displayed in Table 8. The beneficial attributes are toughness index, yield strength, and elastic modulus, whereas density, thermal expansion

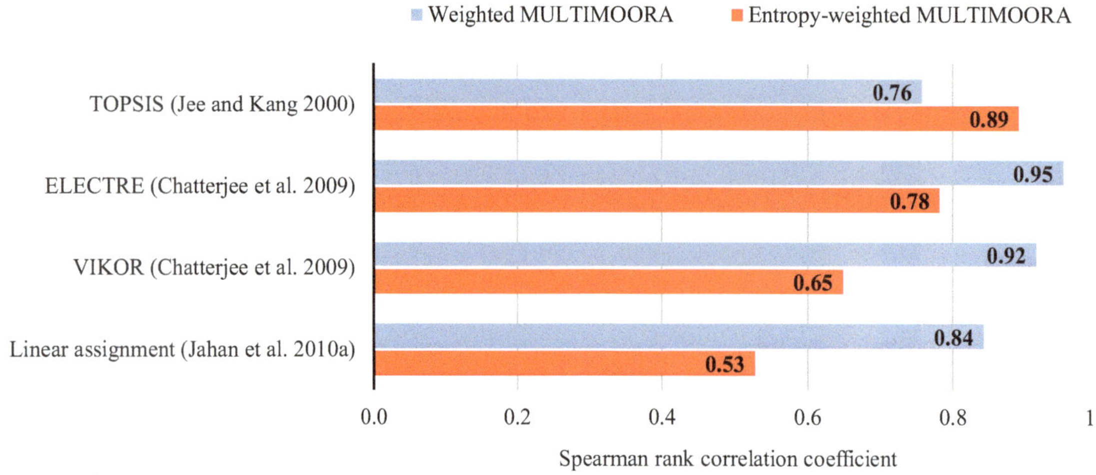

Fig. 1 Correlation between the materials ranks of the proposed model and other methods for Example 1

Table 8 Decision matrix for Example 2 (Manshadi et al. 2007)

Materials	Beneficial and non-beneficial attributes						
	MAX Toughness index (MPa)	MAX Yield strength (MPa)	MAX Elastic modulus (GPa)	MIN Density (g/cm^3)	MIN Thermal expansion (10^{-6}/°C)	MIN Thermal conductivity (cal/s/cm/°C)	MIN Specific heat (cal/g/°C)
Al 2024-T6	75.5	420	74.2	2.80	21.4	0.370	0.16
Al 5052-O	95	91	70	2.68	22.1	0.330	0.16
SS 301-FH	770	1365	189	7.90	16.9	0.040	0.08
SS 310-3AH	187	1120	210	7.90	14.4	0.030	0.08
Ti–6Al–4V	179	875	112	4.43	9.4	0.016	0.09
Inconel 718	239	1190	217	8.51	11.5	0.310	0.07
70Cu–30Zn	273	200	112	8.53	19.9	0.290	0.06

coefficient, thermal conductivity, and specific heat are the non-beneficial attributes. The arrays of the decision matrix were normalized as revealed in Table 9.

Table 10 indicates the values of entropy measure and weights for Example 2. Subjective weights are allocated based on the study of Manshadi et al. (2007). The last row belongs to the general entropy weight w_j^*.

Tables 11 and 12 exhibit the assessment values related to two scenarios of the proposed model and their resultant rankings for Example 2. The final ranks obtained for the materials selection problem of the nitrogen storage tank are presented in the end columns of Tables 11 and 12. Comparison of the subordinate and final ranks reveals that the optimal material A^* is identical ($A^*_{\text{WMF}} = A^*_{\text{EWMF}} =$ SS 301-FH). Tables 11 and 12 show a nearly identical final ranking except for SS 310-3AH and Inconel 718 that have different standings in the weighted and entropy-weighted MULTIMOORA scenarios. Al 5052-O is the worst option for selection in both scenarios.

The final ranks of the weighted and entropy-weighted MULTIMOORA methods for Example 2 were compared with those of the related studies in Table 13. The best material is SS 301-FH in all approaches. The weighted MULTIMOORA ranking is exactly similar to the fuzzy logic (Khabbaz et al. 2009) and the Z-transformation (Fayazbakhsh et al. 2009) rank lists. The GTMA (Rao 2006) ranking shows a direct correspondence with the entropy-weighted MULTIMOORA method.

Figure 2 demonstrates Spearman rank correlation coefficients for Example 2. In the weighted MULTIMOORA method category, the fuzzy logic (Khabbaz et al. 2009) and the Z-transformation (Fayazbakhsh et al. 2009), by Spearman coefficient 1, are exactly correspondent with our proposed approach. The AHP-TOPSIS (Rao and Davim 2008) and the WPM (Manshadi et al. 2007) have the lowest Spearman coefficient value that is 0.89. In the entropy-weighted MULTIMOORA method category, the best value, 1, is for the GTMA (Rao 2006). In this group, the

Table 9 Normalized decision matrix for Example 2

Materials	MAX Toughness index	MAX Yield strength	MAX Elastic modulus	MIN Density	MIN Thermal expansion	MIN Thermal conductivity	MIN Specific heat
Al 2024-T6	0.084	0.179	0.184	0.160	0.472	0.565	0.564
Al 5052-O	0.106	0.039	0.174	0.154	0.487	0.504	0.564
SS 301-FH	0.858	0.581	0.469	0.453	0.373	0.061	0.282
SS 310-3AH	0.208	0.477	0.521	0.453	0.318	0.046	0.282
Ti–6Al–4V	0.199	0.372	0.278	0.254	0.207	0.024	0.317
Inconel 718	0.266	0.506	0.538	0.488	0.254	0.473	0.247
70Cu–30Zn	0.304	0.085	0.278	0.489	0.439	0.443	0.211

(Column group header: Beneficial and non-beneficial attributes)

Table 10 Entropy measure and weighting factors for Example 2

Entropy and weights	MAX Toughness index	MAX Yield strength	MAX Elastic modulus	MIN Density	MIN Thermal expansion	MIN Thermal conductivity	MIN Specific heat
s_j	0.28	0.14	0.05	0.24	0.19	0.05	0.05
E_j	0.855	0.879	0.954	0.953	0.979	0.819	0.964
w_j	0.404	0.338	0.127	0.132	0.058	0.505	0.101
w_j^*	0.571	0.238	0.032	0.159	0.055	0.127	0.026

(Column group header: Beneficial and non-beneficial attributes)

Table 11 Assessment values and rankings of the weighted MULTIMOORA method for Example 2

Materials	Assessment values y_i^w	z_i^w	U_i^w	Rankings y_i^w Rank	z_i^w Rank	U_i^w Rank	Final Rank
Al 2024-T6	−0.127	0.217	0.684	6	7	6	6
Al 5052-O	−0.139	0.210	0.593	7	6	7	7
SS 301-FH	0.148	0.072	1.528	1	1	1	1
SS 310-3AH	−0.034	0.182	1.051	4	4	3	4
Ti–6Al–4V	0.004	0.184	1.243	2	5	2	2
Inconel 718	−0.029	0.166	1.045	3	3	4	3
70Cu–30Zn	−0.122	0.155	0.744	5	2	5	5

Table 12 Assessment values and rankings of the entropy-weighted MULTIMOORA method for Example 2

Materials	Assessment values y_i^{ew}	z_i^{ew}	U_i^{ew}	Rankings y_i^{ew} Rank	z_i^{ew} Rank	U_i^{ew} Rank	Final Rank
Al 2024-T6	−0.042	0.441	0.233	6	7	6	6
Al 5052-O	−0.055	0.429	0.188	7	6	7	7
SS 301-FH	0.535	0.048	1.388	1	1	1	1
SS 310-3AH	0.146	0.370	0.620	3	4	3	3
Ti–6Al–4V	0.148	0.376	0.677	2	5	2	2
Inconel 718	0.132	0.337	0.540	4	3	4	4
70Cu–30Zn	0.039	0.316	0.366	5	2	5	5

Spearman coefficient for the fuzzy logic (Khabbaz et al. 2009) and the Z-transformation (Fayazbakhsh et al. 2009) is 0.96. The AHP-TOPSIS (Rao and Davim 2008) and the WPM (Manshadi et al. 2007), by 0.86, have the lowest correlation with our results. From Fig. 2, it is found that the weighted MULTIMOORA rank is closer to the results of other studies than that of the entropy-weighted MULTIMOORA method. The reason is that except the GTMA (Rao 2006) and the fuzzy logic (Khabbaz et al. 2009), others have utilized nearly identical subjective weights in the derivation of their models. A novel method of assigning subjective weights was employed in the GTMA (Rao 2006). No weighting was considered in the fuzzy logic (Khabbaz et al. 2009).

Conclusion

In the present paper, we extended MULTIMOORA method using entropy weight based on the Shannon information theory to solve materials selection problem. The extended model has two scenarios called the weighted and entropy-

Table 13 Comparison between the materials ranks of the proposed model and other methods for Example 2

Materials	Methods							
	Weighted MULTIMOORA	Entropy-weighted MULTIMOORA	The method of Manshadi et al. (2007)	WPM (Manshadi et al. 2007)	GTMA (Rao 2006)	AHP-TOPSIS (Rao and Davim 2008)	Fuzzy logic (Khabbaz et al. 2009)	Z-transformation (Fayazbakhsh et al. 2009)
Al 2024-T6	6	6	5	5	6	5	6	6
Al 5052-O	7	7	7	6	7	6	7	7
SS 301-FH	1	1	1	1	1	1	1	1
SS 310-3AH	4	3	4	4	3	4	4	4
Ti–6Al–4V	2	2	2	2	2	2	2	2
Inconel 718	3	4	3	3	4	3	3	3
70Cu–30Zn	5	5	6	7	5	7	5	5

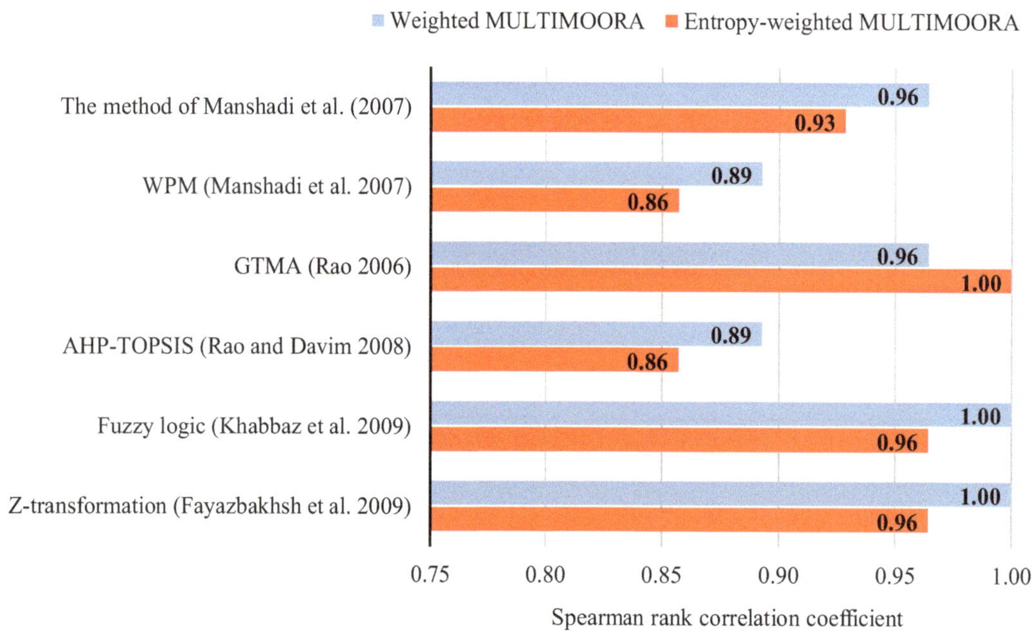

Fig. 2 Correlation between the materials ranks of the proposed model and other methods for Example 2

weighted MULTIMOORA methods. To attach relative importance to attributes, subjective weight was considered in the first scenario whereas the combined subjective and objective weights were used in the second scenario. Subjective weight is obtained straight from decision makers' comments based on their knowledge of materials and their experiences of the engineering design process. However, objective weight is calculated using entropy idea. The two forms of weighting factor can be integrated to produce the general form of Shannon entropy weight. Each of the two scenarios has three subordinate parts. To integrate the subordinate rankings, the dominance theory was exploited. Two practical materials selection examples were discussed to show the effect of the entropy weight on MULTIMOORA ranking. Moreover, the final rankings of the examples were compared with those of other methods.

The comparison between our final ranks and other studies demonstrates close correspondences, especially over the best rank or the optimal material. Spearman rank correlation coefficients obtained for the two examples show that the correlation between the ranks of the weighted MULTIMOORA method and the most of the earlier studies is higher than that of the entropy-weighted MULTIMOORA method. This fact is due to considering subjective weights in the models of the most of the references. Because of readily comprehensible mathematical derivation, the model based on MULTIMOORA method and the entropy concept gives an efficient means for decision making in the field of materials selection. Another strong point of our model is that our final rankings that were calculated by the consolidation of three subordinate ranks are more robust than those of other studies in which a single rank has been reported. The proposed model may have practical limitations in some real-world applications. The data of decision matrix may be presented as uncertain values. In this regard, new developments of the model are required based on fuzzy, interval, green, or other uncertain numbers dependent of the type of vagueness of the data. Moreover, our suggested methodology is to be developed for the case studies in which target-based attributes exist in the decision-making process, such as biomaterials selection problems. If a large number of alternatives and attributes exist in the decision matrix for a practical case, the manual calculation may be exhausting. Thus, the algorithm of this study can be computerized for such cases.

As future research, the extended MULTIMOORA approach can be considered for application in many case studies other than materials selection problem. For instance, decision making over the selection of optimal manufacturing process and the evaluation of failure modes risks can be done using the proposed model. In the field of materials selection, only two typical practical examples were presented in this paper. Other real-world materials selection problems with a number of various alternatives and attributes can be considered. The final rankings of the proposed model for the two examples were compared with a few approaches. The comparison of the present paper results with other MADM methods or expert systems seems to be interesting. As different extensions of the MULTIMOORA method, other concepts for assigning relative importance of attributes can be utilized.

References

Anojkumar L, Ilangkumaran M, Sasirekha V (2014) Comparative analysis of MCDM methods for pipe material selection in sugar industry. Expert Syst Appl 41:2964–2980

Athawale VM, Kumar R, Chakraborty S (2011) Decision making for material selection using the UTA method. Int J Adv Manuf Technol 57:11–22

Bakhoum ES, Brown DC (2013) A hybrid approach using AHP–TOPSIS–entropy methods for sustainable ranking of structural materials. Int J Sustain Eng 6:212–224

Brauers WKM, Ginevičius R (2010) The economy of the Belgian regions tested with MULTIMOORA. J Bus Econ Manag 2:173–209

Brauers WKM, Zavadskas EK (2006) The MOORA method and its application to privatization in a transition economy. Control Cybern 35:445–469

Brauers WKM, Zavadskas EK (2010) Project management by MULTIMOORA as an instrument for transition economies. Technol Econ Dev Econ 16:5–24

Brauers WKM, Zavadskas EK (2011) MULTIMOORA optimization used to decide on a bank loan to buy property. Technol Econ Dev Econ 17:174–188

Brauers WKM, Zavadskas EK (2012) Robustness of MULTIMOORA: a method for multi-objective optimization. Informatica 23:1–25

Brauers WKM, Baležentis A, Baležentis T (2011) MULTIMOORA for the EU Member States updated with fuzzy number theory. Technol Econ Dev Econ 17:259–290

Brauers WKM, Ginevicius R, Podvezko A (2014) Ranking of the Lithuanian Banks during the recession of 2008–2009 by the MULTIMOORA method. Ann Manag Sci 3:1

Chakraborty S (2011) Applications of the MOORA method for decision making in manufacturing environment. Int J Adv Manuf Technol 54:1155–1166. doi:10.1007/s00170-010-2972-0

Chan JWK, Tong TKL (2007) Multi-criteria material selections and end-of-life product strategy: grey relational analysis approach. Mater Des 28:1539–1546

Chatterjee P, Chakraborty S (2012) Material selection using preferential ranking methods. Mater Des 35:384–393

Chatterjee P, Athawale VM, Chakraborty S (2009) Selection of materials using compromise ranking and outranking methods. Mater Des 30:4043–4053

Chauhan A, Vaish R (2013) Pareto optimal microwave dielectric materials. Adv Sci Eng Med 5:149–155

Das D (2012) Selection of materials in engineering design using Ashby's chart and AHP-TOPSIS. Bachelor Dissertation, Jadavpur University

Datta S, Sahu N, Mahapatra S (2013) Robot selection based on grey-MULTIMOORA approach. Grey Syst Theory Appl 3:201–232

Deliktas D, Ustun O (2015) Student selection and assignment methodology based on fuzzy MULTIMOORA and multichoice goal programming. Int Trans Oper Res. doi:10.1111/itor.12185

Dweiri F, Al-Oqla FM (2006) Material selection using analytical hierarchy process. Int J Comput Appl Technol 26:182–189

El-Santawy MF (2014) An IEW-MOORA approach for solving MCDM problems. Comput Inf Syst 18:21–23

Farag MM (2002) Quantitative methods of materials selection. In: Kutz M (ed) Handbook of materials selection. Wiley, New York, pp 3–24

Farzamnia E, Babolghani MB (2014) Group decision-making process for supplier selection using MULTIMOORA technique under fuzzy environment. Kuwait Chapter Arab J Bus Manag Rev 3:203

Fayazbakhsh K, Abedian A (2010) Materials selection for applications in space environment considering outgassing phenomenon. Adv Space Res 45:741–749

Fayazbakhsh K, Abedian A, Manshadi BD, Khabbaz RS (2009) Introducing a novel method for materials selection in mechanical design using Z-transformation in statistics for normalization of material properties. Mater Des 30:4396–4404

Findik F, Turan K (2012) Materials selection for lighter wagon design with a weighted property index method. Mater Des 37:470–477
Huang H, Zhang L, Liu Z, Sutherland JW (2011) Multi-criteria decision making and uncertainty analysis for materials selection in environmentally conscious design. Int J Adv Manuf Technol 52:421–432
Hwang CL, Yoon K (1981) Multiple attribute decision making: methods and applications, a state-of-the-art survey. Springer, Berlin
Jahan A, Edwards KL (2013a) Multi-criteria decision analysis for supporting the selection of engineering materials in product design. Butterworth-Heinemann, Oxford
Jahan A, Edwards KL (2013b) VIKOR method for material selection problems with interval numbers and target-based criteria. Mater Des 47:759–765
Jahan A, Ismail MY, Mustapha F, Sapuan SM (2010a) Material selection based on ordinal data. Mater Des 31:3180–3187
Jahan A, Ismail MY, Sapuan SM, Mustapha F (2010b) Material screening and choosing methods—a review. Mater Des 31:696–705
Jee D-H, Kang K-J (2000) A method for optimal material selection aided with decision making theory. Mater Des 21:199–206
Jiao Q, Lan Y, Guan Z, Li Z (2011) A new material selection approach using PROMETHEE method. In: Proceeding of the 2011 international conference on electronic and mechanical engineering and information technology (EMEIT), IEEE, Heilongjiang, China, pp 2950–2954, 12–14 August 2011
Karande P, Chakraborty S (2012a) Application of multi-objective optimization on the basis of ratio analysis (MOORA) method for materials selection. Mater Des 37:317–324
Karande P, Chakraborty S (2012b) Decision making for supplier selection using the MOORA method. IUP J Oper Manag 11:6–18
Khabbaz RS, Manshadi BD, Abedian A, Mahmudi R (2009) A simplified fuzzy logic approach for materials selection in mechanical engineering design. Mater Des 30:687–697
Lewis G (1990) Selection of engineering materials. Prentice Hall, Englewood Cliffs
Liu H-C, Liu L, Wu J (2013) Material selection using an interval 2-tuple linguistic VIKOR method considering subjective and objective weights. Mater Des 52:158–167
Liu H-C, Fan X-J, Li P, Chen Y-Z (2014a) Evaluating the risk of failure modes with extended MULTIMOORA method under fuzzy environment. Eng Appl Artif Intell 34:168–177
Liu H-C, You J-X, Lu C, Shan M-M (2014b) Application of interval 2-tuple linguistic MULTIMOORA method for health-care waste treatment technology evaluation and selection. Waste Manag 34:2355–2364
Maity SR, Chatterjee P, Chakraborty S (2012) Cutting tool material selection using grey complex proportional assessment method. Mater Des 36:372–378
Maniya K, Bhatt MG (2010) A selection of material using a novel type decision-making method: preference selection index method. Mater Des 31:1785–1789
Manshadi BD, Mahmudi H, Abedian A, Mahmudi R (2007) A novel method for materials selection in mechanical design: combination of non-linear normalization and a modified digital logic method. Mater Des 28:8–15
Mayyas A, Shen Q, Mayyas A, Shan D, Qattawi A, Omar M (2011) Using quality function deployment and analytical hierarchy process for material selection of body-in-white. Mater Des 32:2771–2782
Mishra S, Sahu AK, Datta S, Mahapatra SS (2015) Application of fuzzy integrated MULTIMOORA method towards supplier/partner selection in agile supply chain. Int J Oper Res 22:466–514
Özçelik G, Aydogan EK, Gencer C (2014) A hybrid moora-fuzzy algorithm for special education and rehabilitation center selection. J Mil Inf Sci 2:53–62
Peng A-H, Xiao X-M (2013) Material selection using PROMETHEE combined with analytic network process under hybrid environment. Mater Des 47:643–652
Prasad K (2013) Selection of materials, machining centers and advanced machining processes using quality function deployment technique. Master dissertation, Jadavpur University
Prasad K, Chakraborty S (2013) A quality function deployment-based model for materials selection. Mater Des 49:525–535
Rached-Paoli G, Baunda M (2014) Analysis on selection of project by SDV-MOORA approach. Scientia Guaianae 5:6–9
Rao RV (2006) A material selection model using graph theory and matrix approach. Mater Sci Eng A 431:248–255
Rao RV, Davim JP (2008) A decision-making framework model for material selection using a combined multiple attribute decision-making method. Int J Adv Manuf Technol 35:751–760
Shanian A, Savadogo O (2009) A methodological concept for material selection of highly sensitive components based on multiple criteria decision analysis. Expert Syst Appl 36:1362–1370
Shannon CE (1948) A mathematical theory of communication. Bell Syst Tech J 27:379–423
Torrez JB, Milani AS, Schuetze DDC (2012) Material selection for light-weight naval crafts using the modified digital logic method: dealing with ranking abnormalities. Int J Mater Struct Integr 6:241–256
Voogd H (1983) Multicriteria evaluation for urban and regional planning. Pion Ltd., London
Waterman NA, Ashby F (1991) CRC-Elsevier materials selector, vol 2. CRC Press, Boca Raton
Zhao R, Neighbour G, Deutz P, McGuire M (2012) Materials selection for cleaner production: an environmental evaluation approach. Mater Des 37:429–434

3

A discrete particle swarm optimization algorithm with local search for a production-based two-echelon single-vendor multiple-buyer supply chain

Mehdi Seifbarghy[1] · Masoud Mirzaei Kalani[2] · Mojtaba Hemmati[2]

Abstract This paper formulates a two-echelon single-producer multi-buyer supply chain model, while a single product is produced and transported to the buyers by the producer. The producer and the buyers apply vendor-managed inventory mode of operation. It is assumed that the producer applies economic production quantity policy, which implies a constant production rate at the producer. The operational parameters of each buyer are sales quantity, sales price and production rate. Channel profit of the supply chain and contract price between the producer and each buyer is determined based on the values of the operational parameters. Since the model belongs to non-linear integer programs, we use a discrete particle swarm optimization algorithm (DPSO) to solve the addressed problem; however, the performance of the DPSO is compared utilizing two well-known heuristics, namely genetic algorithm and simulated annealing. A number of examples are provided to verify the model and assess the performance of the proposed heuristics. Experimental results indicate that DPSO outperforms the rival heuristics, with respect to some comparison metrics.

✉ Mehdi Seifbarghy
seifbar@yahoo.com; m.seifbarghy@alzahra.ac.ir

Masoud Mirzaei Kalani
masood.mirzaie@gmail.com

Mojtaba Hemmati
mojtaba_h86@yahoo.com

[1] Department of Industrial Engineering, Alzahra University, Tehran, Iran

[2] Faculty of Industrial and Mechanical Engineering, Qazvin Branch, Islamic Azad University, Qazvin, Iran

Keywords Vendor-managed inventory · Economic production quantity · Supply chain · Particle swarm optimization

Introduction

A supply chain consists of a number of organizations with materials, information and cash flows among them. Considering the first and last organizations as supplier and customer, respectively, the chain's objective is satisfying customer requirements with optimal operational cost. Vendor-managed inventory (VMI) as a modern IT-based partnership technique has been of great attention in recent years. In a VMI partnership, the supplier, usually the manufacturer but sometimes a reseller or distributor, makes the main inventory replenishment decisions for the consuming organization in such a way that the vendor monitors the buyer's inventory levels (physically or via electronic messaging) and makes periodic replenishment decisions.

This paper is an extension to Nachiappan and Jawahar (2007) in which a two-echelon single-producer multi-buyer supply chain (TSPMBSC) model while the vendor applies economic production quantity (EPQ) instead of economic order quantity (EOQ) is formulated. The producer and buyers apply VMI mode of operation. The production rate of the producer is assumed to be restricted. As an EOQ inventory control system, the production is done during a specific part of the replenishment cycle time. The buyers are assumed to employ the well-known EOQ inventory control system. The operational parameters are sales quantity, sales price and the production rate for each buyer which should be determined at the producer's location. Channel profit of the supply chain and contract price between the producer and buyers is determined based on the optimal values of the addressed operational parameters. A mathematical programming model is developed to find out the optimal values of the operational parameters. The model has a nonlinear objective function involving several integer variables and three different sets of linear constraints; it belongs to nonlinear integer programs (NIP). Considering Govindan (2013) which researches on VMI are classified into three categories of modeling, simulation, and case studies, we can conclude that this paper falls into the modeling category.

Costa and Oliveira (2001) addressed that the evolutionary strategies such as genetic algorithm (GA) and simulated annealing algorithm (SA) are emerging as the best algorithms for solving NIP problems. GA and SA could be useful for this NIP problem to provide near to optimal solutions.

The revenue sharing and the partnership among members of the supply chain are the major issues for the success of a supply chain. The net revenue is addressed as channel profit '*Pc*' which depends on sales quantity. Sales quantity is influenced by sales price (Waller et al. 2001). However, the formulations would show that it depends on both sales quantity and production rate while the vendor is also the producer. The relationship between sales quantity and sales price could be assumed to behave linearly (Lau and Lau 2003). It is generally believed that the pricing acceptable (fair) to the partners involved is an important factor to make constant relations in VMI, and that it requires acceptable revenue sharing that would satisfy both the vendor and the buyer (Grieger 2003). This reveals that the revenue sharing between the vendor and the buyer plays a vital role in determining the contract price.

The organization of the rest of the paper is as follows: Sect. 2 is on the literature review, problem description and modeling is given in Sect. 3. Section 4 gives the proposed heuristics to solve the problem. In Sect. 5, a number of numerical examples of different sizes are presented and solved to measure the accuracy of the proposed heuristics. Section 6 gives the research conclusions and ideas for further research.

Literature review

The literature on two-echelon supply chains is rich enough. Bhattacharjee and Ramesh (2000) developed two efficient heuristics to derive the optimal price and ordering policies to maximize the net profit of the retailer for a multi-period inventory and pricing model. Lu (1995) pointed out that future researches should consider the buyer's point of view and there should be a minimal acceptable profit level to both the vendor and buyers; this made a suitable base for the well-known concept which is revenue sharing. Maloni and Benton (1997) stated that the major focus of revenue sharing is to share the revenues/profits generated based on the assignments and responsibilities to avoid the conflict between supply chain partners. Yao and Chiou (2004) considered the single-vendor and multi-buyer model proposed by Lu (1995) and identified that the vendor's optimal annual cost function was a piecewise convex curve with respect to the vendors' production setup interval; they suggested that a search algorithm can be developed to obtain an optimal solution for a sub-problem. They also proposed a search algorithm and demonstrated that their algorithm reached a better result than Lu's search procedure.

Nachiappan and Jawahar (2007) formulated an integrated inventory model of a two-echelon single-vendor multiple buyers (TSVMBSC) under the VMI mode of operation to maximize the channel profit and to share the profit among the members involved assuming that both vendors and buyers follow EOQ conditions. The given model in this paper is an extension to the model given by Nachiappan and Jawahar (2007) assuming that the entrance rate of products to the vendor's location is bounded (i.e., EPQ conditions); in the new formulation, the optimal production rate for each buyer in the vendor's (i.e., producer's) location is determined as well as the optimal sales quantities and sales prices.

Zhang et al. (2007) presented an integrated VMI model for a single vendor and multiple buyers, where the vendor purchases and processes raw materials and then delivers finished items to the buyers. A joint relevant cost model is developed with constant production and demand rates under the assumption that buyers' ordering cycles may be different and that each buyer can replenish more than once in one production cycle. The main point of this research is that demand rate at all buyers is constant while it is determined as a function of the sales price in the current research. Yao et al. (2007) developed an analytical model that explores how important supply chain parameters affect the cost savings to be realized from collaborative initiatives as VMI. Van der Vlist et al. (2007) argue on the conclusions drawn from Yao et al. (2007). They express that the model ignores the costs of shipments from the supplier to the buyer and plans the incoming and outgoing flows at the supplier in a manner that overstates the inventory needed.

Toptal and Çetinkaya (2008) aimed to develop analytical and numerical results representing the system-wide cost improvement rates which are due to coordination. Revisiting a few basic researches, they consider generalized replenishment costs under centralized decision making. This research analyzes (1) how the counterpart centralized and decentralized solutions differ from each other, (2) under what circumstances their implications are similar, and (3) the effect of generalized replenishment costs of the system-wide cost improvement rates which are subject to coordination. Wang (2009) studied a decentralized supply chain consisting of a single manufacturer and a single distributor for a short lifecycle product with random yield and uncertain demand as in the semiconductor industry. Two scenarios for handling the business are considered. One scenario is the traditional supply chain arrangement, where the distributor is fully responsible for the inventory decision, whereas the manufacturer is fully responsible for the production decision. The other scenario is the VMI arrangement, where the manufacturer is fully responsible for the entire production and inventory decisions in the supply chain. The optimal production and inventory decisions under both scenarios are compared.

Yu et al. (2009a) discussed how the manufacturer (vendor) can take advantage of the information received from retailers for increasing his own profit using a Stackelberg game in a VMI system. The manufacturer produces a finished product and supplies it at the same wholesale price to multiple retailers. The retailers sell the product in independent markets at retail prices. Solution procedures are developed to find the Stackelberg game equilibrium which each enterprise is not interested in deviating from Yu et al. (2009b) investigated how a manufacturer and its retailers cooperate each other to find their individual optimal net profits considering product marketing (advertising and pricing) and inventory policies in an information-asymmetric VMI supply chain. The manufacturer produces and gives a single product at the same wholesale price to multiple retailers who sell the product in their independent markets at retail prices. The manufacturer determines its wholesale price, advertising investment, replenishment cycles for the raw materials and finished product, and backorder quantity to maximize the profit. Retailers in turn consider the replenishment policies and the manufacturer's promotion policies and determine the optimal retail prices and advertisement investments to maximize their profits.

Zavanella and Zanoni (2009) investigated the way how a particular VMI policy, known as Consignment Stock (CS), may represent a successful strategy for both the buyer and the supplier. The most radical application of CS may lead to the suppression of the vendor inventory, as this actor uses the buyer's store to stock its finished products. As a counterpart, the vendor will guarantee that the quantity stored in the buyer's store will be kept between a maximum and a minimum level, also supporting the additional costs eventually induced by stock-out conditions. The buyer will pick up from its store the quantity of material needed to meet its production plans and the material itself will be paid to the buyer according to the agreement signed. Wong et al. (2009) studied on how a sales rebate contract helped to achieve supply chain coordination. For this purpose, a model in the context of a two-echelon supply chain with a single supplier serving multiple retailers in VMI partnership is proposed. VMI facilitates the application of the sales rebate contract since information sharing in VMI partnership lets the supplier to obtain actual sales data in a timely manner and determine the rebate for retailers. The proposed model indicates that the supplier gains more profit with competing retailers than without as competition among the retailers lowers the prices and correspondingly increases demand. Bichescu and Fry (2009) analyzed decentralized supply chains, which followed continuous review (Q, R) inventory policies considering VMI agreements where the supplier chooses the order quantity Q, and the retailer chooses the reorder point R. The effect of divisions of channel power on supply chain and individual agent performance is investigated by examining different game theoretic models. The results showed that VMI can result in considerable supply chain savings rather than traditional relationships; furthermore, the greatest system benefits from VMI arise in asymmetric channel power relationships.

Almehdawe and Mantin (2010) consider a supply chain consisting of a single capacitated manufacturer and multiple retailers. A Stackelberg game VMI framework under two scenarios is utilized. Initially, the traditional approach wherein the manufacturer is the leader is considered; in the second, one of the retailers acts as the dominant player of the supply chain. Darwish and Odah (2010) developed a model for a supply chain with a single vendor and multiple retailers under the VMI mode of operation. The developed model can easily describe supply chains with capacity constraints by selecting high penalty cost. Theorems are given to tackle the complexity of the model. Furthermore, an efficient algorithm is devised to find the global optimal solution. Wang et al. (2010) investigate a recent paper by Yao et al. (2007) and a critique by Van der Vlist et al. (2007). Both researches presented interesting arguments to show their valuable findings. However, their finding on the buyer's order sizes seems to conflict with each other. Revisiting both papers, they come to the conclusion that both papers are valid within the scopes and assumptions of their own studies. Guan and Zhao (2010) considered a single-vendor and a single-buyer supply chain and study contracts for a VMI program. They design a revenue sharing contract for vendor with ownership scenario, and a franchising contract for retailer with ownership scenario. Based on continuous review (R, Q) policy, without consideration of order policy and related costs at the vendor site, it is indicated that one contract can perform satisfactorily while the other one is a perfect contract. Considering order policy and related costs at the vendor site, it is indicated that one contract can perform satisfactorily while the performance of the other one depends on the system parameters. Bookbinder et al. (2010) consider a vendor, which manufactures a single product sold to a retailer. Three scenarios are studied: independent decision making in which there is no agreement between the parties; VMI, whereby the vendor initiates orders on behalf of the retailer; and central decision making in which both vendor and retailer are controlled by the same corporate entity. Optimal solutions are obtained analytically for the retailer's order quantity, the vendor's production quantity, the parties' individual and total costs in the three scenarios. Those situations in which VMI is beneficial are recognized. Razmi et al. (2010) considered a buyer–supplier supply chain and compared the performance of the traditional and VMI system using the total inventory cost of the supply chain as the performance measure. The concept of extent point is introduced in which the difference between the total cost of both traditional and VMI systems is minimal. It is applied to investigate how increasing or reducing the key parameters changes the total cost of the two systems with respect to each other. Goh and Ponnambalam (2010) proposed a mathematical model to determine the optimal sales quantity, optima sales price, optimal channel profit and contract price between the vendor and buyer in TSVMBSC under the VMI mode of operation. All the parameters depend on the understanding of the revenue sharing between the vendor and buyers. A particle swarm optimization (PSO) was proposed to solve the problem. The solutions obtained from PSO were compared with the previous results reported in the literature. Pasandideh et al. (2010) developed a model for a two-level supply chain consisting of a single supplier and a single retailer studying the inventory management practices before and after implementation of VMI. This research explores the effect of important supply chain parameters on the cost savings realized from collaborative initiatives. The results indicate that the VMI implementation of EOQ model when unsatisfied demand is backlogged sometimes has the ability to reduce total costs of supply chains.

Pasandideh et al. (2011) developed an EOQ model for a two-level supply chain consisting of one supplier and one retailer in which unsatisfied demands are backordered, the supplier's storage is constrained and there is an upper bound on the number of orders. They assume that the supplier utilizes the retailer's information in decision making on the replenishments and supplies orders to the retailer according to (R, Q) policy. A GA is proposed to find the order quantities and the maximum backorder levels, so that the total inventory cost of the supply chain is minimized. Shao et al. (2011) studied inventory and pricing policies in a non-cooperative supply chain with one supplier and several retailers under an information-asymmetric VMI environment. The supplier produces a product at the wholesale price and gives to the retailers. The retailers distribute the product in markets at retail selling prices. The demand rate for each independent market is a non-decreasing concave function of the marketing expenditures of both local retailers and the manufacturer, but a non-increasing and convex function of the retail selling prices. Wholesale price, marketing expenditure for supplier and retailers, replenishment cycles for the product and backorder quantity are determined in such a way as to maximize the total profit. Sana et al. (2011) present an integrated production-inventory model that is presented for supplier, manufacturer and retailer supply chain, considering perfect and imperfect quality items. This model considers the impact of business strategies such as the optimal order size of raw materials, production rate and unit production cost, and idle times in different sectors on the collaborating marketing system. An analytical method is employed to optimize the production rate and raw material order size for maximum expected average profit. An example is illustrated to study the behavior and application of the model. Pal et al. (2012a, b) present a production inventory model for various types of items where multiple suppliers, a manufacturer and the multiple non-competing retailers are the members of the supply chain. And each supplier supplies only one type of raw material to the manufacturer. The manufacturer produces a finished item by the combination of a certain percentage of the various types of raw materials. The manufacturer produces also multi-items and delivers them according to the demand of the different retailers. Finally, an integrated profit of the supply chain is optimized by optimal ordering lot sizes of the raw materials. A numerical example is provided to justify the proposed model.

Pal et al. (2012a, b) develop a multi-echelon supply chain model for multiple markets with different selling seasons. Here, two suppliers are involved to supply the raw materials to the manufacturer where the main supplier may face supply disruption after a random time and the secondary supplier is perfectly reliable but more expensive than the main supplier. In their article, the manufacturer produces a random proportion of defective items which are reworked after regular production and are sold in a lot to another market just after completion of rework. The retailer sells the finished products in different markets according to seasons. Finally, an integrated expected cost per unit product of the chain is minimized analytically by considering the lot-size ordered as a decision variable. An appropriate numerical example is also provided to justify the proposed model. Goh et al. (2012) solved TSVMBSC model proposed by Nachiappan and Jawahar (2007) utilizing PSO and a hybrid of GA and artificial immune system (GA–AIS). These two algorithms are evaluated for their solution quality in the addressed research. Cárdenas-Barrón et al. (2012)

presented an alternative heuristic algorithm to solve the vendor management inventory system with multi-product and multi-constraint based on an EOQ model with authorized stock out. Stock-out cost is considered linear and fixed. Since the problem is a nonlinear integer programming, a heuristic algorithm is proposed to solve the problem.

Sadeghi et al. (2013) studied a multi-vendor multi-retailer single-warehouse supply chain under the VMI mode of operation with constrained space and annual number of orders for the warehouse. The objective was to find the order quantities along with the number of shipments received by retailers and vendors in such a way as to minimize the total inventory cost. Nia et al. (2013) developed a multi-product EOQ model under a VMI policy in a single-vendor single-buyer supply chain. Unsatisfied demands are backordered. A few constraints such as storage capacity, number of deliveries and order quantity are considered in the given model. Demand, available storage and total order quantity are considered as fuzzy numbers. An ant colony optimization algorithm along with GA is utilized to find a near-optimum solution. AriaNezhad et al. (2013) attempts to develop the retailer's inventory model with the effect of order cancellations during the advance sales period. The retailer announces a price discount program during advance sales period to promote his sales and also offers trade credit financing during the sales periods. The retailer availing trade credit period from his supplier offers a permissible delay period to his customers. The customer who gets an item is allowed to pay on or before the permissible delay period which is accounted from the buying time rather than from the start period of inventory sales. This accounts for significant changes in the calculations of interest payable and interest earned by the retailer. The retailer's total cost is minimized so as to find out the optimal replenishment cycle time and price discount policies through a solution procedure. The results derived in mathematical theorems are implemented in numerical examples, and sensitivity analyses on several inventory parameters are obtained.

Diabat (2014) considered a two-echelon single-vendor multi-buyer supply chain network operated under VMI policy and found the optimal sales. Hybrid genetic/simulated annealing algorithm is developed to deal with the problem. Rad et al. (2014) considered a two-echelon supply chain consisting of a single vendor and two buyers. The vendor gives a single product to both buyers at a finite production rate. A mathematical model for the integrated VMI policy is developed. Furthermore, solution algorithms are proposed to determine the optimal lot size and total inventory cost of the supply chain. The effect of key parameters such as buyer's demand and vendor's holding cost on lot size variation is also studied. Results show that greater reduction in the total cost of the supply chain can be obtained using VMI. Verma et al. (2014) proposed an alternative replenishment scheme allowing for different replenishment cycles for each retailer in the single-vendor multi-retailer supply chain under VMI partnership. Taleizadeh and Noori-daryan (2014) considered a decentralized three-layer supply chain including a supplier, a producer and arbitrary number of retailers. Retailers order from the producer who is replenished by the supplier. Demand is assumed to be price sensitive. The paper optimizes the total cost of the supply chain network integrating decision-making policy using Stackelberg–Nash equilibrium. The decision variables of the model are the supplier's price, the producer's price and the number of shipments received by the supplier and the producer. Pasandideh et al. (2014a) studied single-vendor single-buyer supply chain system under VMI working condition. The multiproduct EPQ model considering backordering subject to the constraints of storage capacity, number of orders, and available budget was considered. The near optimal order quantities along with the maximum backorder levels of the products in a cycle are determined so that the total VMI inventory cost is minimized. A GA-based heuristic is proposed to solve the problem. Pasandideh et al. (2014b) present an integrated vendor-managed inventory model for a two-echelon supply chain organized as a single capacitated manufacturer at the first echelon and multiple retailers at the second echelon. Manufacturer produces different products whose demands are assumed decreasing functions of retail prices. A fair profit contract is designed for the manufacturer and the retailers and the problem is formulated into a bi-objective non-linear mathematical model. The lexicographic max–min approach is utilized to obtain a fair non-dominated solution.

Sana (2014) develops a production-inventory model of a two-stage supply chain consisting of one manufacturer and one retailer to study production lot size/order quantity, reorder point sales teams' initiatives where the demand of the end customers is dependent on random variable and sales teams' initiatives simultaneously. The manufacturer produces the order quantity of the retailer at one lot in which the procurement cost per unit quantity follows a realistic convex function of production lot size. In the chain, the cost of sales team's initiatives/promotion efforts and wholesale price of the manufacturer are negotiated at the points such that their optimum profits reached nearer to their target profits. This study suggests to the management of firms to determine the optimal order quantity/production quantity, reorder point and sales teams' initiatives/promotional effort to achieve their maximum profits. An analytical method is applied to determine the optimal values of the decision variables. Finally, numerical examples with its graphical presentation and sensitivity analysis of the key parameters are presented to illustrate more insights of the model. In Sana et al. (2014), the replenishment size/production lot size problem both for perfect and imperfect quality products studied in their paper is motivated by the optimal strategy in a three-layer supply chain consisting of multiple suppliers, manufacturers and retailers. And each manufacturer produces each product with a combination of several raw materials which are supplied by each supplier. The defective products at suppliers and manufacturers are sent back to the respective upstream members at lower price than the respective purchasing price. Finally, the expected average profits of suppliers, manufacturers and retailers are formulated by trading off setup costs, purchasing costs, screening costs, production costs, inventory costs and selling prices. The objective of this chain is to compare between the collaborating system and Stakelberg game structure so that the expected average profit of the chain is maximized. In a numerical illustration, the optimal solution of the collaborating system shows a better optimal solution than the approach by Stakelberg. Thangam (2014) in their paper attempts to develop the retailer's inventory model with the effect of order cancellations during the advance sales period. The retailer announces a price discount program during advance sales period to promote his sales and also offers trade credit financing during the sales periods. The retailer availing trade credit period from his supplier offers a permissible delay period to his customers. The customer who gets an item is allowed to pay on or before the permissible delay period which is accounted from the buying time rather than from the start period of inventory sales. This accounts for significant changes in the calculations of interest payable and interest earned by the retailer. The retailer's total cost is minimized so as to find out the optimal replenishment cycle time and price discount policies through a solution procedure. The results derived in mathematical theorems are implemented in numerical examples, and sensitivity analyses on several inventory parameters are obtained.

Notation and modeling

The major notations used in this paper are as follows:

n	Number of buyers
a_j	Intercept of the demand curve of buyer j
b_j	Slope of the demand curve of buyer j
Hb_j	Inventory holding cost of buyer j at the independent mode (without VMI implementation)
Hs	Inventory holding cost of the vendor (producer) at the independent mode
$H_{j\text{VMI}}$	Resultant inventory holding cost of the integrated system of vendor and buyer j
Sb_j	Ordering (setup) cost of buyer j at the independent mode
Ss	Setup cost of the vendor per order at the independent mode
$S_{j\text{VMI}}$	Continuously monitoring the stock status of buyer j in VMI mode
P	Total production rate of the vendor (producer)
θ_j	Flow cost per unit from producer to buyer j
υ_j	Transportation cost per unit delivered from vendor to buyer j
δ	Production cost per unit made by the vendor (producer)
PD_j	Production and distribution cost of products to buyer j
PR_j	Revenue share ratio between vendor and buyer j
Q_j	Replenishment quantity for each buyer j
W	Contract price between a vendor and a buyer
W_j	Contract price between vendor and buyer j
P_j	Production rate for buyer j at the vendor's location
y_j	Sales quantity of buyer j
$P(y_j)$	Sales price of the product by buyer j corresponding to sales quantity 'y_j'
$P(y)$	Sales price of the product
$y_{j_{\min}}$	Minimum expected sales quantity of buyer j
$y_{j_{\max}}$	Maximum expected sales quantity of buyer j

This paper investigates a TSPMBSC model operating under VMI mode.

Description of the demand curve and contract price

There are a lot of examples in practice in which each producer (vendor) has its own set of direct outlets (distribution centers/retailers addressed here as buyers). The major parameters of the corresponding models are: sales quantity 'y', the sales price at buyer's market '$P(y)$', the contract price between the vendor and the buyer 'W' and the production rate for each buyer at vendor location. The sales quantity of the product at each location is highly influenced by its sales price and it depends on the factors such as the necessity of the commodity, the purchasing power of the customers, and the nature of the product (being perishable or storable). The general observation is that the higher the sales price, the lower is the sales quantity and vice versa. The relation between '$P(y)$' and 'y' may be assumed to behave linearly and is given as (Nachiappan and Jawahar 2007):

$$P(y) = a - by \tag{1}$$

where a and b represent the intercept and slope of the demand curve, respectively, as indicated in Fig. 1. Besides, sales quantity lies between a specific range between $y_{j_{\min}}$ and $y_{j_{\max}}$ and the validity of the linear demand assumption

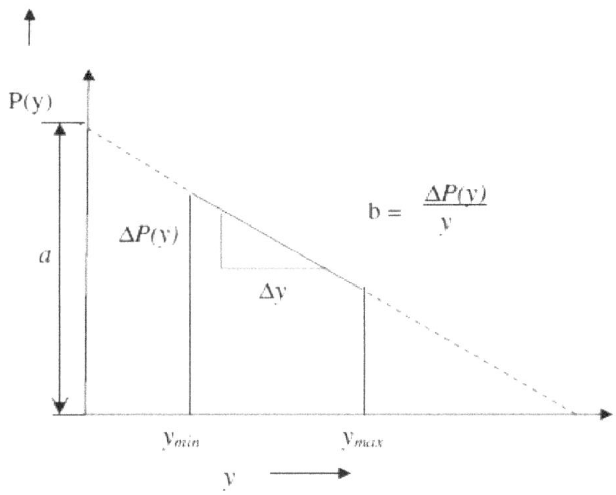

Fig. 1 Relation between the sales price and the sales quantity

function holds very well within this range. Since the buyers are not necessarily identical, the demand function of buyer j may be stated as in (2)–(3)

$$P(y_j) = a_j - b_j y_j \tag{2}$$

$$\text{s.t. } y_{j\min} \leq y_j \leq y_{j\max} \tag{3}$$

A parameter which plays an important role on the profits of the both vendor and buyer(s) is the contract price. It is a price which is mutually agreed between the vendor and the buyer(s). Usually, it is assumed a value between the cost of manufacturing and the sales price. The nature of the product, the demand and the logistic cost play a critical role on determination of the contract price value. The commodities which have a good reputation and higher demand are usually fast moving and are involved with lower risk; in these circumstances, the buyer accepts the contract price closer to sales price. However, in other cases where the product is new and the demand is not yet stabilized, the contract price is expected being settled at a lower level, closer to the cost of manufacturing. In Nachiappan and Jawahar (2007), the contract price is a variable which is dependent on location, the competitiveness of the products, the production and the operational costs between vendor and buyer(s). The contract price between vendor and buyer j is addressed by W_j.

Vendor operations and costs

Disney and Towill (2002) state that in VMI mode of cooperation among the members of a vendor–buyers chain, the vendor has more responsibility than the buyers and acts as a leader. The vendor monitors, manages and replenishes the inventory of all members (Achabal et al. 2000). The associated costs include production cost, distribution cost, order cost and stock holding cost. Production cost is derived from the expenses spent for producing a single unit 'δ' and the aggregate demand 'y' (i.e., $y = \sum_{j=1}^{n} y_j$). Therefore, the total production cost can be stated as δy. The distribution cost is the multiplication of flow and transportation resource cost. The flow cost is the direct mileage and the carrier contract cost per unit of buyer j 'θ_j' and the transportation resource cost is the indirect cost such as mode of transport, human router cost and administrative costs and termed as 'υ_j' per unit demand for the buyer j (Dong and Xu 2002). Therefore, the distribution cost can be stated as '$\theta_j y_j \upsilon_j y_j$'. In this paper, it is assumed that the products to all locations are delivered by road and the value of 'υ_j' is taken as 0.5 per unit as Dong and Xu (2002) consider. Therefore, the production and distribution costs 'PD_j' to the vendor for meeting sales 'y_j' of buyer j can be given by (4)

$$PD_j = \delta y_j + 0.5\theta_j y_j^2 \tag{4}$$

The vendor monitors the stock status and replenishes the stock. The buyer does not initiate orders. Therefore, the order cost per replenishment '$S_{j\mathrm{VMI}}$' associated with continuously monitoring the stock status is assumed as sum of the order cost of vendor 'Ss' and order cost of buyer j 'Sb_j' (Nachiappan and Jawahar 2007) and is given as in (5)

$$S_{j\mathrm{VMI}} = Ss + Sb_j \tag{5}$$

So, the cost involved with replenishing the batches 'Q_j' of demand of the buyer j 'y_j' can be stated as '$y_j(Ss + Sb_j)/Q_j$'. Nachiappan and Jawahar (2007) give this result in case where the vendor's production rate is infinite. However, this is also valid while the production rate is finite.

The inventory is held at both the vendor and the buyer(s) locations; the cost of holding one unit per unit time at vendor and buyer j locations can be represented by 'Hs' and 'Hb_j', respectively. The vendor accumulates inventory before delivery to buyer 'j'. As EPQ model, the vendor holds an average inventory of '$Q_j(1 - \frac{y_j}{P_j})/2$' to replenish buyer j. The inventory held at the vendor to replenish buyer j is given to the buyer. The average inventory at the buyer location turns out to be '$Q_j(1 - \frac{y_j}{P_j})/2$'; this is why the members use the VMI mode of cooperation. Therefore, in VMI mode, the cost of holding inventory '$H_{j\mathrm{VMI}}$' becomes the sum of the inventory holding cost at vendor and buyer (Nachiappan and Jawahar 2007); it can be given as in (6).

$$H_{j\mathrm{VMI}} = Hs + Hb_j \tag{6}$$

The sum of the order cost and average inventory holding cost of the vendor for buyer 'j' namely 'OSM_j', thus can be stated as in (7):

$$\mathrm{OSM}_j = (Ss + Sb_j)y_j/Q_j + Q_j(Hs + Hb_j)(1 - y_j/P_j)/2 \tag{7}$$

Since the vendor produces for each buyer, assuming a common cycle time 'T' for different buyers, the common cycle time can be indicated as $T = \frac{Q_j}{y_j}$, $\forall j = 1, \ldots, n$ (Silver et al. 1998). The sum of order and average inventory holding costs of the vendor for all buyers 'OSM' can be stated as in (8)

$$\text{OSM} = \sum_{j=1}^{n} [(Ss + Sb_j)/T + (Hs + Hb_j) \cdot T \cdot y_j(1 - y_j/P_j)/2], \quad (8)$$

where 'T' is computed as:

$$T = \sqrt{\frac{2\sum_{j=1}^{n}(Ss + Sb_j)}{\sum_{j=1}^{n} y_j(Hs + Hb_j)(1 - y_j/P_j)}} \quad (9)$$

The profit of the vendor when supplying the product to buyer j 'PV_j' can be obtained from the difference between revenue to the vendor (W_jy_j) and the total cost involved 'PD_j + OSM$_j$'. Therefore, the total profit to the vendor 'PV' by supplying its products to all the buyers can be obtained from (10)

$$PV = \sum_{j=1}^{n} \{W_jy_j - (\delta y_j + 0.5\theta_j y_j^2) - [(Ss + Sb_j)/T + (Hs + Hb_j) \cdot T \cdot y_j(1 - y_j/P_j)/2]\} \quad (10)$$

Buyer operations and costs

As Nachiappan and Jawahar (2007) declare, the costs associated with the buyers in VMI mode are the sales price and the contract price. The sales price for each buyer is determined using Eq. (2). The acceptable contract prices that would satisfy both the vendor and the buyer are derived from the revenue share ratio 'PR_j'. Thus, the profit of buyer j 'Pb_j' in VMI mode is equal to the difference between the sales revenue and the cost of purchase as in (11).

$$Pb_j = P(y_j)y_j - W_jy_j = (a_j - b_jy_j)y_j - W_jy_j \quad (11)$$

For a pre-specified value of revenue share ratio '$PR_j = PV_j/Pb_j$' between the vendor and buyer j, the contract price can be stated as in (12).

$$W_j = \frac{a_jy_jPR_j - b_jy_j^2PR_j + \delta y_j + 0.5\theta_j y_j^2 + [(Ss + Sb_j)/T + (Hs + Hb_j) \cdot T \cdot y_j(1 - y_j/P_j)/2]}{(1 + PR_j)y_j}, \quad (12)$$

where 'T' is computed as in (9).

Objective function

The objective function is considered as the maximization of channel profit of the supply chain. The mathematical expression of channel profit 'PC' can be stated as in (13).

$$PC = PV + \sum_{j=1}^{n} Pb_j = \sum_{j=1}^{n} \{a_jy_j - b_jy_j^2 - (\delta y_j + 0.5\theta_j y_j^2) - [(Ss + Sb_j)/T + (Hs + Hb_j) \cdot T \cdot y_j(1 - y_j/P_j)/2]\} \quad (13)$$

Mathematical programming model

The optimal or near optimal sales quantity and production rate for buyer j namely '$y_{j_{opt}}$' and '$P_{j_{opt}}$' are obtainable from the following mathematical model which maximizes the channel profit namely 'PC'.

$$\text{Max}\, PC = \sum_{j=1}^{n} \{a_jy_j - b_jy_j^2 - (\delta y_j + 0.5\theta_j y_j^2) - [(Ss + Sb_j)/T + (Hs + Hb_j) \cdot T \cdot y_j(1 - y_j/P_j)/2]\} \quad (14)$$

$$\text{s.t}: y_{j_{min}} \le y_j \le y_{j_{max}}, \quad \forall j = 1, \ldots, n \quad (15)$$

$$\sum_{j=1}^{n} P_j = P \quad (16)$$

$$y_j \le P_j, \quad \forall j = 1, \ldots, n \quad (17)$$

$$y_j \ge 0, \quad P_j \ge 0, \quad \forall j = 1, \ldots, n \quad (18)$$

Constraint (15) gives the valid upper and lower bounds of the sales quantity for the buyers. Constraint (16) guarantees that the sum of the buyer's production rates should be equal to the total production rate. Constraint (17) guarantees that the demand rate be less than the production rate for all the buyers as in EPQ model. Constraint (18) represents that the decision variables of the models should be non-negative.The optimal sales price '$P(y_{j_{opt}})$' can be obtained from (19).

$$P(y_{j_{opt}}) = a_j - b_jy_{j_{opt}} \quad (19)$$

The acceptable contract price '$W_{j_{opt}}$' is yielded by substituting the optimal sales quantity '$y_{j_{opt}}$' in Eq. (12); the result is given as in (20).

$$W_{j_{opt}} = \frac{a_j y_{j_{opt}} PR_j - b_j y_{j_{opt}}^2 PR_j + \delta y_{j_{opt}} + 0.5\theta_j y_{j_{opt}}^2 + [(Ss + Sb_j)/T_{opt} + (Hs + Hb_j) \cdot T_{opt} \cdot y_{j_{opt}}(1 - y_{j_{opt}}/P_{j_{opt}})/2]}{(1 + PR_j) y_{j_{opt}}}, \quad (20)$$

where

$$T_{opt} = \sqrt{\frac{2\sum_{j=1}^{n}(Ss + Sb_j)}{\sum_{j=1}^{n} y_{j_{opt}}(Hs + Hb_j)(1 - y_{j_{opt}}/P_{j_{opt}})}} \quad (21)$$

Proposed heuristics

Particle swarm optimization

Particle swarm optimization (PSO) was introduced by Kennedy and Eberhart (1995) as a population-based search algorithm. PSO is motivated from the simulation of social behavior of bird flocking. PSO uses a population of particles that fly through the search space to reach an optimum. Optimization with particle swarms has two major ingredients, the particle dynamics and the particle information exchange. The particle dynamics are derived from swarm simulations in computer graphics, and the information exchange component is inspired by social networks. These ingredients combine to make PSO a robust and efficient optimizer of real-valued objective functions (although PSO has also been successfully applied to combinatorial and discrete problems). PSO is accepted as a computational intelligent technique; the major difference between PSO and other well-known heuristics such as GA and SA is that the society members are aware of the other members' situation or at least of the best member and consider the obtained information in their decision making. Since the members can remember their best situation during the algorithm operations and always try to include this in their decision making, they can compensate immediately in case of a bad decision making. Each member can search the corresponding neighboring boundary without being worried about worsening the situation. The degree of being influenced by other members of the population is determined by a coefficient called learning coefficient.

PSO is similar to GA in that the system is initialized with a population of random solutions (called particle position); however, unlike GA, each potential solution is also assigned a randomized velocity and does not necessarily need to be encoded. Each individual or potential solution (i.e., particle) flies in the problem dimensional space with a velocity which is dynamically adjusted according to the flying experiences of its own and its colleagues. Each particle is affected by three factors: its own velocity, the best position it has achieved so far called '*p*best' and the overall best position achieved by all particles called '*g*best'. A particle changes its velocity based on the three addressed factors. Denoted by *np* the number of particles in the population (here, we assume $p = 2n$), Let $X_i^t = [x_{i,1}^t, x_{i,2}^t, \ldots, x_{i,2n}^t]$ representing the position value of particle i with respect to dimension j ($j = 1, 2, \ldots, 2n$) at iteration t. We define the velocity of each particle as $V_i^t = [v_{i,1}^t, v_{i,2}^t, \ldots, v_{i,2n}^t]$ while each member of v_i^t corresponds to each member of X_i^t. Let $Pb_i^t = [pb_{i,1}^t, pb_{i,2}^t, \ldots, pb_{i,2n}^t]$ be the best solution which particle i has obtained by iteration t, and let $P_g^t = [p_{g,1}^t, p_{g,2}^t, \ldots, p_{g,2n}^t]$ be the best solution obtained by iteration t.

Solution representation is one of the important steps while designing a PSO-based heuristic. The decision variables can be very good guidelines in this regard. In this paper, the solutions are represented as a string of $2n$ characters in which the first n characters represent the buyers' sales values and the second n characters represent the production rates of the vendor as $\{(y_1, \ldots, y_n, p_1, \ldots, p_n) | y_{min} \le y_i \le y_{max}, y_i \le p_i, \sum p_i \le P\}$. Imagine that there are three buyers whose sales values are uniformly distributed as $y_1 \sim U[1600, 4800]$, $y_2 \sim U[700, 1400]$, $y_3 \sim U[1200, 3600]$ and the production capacity of the vendor is as $P = 18{,}000$. As the constraints of the model we should have $y_i \le p_i$, $\sum_{i=1}^{3} p_i \le 18{,}000$; the particle length should be $2n = 6$. Three random numbers should be generated corresponding to y_1, y_2, y_3 noting that $y_{min} \le y_i \le y_{max}$. As the solutions are continuous, they will convert to the discrete solutions by random number generation in order to be usable in the problem. Table 1 illustrates a sample vector of particles X_i^t used by PSO algorithm.

Table 1 Solution representation of X_i^t

Dimension, j						
Location	y_1	y_2	y_3	p_1	p_2	p_3
x_{ij}^t	4620	1333	1505	5801	2296	2640

We give a brief outline of the algorithm as:

Step 1 Initialization

- Set $t = 0$, $np = 2n$.
- Generate np particles randomly as explained before and make the vector of particles as $X_i^0 = [y_1^0, \ldots, y_n^0, p_1^0, \ldots, p_n^0]$ where $y_{\min} \le y_i \le y_{\max}$, $y_i \le p_i$ and $\sum_{pi} \le P$; the continuous values of the positions are generated randomly.
- Generate the initial velocities for each particle randomly, i.e., $\{V_i^0,\quad i = 1, 2, \ldots, NP\}$ where $V_i^0 = [v_{i,1}^0, v_{i,2}^0, \ldots, v_{i,2n}^0]$. Initial velocities are generated using the formula $v_{ij}^0 = v_{\min} + (v_{\max} - v_{\min}) \times r$ where $v_{\min} = -4$, $v_{\max} = 4$ and r is a uniform random number between [0, 1]. Continuous velocity values are restricted as $v_{ij}^t = [v_{\min}, v_{\max}] = [-4, 4]$, where $v_{\min} = -v_{\max}$.
- Evaluate each particle in the swarm using the objective function $f_i^0(X_i^0)$ for $i = 1, 2, \ldots, np$.
- For each particle in the swarm, set $Pb_i^0 = X_i^0$, where $P_i^0 = [pb_{i,1}^0 = x_{i,1}^0, pb_{i,2}^0 = x_{i,2}^0, \ldots, pb_{i,2n}^0 = x_{i,2n}^0]$ together with its best fitness value, f_i^{pb} for $i = 1, 2, \ldots, np$.
- Find the best fitness value among the whole swarm such that $f_l = \min\{f_i^0\}$ for $i = 1, 2, \ldots, np$ together with its corresponding positions X_l^0. Set global best to $G^0 = X_l^0$ such that $G^0 = [g_1 = x_{l,1}, \ldots, g_{DT} = x_{l,2n}]$ with its fitness value $f^{gb} = f_l$.

Step 2 Update iteration counter, i.e., $t = t + 1$

Step 3 Update inertia weight, i.e., $w^t = w^{t-1} \times \beta$ where β is a decrement factor

Step 4 Update velocity, i.e., $v_{ij}^t = w^{t-1} v_{ij}^{t-1} + c_1 r_1 (pb_{ij}^{t-1} - x_{ij}^{t-1}) + c_2 r_2 (g_j^{t-1} - x_{ij}^{t-1})$, where c_1 and c_2 are acceleration coefficients and r_1 and r_2 are uniform random numbers between [0, 1]

Step 5 Update positions, i.e., $x_{ij}^t = x_{ij}^{t-1} + v_{ij}^t$

Step 6 If x_{ij}^t is continuous, round it to the nearest integer number in accordance with the model's constraints

Step 7 Update personal best; each particle is evaluated using the permutation to see if the personal best will improve. If $f_i^t < f_i^{pb}$ for $i = 1, 2, \ldots, np$, then personal best is updated as $P_i^t = X_i^t$ and $f_i^{pb} = f_i^t$

Step 8 Update global best. Find the minimum value of personal best. That is, $f_l^t = \min\{f_i^{pb}\}$, for $i = 1, 2, \ldots, np$; $l \in \{i; i = 1, 2, \ldots, np\}$. If $f_i^t < f_i^{gb}$, then the global best is updated as $G^t = X_l^t$ and $f^{gb} = f_l^t$

Step 9 Stopping criterion. If the number of iterations exceeds the maximum number of iterations, or maximum CPU time, then stop; otherwise go to step 2.

Genetic algorithm

The proposed mathematical model for determining '$y_{j_{\rm opt}}$' and '$P_{j_{\rm opt}}$' belongs to nonlinear integer programming (NIP) problem. A GA-based heuristic is proposed to evolve an optimal or near optimal sales quantity and production rate for buyer j, i.e., '$y_{j_{\rm opt}}$' and '$P_{j_{\rm opt}}$' to maximize the channel profit. The optimal sales price '$P(y_{j_{\rm opt}})$' and contract price '$W_{j_{\rm opt}}$' are derived subsequently from the '$y_{j_{\rm opt}}$'and '$P_{j_{\rm opt}}$'. GA as a population-based algorithm is a class of evolutionary algorithms. It is a generic optimization method, which can be applied to almost every problem. The feasible solutions of the problem are usually represented as strings of binary or real numbers called chromosomes. Each chromosome has a fitness value, which corresponds to the objective function value of the model. Initially, there is a population of chromosomes randomly generated; then, a number of chromosomes are selected as parents for mating to produce new chromosomes (i.e., solutions) called offspring. The mating of parents is carried out applying a few GA operators, such as crossover and mutation. The selection of parents and producing offspring are repeated until a stopping rule (e.g., elapsing a certain number of iterations) is satisfied (Goldberg 1989). Before giving a general outline of the proposed GA-based heuristic, a few additional notations are defined:

Pop_size Size of the population of solutions that remains constant during the algorithm performance.

Max_iteration Number of generations produced until the algorithm stops.

p_c Crossover rate, which is the probability of selecting a chromosome in each generation for crossover.

p_m Mutation rate, which is the probability of selecting a gene or bit inside a chromosome for mutating.

Fitness_function Fitness function value, which exactly corresponds to the objective function value in this paper.

We give a brief outline of the algorithm as:

Step 1 Initialization

- Set Pop_size, Max_iteration, p_c and p_m.

Step 2 Randomly generate the initial population

Step 3 Repeat until *Max_iteration*:

Step 3.1 Perform the reproduction operator according to the roulette wheel rule to make a new population

Step 3.2 Select the parent chromosomes from the obtained population, each with probability p_c

Step 3.3 Crossover:

(a) Determine the pairs of parents among the parent chromosomes.

(b) Apply the crossover operator to produce two offspring corresponding to each pair.

(c) Replace each offspring in the population instead of the parents.

Step 3.4 Apply the mutation operator on the population with probability p_m

Step 3.5 Calculate *Fitness_function* for each chromosome and save the best value in *bv* (best value)

Step 4 Print *bv*

Each chromosome consists of $2n$ genes. The first n genes represent the sales quantities of the buyers and the second n genes represent the production rates of them. As an example, the chromosome [1 1 3 4 5 7 9 10] indicates that there are four buyers whose sales quantities are 1, 1, 3 and 4, respectively, and production rates are 5, 7, 9 and 10, respectively.

The *Pop_size*, p_c and p_m are determined through the try and error method while *Max_iteration* is assumed equal to 2000.

Simulated annealing

SA proposed by Kirkpatrick et al. (1983) is a stochastic and neighborhood-based search algorithm motivated from an analogy between the simulation of the annealing of solids and the strategy of solving combinatorial optimization problems. SA has been widely applied to solve combinatorial optimization problems as Yao (1995) declares. It is inspired by the physical process of heating a substance and then slowly cooling it, until a strong crystalline structure to be formed. This process is simulated through gradually lowering an initial temperature until the system reaches an equilibrium point so that no more changes occur. Generally, details of SA proposed are as follows:

Algorithm: simulated annealing

Initialize parameters T_0, N, K, α, T_f
Initialized counter $n = 0$, $k = 0$
Do (outside loop)
Set $n = 0$
Generate initial solution X^0 : Set $X^{\text{Best}} = X^0$
Do (inside loop)
Generate neighboring solution X^{n-1} by operation $(X_n \rightarrow X^{n-1})$
If $f(X^{n+1}) \prec f(X^n)$ then
$X^n = X^{n+1}$; Set $n = n + 1$
Else
Generate random Rand $\rightarrow u(0, 1)$
If Rand $< e^{-\Delta F/T_k}$ then $X^n = X^{n+1}$; set $n = n + 1$
End if
Update X^{Best}
Loop until $(n \leq N)$
$T_{r+1} = \alpha T_r$
Loop until frozen

Computational experiments

Nachiappan and Jawahar (2007) analyzed their proposed model and methodology by a case study carried out at the SNP dairy company located in Madurai, India. The dairy manufacturer (vendor) supplies its product (milk packets) to the customers at different locations (buyers). Since the structure of our proposed model is near to that of Nachiappan and Jawahar (2007), we have provided a few numerical problems inspiring from those of Nachiappan and Jawahar (2007). The numerical problems are given in three categories while considering three, five, and eight buyers in the model. Since the number of parameters is too much, the buyer-related parameters are considered fixed, while the vendor-related parameters are changed in order to do the sensitivity analysis. The values of the buyer-related parameters are given in Table 2. Five problems are selected from each category as small, medium and large size

Table 2 Values of the buyer-related parameters

	Buyer-related data for $n = 3$			Buyer-related data for $n = 5$					Buyer-related data for $n = 8$							
j	1	2	3	1	2	3	4	5	1	2	3	4	5	6	7	8
Hb_j	8	10	10	8	10	10	6	7	8	10	10	6	7	12	13	14
Sb_j	24	11	29	24	11	29	14	25	24	11	29	14	25	12	30	22
a_j	31	35	37	31	35	37	32	39	31	35	37	32	39	33	36	38
b_j	0.008	0.004	0.006	0.008	0.004	0.006	0.003	0.004	0.008	0.004	0.006	0.003	0.004	0.005	0.007	0.006
$y_{j\min}$	1600	700	1200	1600	700	1200	1500	900	1600	700	1200	1500	900	700	800	1200
$y_{j\max}$	4800	1400	3600	4800	1400	3600	3000	2700	4800	1400	3600	3000	2700	3500	4900	3000
θ_j	0.004	0.008	0.005	0.004	0.008	0.005	0.005	0.007	0.004	0.008	0.005	0.005	0.007	0.005	0.007	0.006

Table 3 Values of the vendor-related parameters

Level	Hs	Ss	δ	P
Low (−1)	3	5	5	18,000
Up (+1)	15	40	10	27,000

problems. The values of the vendor-related parameters are given in Table 3.

In the rest of this section, we are going to compare the proposed GA, discrete particle swarm optimization algorithm (DPSO), and SA for TSVMBSC problem. We have also used LINGO solver to assess the performance of the proposed heuristics. All the heuristics are coded in Matlab7.0 software and run on a PC with 1.67 GHz processor (Intel Pentium 4), 256 MG memory and windows XP Professional Operating System.

We have used relative percentage index (RPI) to assess the performance of the proposed heuristics. This index is one of the well-known indexes in this regard for single objective problems. We have solved a number of instances for each numerical problem. RPI can be computed by Eq. (22) in which Max_{sol} and $\text{Worst}_{\text{sol}}$ represent the best and worst objective function values obtained from solving the instances of each numerical problem while solving it by different heuristics; $A\,\text{lg}_{\text{sol}}$ represents the objective function value for each instance of a numerical problem.

$$\text{RPI} = \frac{\text{Max}_{\text{sol}} - A\,\text{lg}_{\text{sol}}}{\text{Max}_{\text{sol}} - \text{Worst}_{\text{sol}}} \tag{22}$$

RPI can take values between 0 and 1. Clearly, lower values of RPI are preferred. Table 4 gives the RPI values for each numerical problem while solving it by each heuristic as well as using LINGO solver. We have considered the number of instances for each numerical problem equal to five times; the average of the obtained objective function values from solving the five instances is considered as $\overline{\text{RPI}}$ for each numerical problem with respect to each heuristic. The CPU times are considered the same for the heuristics; however, we have reported the CPU times when each algorithm reached the best corresponding solution.

As it is clear from Table 4, the $\overline{\text{RPI}}$ for DPSO is less than that of other heuristics; however, the average CPU time of LINGO solver is less than that of other heuristics. We have also used statistical t test at significance level $\alpha = 0.05$ to compare each heuristic with the other considering H_0: $D = 0$ against H_1: $D > 0$ in which D represents the difference between the average of $\overline{\text{RPI}}$ of the first heuristic and that of the second. Therefore, hypothesis H_0 can be rejected if and only if $t = \frac{\bar{D}}{S_D/\sqrt{n}} > t_{\alpha;n-1}$. Table 5 illustrates the results.

Figure 2 indicates the average value of LSD with confidence interval 95 % for various heuristics. It is clear from Fig. 2 that DPSO is superior compared with other heuristics and LINGO solver.

Conclusions and suggestion

This paper presents a TSPMBSC model under the VMI mode of operation. It is the extension of Nachiappan and Jawahar (2007) for the case where the vendor (producer) replenishes orders as EPQ, i.e., the product gradually enters into the vendor's location. The final model can be stated as a mathematical programming model with the objective function of channel profit and the two decision variables of sales quantity and production rate. Thereafter, the optimal values of the decision variables are determined. The aforementioned problem is NP-hard which means too difficult to be solved during a logical amount of time. We presented a DPSO-based heuristic to solve the problem. To prove the efficiency of the proposed heuristic, two distinct kinds of heuristics were used, including innovative searching method of the GA, and SA; however, LINGO solver was also used. The heuristics applied to solve a set of small, medium and large

Table 4 Average relative percentage deviation ($\overline{RPI}$) and average CPU time for algorithms

Problem size	Comparative algorithms							
	GA		SA		DPSO		LINGO	
	$\overline{RPI}$	CPU time	$\overline{RPI}$	CPU time	$\overline{RPI}$	CPU time	$\overline{RPI}$	CPU time
Small problem								
PS1	0.0000	0.30	0.0000	0.27	0.0000	0.26	0.0000	0.27
PS2	0.2000	2.08	0.2000	1.87	0.1000	1.59	0.0000	1.62
PS3	0.2000	1.87	0.2000	1.69	0.2000	1.39	0.0000	1.41
PS4	0.4000	5.74	0.2000	4.74	0.2000	4.37	0.0000	4.33
PS5	0.2000	6.23	0.1333	6.16	0.0667	5.20	0.0000	5.15
Medium problems								
PM1	0.8667	16.33	0.7000	14.26	0.6000	15.52	0.5000	15.91
PM2	0.3030	37.21	0.2017	32.02	0.1517	34.65	0.1617	33.12
PM3	0.7750	37.18	0.5002	32.24	0.4379	31.87	0.5128	30.91
PM4	0.3715	32.67	0.3004	29.65	0.2307	24.42	0.5081	24.42
PM5	0.4806	35.55	0.4444	26.80	0.5833	27.82	0.8056	28.43
Large problems								
PL1	0.1044	478.28	0.3116	478.28	0.2889	404.75	0.3864	389.78
PL2	0.3667	448.07	0.3833	448.07	0.3750	437.87	0.6333	433.87
PL3	0.5331	352.03	0.6662	352.03	0.5996	309.00	0.9975	293.93
PL4	0.0333	539.38	0.1833	539.38	0.1778	513.78	0.2833	513.78
PL5	0.7273	935.95	0.8473	935.95	0.6909	832.05	0.9818	832.05
Average	0.371	195.26	0.351	193.56	0.313	186.30	0.385	173.93

Table 5 Results from t test for the $\overline{RPI}$

Algorithms	t	P value
DPSO: LINGO	−0.93	0.717
DPSO: GA	−1.89	0.853
DPSO: SA	−1.48	0.802
GA: SA	1.40	0.173*
SA: LINGO	−0.67	0.476
GA: LINGO	1.90	0.113*

* Means that the difference is significant, i.e., P value $<\alpha$

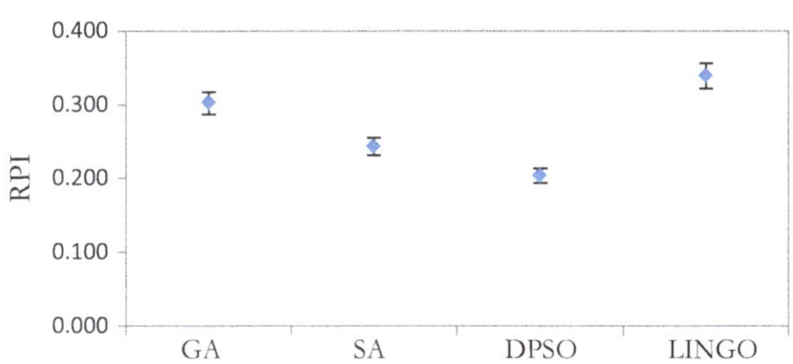

Fig. 2 Plot of $\overline{RPI}$ for the type of algorithm factor

size problems. The results indicated that the DPSO excels compared to the other rival heuristics.

Though the model considered in this paper is restricted to two-echelons, further analysis is required to study the performance under multi-echelon supply chains. Besides, demand can be lost or backordered while it is stochastic.

References

Achabal DD, Mcintyre SH, Smith SA, Kalyanam K (2000) A decision support system for VMI. J Retail 76(4):430–454

Almehdawe E, Mantin B (2010) Vendor managed inventory with a capacitated manufacturer and multiple retailers: retailer versus manufacturer leadership. Int J Prod Econ 128(1):292–302

AriaNezhad MG, Makuieand A, Khayatmoghadam S (2013) Developing and solving two-echelon inventory system for perishable items in a supply chain: case study (Mashhad Behrouz Company). J Ind Eng Int. doi:10.1186/2251-712X-9-39

Bhattacharjee S, Ramesh R (2000) A multi-period profit maximizing model for retail supply chain management: an integration of demand and supply side mechanism. Eur J Oper Res 122:584–601

Bichescu BC, Fry MJ (2009) Vendor-managed inventory and the effect of channel power. OR Spectr 31(1):195–228

Bookbinder JH, Gümüş M, Jewkes EM (2010) Calculating the benefits of vendor managed inventory in a manufacturer-retailer system. Int J Prod Res 48(19):5549–5571

Cárdenas-Barrón LE, Treviño-Garza G, Wee HM (2012) A simple and better algorithm to solve the vendor managed inventory control system of multi-product multi-constraint economic order quantity model. Expert Syst Appl 39(3):3888–3895
Costa L, Oliveira P (2001) Evolutionary algorithms approach to the solution of mixed integer nonlinear programming problems. Comput Chem Eng 25:257–266
Darwish MA, Odah OM (2010) Vendor managed inventory model for single-vendor multi-retailer supply chains. Eur J Oper Res 204(3):473–484
Diabat A (2014) Hybrid algorithm for a vendor managed inventory system in a two-echelon supply chain. Eur J Oper Res 238(1):114–121
Disney SM, Towill DR (2002) A procedure for optimization of the dynamic response of a vendor managed inventory systems. Comput Ind Eng 43:27–58
Dong Y, Xu K (2002) A supply chain model of vendor managed inventory. Transp Res Part E Logist Transp Rev 38(2):75–95
Goh S-A, Ponnambalam SG (2010) A particle swarm optimization algorithm for optimal operating parameters of VMI systems in a two-echelon supply chain. Swarm Evolut Memet Comput 6466:440–447
Goh S-A, Ponnambalam SG, Jawahar N (2012) Evolutionary algorithms for optimal operating parameters of vendor managed inventory systems in a two-echelon supply chain. Adv Eng Softw 52:47–54
Goldberg DE (1989) Genetic algorithms in search, optimisation and machine learning. Addision Wesley, Reading
Govindan K (2013) Vendor-managed inventory: a review based on dimensions. Int J Prod Res 51(13):3808–3835
Grieger M (2003) Electronic marketplaces: a literature review and a call for supply chain management research. Eur J Oper Res 144:280–294
Guan R, Zhao X (2010) On contracts for VMI program with continuous review (r, Q) policy. Eur J Oper Res 207(2):656–667
Kennedy J, Eberhart RC (1995) Particle swarm optimization. In: Proceedings of IEEE international conference on neural networks, IV, pp 1942–1948
Kirkpatrick S, Gelatt CD, Vecchi MP (1983) Optimization by simulated annealing. Science New Ser 220:671–680
Lau AHL, Lau HS (2003) Effects of a demand curve's shape on the optimal solutions of a multi echelon inventory/pricing model. Eur J Oper Res 147:530–548
Lu L (1995) A one-vendor multi-buyer integrated inventory model. Eur J Oper Res 81(2):312–323
Maloni MJ, Benton WC (1997) Supply chain partnership: opportunities for operations research. Eur J Oper Res 101:419–429
Nachiappan SP, Jawahar N (2007) A genetic algorithm for optimal operating parameters of VMI system in a two-echelon supply chain. Eur J Oper Res 182(3):1433–1452
Nia AR, Far MH, Niaki STA (2013) A fuzzy vendor managed inventory of multi-item economic order quantity model under shortage: an ant colony optimization algorithm. Int J Prod Econ 155:259–271
Pal B, Sana SS, Chaudhuri K (2012a) A three layer multi-item production-inventory model for multiple suppliers and retailers. Econ Model 29(6):2704–2710
Pal B, Sana SS, Chaudhuri K (2012b) A multi-echelon supply chain model for reworkable items in multiple-markets with supply disruption. Econ Model 29(5):1891–1898
Pasandideh SHR, Niaki STA, Nia AR (2010) An investigation of vendor-managed inventory application in supply chain: the EOQ model with shortage. Int J Adv Manuf Technol 49(1–4):329–339
Pasandideh SHR, Niaki STA, Nia AR (2011) A genetic algorithm for vendor managed inventory control system of multi-product multi-constraint economic order quantity model. Expert Syst Appl 38(3):2708–2716
Pasandideh SHR, Niaki STA, Far MH (2014a) Optimization of vendor managed inventory of multiproduct EPQ model with multiple constraints using genetic algorithm. Int J Adv Manuf Technol 71(1–4):365–376
Pasandideh SHR, Niaki STA, Niknamfar AM (2014b) Lexicographic max–min approach for an integrated vendor-managed inventory problem. Knowl Based Syst 59:58–65
Rad RH, Razmi J, Sangari MS, Ebrahimi ZF (2014) Optimizing an integrated vendor-managed inventory system for a single-vendor two-buyer supply chain with determining weighting factor for vendor's ordering cost. Int J Prod Econ 153:295–308
Razmi J, Rad RH, Sangari MS (2010) Developing a two-echelon mathematical model for a vendor-managed inventory (VMI) system. Int J Adv Manuf Technol 48(5–8):773–783
Sadeghi J, Mousavi SM, Niaki STA, Sadeghi S (2013) Optimizing a multi-vendor multi-retailer vendor managed inventory problem: two tuned meta-heuristic algorithms. Knowl Based Syst 50:159–170
Sana SS (2014) Optimal production lot size and reorder point of a two-stage supply chain while random demand is sensitive with sales teams' initiatives. Int J Syst Sci. doi:10.1080/00207721.2014.886748
Sana SS, Chedid JA, Navarro KS (2011) A production-inventory model of imperfect quality products in a three-layer supply chain. Decis Support Syst 50(2):539–547
Sana SS, Chedid IA, Navarro KS (2014) A three layer supply chain model with multiple suppliers, manufacturers and retailers for multiple items. Appl Math Comput 229:139–150
Shao H, Li Y, Zhao D (2011) An optimal dicisional model in two-echelon supply chain. Procedia Eng 15:4282–4286
Silver AE, Pyke DF, Peterson R (1998) Inventory management and production planning and scheduling. Wiley, London
Taleizadeh AA, Noori-daryan M (2014) Pricing, manufacturing and inventory policies for raw material in a three-level supply chain. Int J Syst Sci. doi:10.1080/00207721.2014.909544
Thangam A (2014) Retailer's inventory system in a two-level trade credit financing with selling price discount and partial order cancellations. J Ind Eng Int. doi:10.1186/2251-712X-10-3
Toptal A, Çetinkaya S (2008) Quantifying the value of buyer-vendor coordination: analytical and numerical results under different replenishment cost structures. Eur J Oper Res 187(3):785–805
Van der Vlist P, Kuik R, Verheijen B (2007) Note on supply chain integration in vendor managed inventory. Decis Support Syst 44(1):360–365
Verma NK, Chakraborty A, Chatterjee AK (2014) Joint replenishment of multi retailer with variable replenishment cycle under VMI. Eur J Oper Res 233(3):787–789
Waller M, Johnson ME, Davis T (2001) Vendor managed inventory in the retail supply chain. J Bus Logist 20(1):183–203
Wang CX (2009) Random yield and uncertain demand in decentralized supply chains under the traditional and VMI arrangements. Int J Prod Res 47(7):1955–1968
Wang W-T, Wee H-M, Tsao H-SJ (2010) Revisiting the note on supply chain integration in vendor-managed inventory. Decis Support Syst 48(2):419–420
Wong WK, Qi J, Leung SYS (2009) Coordinating supply chains with sales rebate contracts and vendor-managed inventory. Int J Prod Econ 120(1):151–161
Yao X (1995) A new simulated annealing algorithm. Int J Comput Math 56:161–168
Yao MJ, Chiou CC (2004) On a replenishment coordination model in an integrated supply chain with one vendor and multiple buyers. Eur J Oper Res 159:406–419

Yao Y, Evers PT, Dresner ME (2007) Supply chain integration in vendor-managed inventory. Decis Support Syst 43:663–674

Yu Y, Chu F, Chen H (2009a) A Stackelberg game and its improvement in a VMI system with a manufacturing vendor. Eur J Oper Res 192(3):929–948

Yu Y, Huang GQ, Liang L (2009b) Stackelberg game-theoretic model for optimizing advertising, pricing and inventory policies in vendor managed inventory (VMI) production supply chains. Comput Ind Eng 57(1):368–382

Zavanella L, Zanoni S (2009) A one-vendor multi-buyer integrated production-inventory model: the 'Consignment Stock' case. Int J Prod Econ 118(1):225–232

Zhang T, Liang L, Yu Y, Yu Y (2007) An integrated vendor-managed inventory model for a two-echelon system with order cost reduction. Int J Prod Econ 109:241–253

4

Performance measurement in horizontal LSP cooperation as a field of conflict: the preventive role of collaborative processes

Carl Marcus Wallenburg[1] · **Thorsten Schäffler**[1]

Abstract Engaging in horizontal cooperation is a prevalent strategy of logistics service providers (LSPs) to deal with the challenges they are confronted with by today's supply chain environment (e.g., complex and global supply chains, increased competition). These horizontal LSP alliances, for example, for providing a national less-than-truck-load network, are complex in nature and characterized by coopetition. This situation provides a fertile soil for conflict. The literature outlines that conflict, the experience that goals or interests are in opposition, emerges in areas of essential activities and that performance measurement is such an area. Yet, when these differences meet, they also become apparent and can potentially be eased. Against this background, we develop a theoretical model and test it with survey data from 193 horizontal LSP alliances. The results show that, in general, applying collaborative processes in the performance measurement reduces overall conflict in horizontal LSP cooperations. Further, based on resource dependency theory and social contract theory, this research identifies the power structure of the alliance as a relevant contingency factor and shows that joint action in PM is effective in reducing conflict in symmetrical power cooperations, while information sharing as less intense form of collaboration is effective in reducing conflict in asymmetrical power cooperations.

✉ Carl Marcus Wallenburg
wallenburg@whu.edu

1 Kühne-Foundation Chair of Logistics and Services Management, WHU – Otto Beisheim School of Management, Erkrather Straße 224a, Düsseldorf, Germany

Keywords Alliance management · Logistics service provider · Performance measurement · Conflict · Collaborative processes

1 Introduction

Horizontal cooperation is a prevalent strategy for logistics service providers (LSPs) to deal with the extensive challenges of today's supply chain environment (e.g., complex and global supply chains, increased competition, rising requirements by customers) [1, 2]. Examples are multi-company networks for less-than-truck-load transportation. Today, more than half of all small, medium, and large LSPs are engaged in at least one such horizontal alliance with one or multiple other LSPs [3]. This allows them to strengthen their competitive position through improved service scope and quality, increased productivity, and access to new markets [1, 3, 4].

Horizontal LSP alliances are complex in nature [5] and entail various strategically relevant decisions on which opinions will widely differ. Examples include the question how to develop the scope of activities or in which way to support common activities with IT systems (e.g., whether to offer joint mobile apps at the customer interfaces). In addition, the alliances are influenced by the ambiguity of cooperation and competition often referred to as coopetition [6, 7], where members are simultaneously partners and potential competitors in the marketplace. Overall, this provides fertile ground for conflicts to emerge with the tendency to reduce performance [8].

In this respect, prior research has focused on differing conflict types and their outcomes (e.g., [9]), and on post-manifestation topics such as conflict management and resolution strategies (e.g., [10, 11]. It is clear that conflicts arise due to very different reasons (for example, "from incompatible goals, resource allocation disagreements, opportunistic behavior, knowledge imitation, and competition in downstream markets" [12, p. 291]), at different hierarchical levels and in different areas of the cooperation [13]. To complement this, our research views the areas of activities as possible source of and/or possible remedy to conflict, a domain that has been neglected in prior research. We concentrate on performance measurement (PM) as one key area of any organization and as such also of a horizontal alliance [14], and will show that the way PM is conducted strongly influences the level of conflict in a horizontal alliance.

While the first impulse may be to regard PM as a subordinate, operational activity within a horizontal alliance, PM also has a truly strategic nature in communicating and implementing strategies [15]. Consequently, Melnyk et al. [15] argue that PM needs to be co-created simultaneously with the strategy. Besides the operational measurement, PM not only entails the definition of metrics and targets for the cooperation [15, 16], but also establishing—and over time adapting [15]—what performance is (i.e., the objectives of the alliance) and how these objectives are going to be reached (i.e., the underlying performance drivers). All of these are fields where the potentially different goals and objectives, but also the different management approaches of the alliance members meet and clash [17]. As such, PM is an area prone to conflict, both relating to the question *what* to measure (e.g., speed vs. punctuality of delivery) and *how* to measure [18].

When such differences in opinion or interest meet in PM, they also become apparent and can potentially be eased, which, in turn, substantially lowers overall potential for tension and conflict within the alliance. In this respect, we build on the collaborative processes of joint action and information sharing, which, in general, have been identified as methods for effectively preventing the manifestation of conflicts [19, 20]. We show that in the context of horizontal LSP alliances collaborative activities in PM can effectively offset the conflict potentials inherent to the alliance.

At the same time, extant literature emphasizes that the emergence and resolution of conflicts also relate to the specific context in which the conflicts develop [21, 22]. Here, the power structure has been identified to influence actions and perceptions in cooperating [23] and to be decisive in influencing whether conflict potential manifests itself into actual conflict [24].

This research extends prior research by outlining how the choice of the specific collaborative approach in PM needs to account for the alliance context it is going to be applied in. In order to do so, we draw upon resource dependency theory (RDT) [25] and social contract theory (SCT) [26]. Based on SCT, we conclude that collaborative processes in PM will shape the expectations of the alliance members and, through this, will—according to SCT—create implicit social contracts [27]. Depending on the alliance context—we follow Dant and Schul [28] in distinguishing symmetrical and asymmetrical power relations—these mechanisms have the potential to create tension by running counter to the pursuit of autonomy that companies exhibit according to RDT.

Overall, this research addresses two research questions: (RQ 1) "How effective are joint action in PM and information sharing in PM as collaborative processes in reducing conflict in horizontal LSP alliances?" and (RQ 2) "To what extent does the effectiveness of these two collaborative processes in PM depend on the power structure inherent to the alliance?" Besides developing a conceptual model based on RDT and SCT, the contribution of this research lies in testing the corresponding hypotheses using structural equation modeling based on survey data from 193 logistics service providers (LSPs) engaged in a horizontal alliance.

2 Conceptual framework

2.1 Conflict types and their outcomes

According to the established understanding, conflicts are viewed in this research as "the experience between or among parties that their goals or interest are incompatible or in opposition" [19, p. 1224]. Inter-organizational relationships, where parties with different corporate cultures, different regions, different mindsets, and different ways of doing things interact intensively, provide fertile ground for such conflicts [18, 29]—a situation typical for horizontal LSP alliances [3]. Further factors that create tension and almost inevitably lead to conflicts are mutual dependencies that require continuous interactions [30], different viewpoints concerning temporal aspects (i.e., short-term vs. long-term) [31], diverging goals, and, in the specific case of horizontal cooperation, competitive tension [31, 32] among members.

Viewing conflict from one perspective only "can obscure important differences among different types of [...] conflict" [10, p. 213]. We agree and focus on the two most prominent established dimensions of conflict, cognitive and affective conflicts [33], as dependent variables in our research as these have been found to have differing implications [33, 34].[1]

[1] Going back to [35], the literature sometimes views process conflict as an additional third type of conflict. Here, some authors argue that process conflict is not sufficiently distinct as it entails both task-related elements and person-related aspects. Further, Jehn et al. [36] show that task conflict and process conflict are highly correlated.

Cognitive conflict, alternatively termed task conflict by [35], refers to disagreements between alliance members around task-related issues [19, 37] that revolve around the question of "how best to accomplish an organization's objectives" [33, p. 127]. *Affective conflict,* alternatively termed relationship conflict by [35], in contrast, focuses on personal elements and, therefore, tends to be more emotional in nature [38].

Conflict can both be functional and dysfunctional [39], and while research agrees about the dysfunctional nature of affective conflict (e.g., [9]), research on cognitive conflict has produced differing results. While most researchers suggest that cognitive conflict is generally harmful to cooperation (e.g., [34, 40]), some view the effect as contingent to the routineness of the task [10, 41]. Here, the argumentation is that in complex tasks such as innovation and strategic decision-making that require out-of-the-box thinking, cognitive conflict is positive as disagreements and the associated friction create new ideas and trigger improvement of the task [10, 21, 41]. In contrast, in routine tasks that are characterized by a low level of variability and high repetitiveness [42], dissent is counterproductive as its management is time consuming and frustrating [41]. This is consistent with the empirical findings for horizontal LSP alliances, where [43] show that conflict only can be positive in the non-routine field of innovation generation, provided that functionality of conflict is high, and otherwise is negative. Knowing that conflicts heavily influence key outcomes of horizontal cooperation, it is important to understand the formation of conflicts and how they can be counteracted before they manifest themselves and show their negative sides.

2.2 Power structure in horizontal alliances

The literature has shown that the way in which conflict emerges and manifests itself depends on the business context of the cooperation [21, 22] as this context determines how "a firm's practices, procedures and processes are shaped and constrained" [44]. One central contextual variable that affects conflict is the power structure specific to the relationship [21, 24, 28].

Within RDT, power is viewed as the complement to dependency following the rationale that a party A has power over a party B to the degree that B depends on A [45]. Correspondingly, the literature distinguishes symmetrical power relations and asymmetrical power relations [28].

Also from a SCT perspective symmetrical power relations and asymmetrical power relations differ substantially. SCT provides a legitimacy perspective where legitimization of activities within a relationship is provided or denied via social contracts [26, 46]. These contracts are "norms, assumptions, and beliefs that [alliance members] conceive as fair and appropriate [i.e., legitimate] for parties involved in [alliance] relationships" [47, p. 67] and are embedded in any inter-firm relationship. Whenever any activities in the alliance run counter to the social contracts, this will create tension and evoke reactance by alliance members [26].

In a *symmetrical power relation*, the individual company depends on the cooperation to the same degree to which the cooperation depends on this individual company and its contributions [48]. In such a balanced power situation, the alliance members meet as equals [49, 50] and the degree of mutuality can be expected to be high as the members need each other to reach their individual as well as their joint goals [12]. With respect to RDT, this is a situation where every LSP gives up some autonomy, or as [51] put it, its "freedom to make its own decisions [...] without [...] regard to the demands or expectations of [...] partners," in order to secure resources necessary for its own success via alliance membership.

An *asymmetrical power relation*, in contrast, signifies imbalance regarding the individual company dependence on the cooperation. From the RDT perspective, it implies that some LSPs (the ones with high dependence) gave up much autonomy by entering the alliance, while other (the ones with low dependence) gave up less autonomy. This asymmetrical setup enables some parties to exert power over the other members [32, 50], and a more hierarchical and authoritarian behavior can be expected, reflected in aspects like unilateral decision-making [52].

2.3 Performance measurement as a field of conflict

One key area in an alliance, in which conflicts may arise due to its importance [12], is performance measurement (PM) [14]. While PM involves many operational aspects and may be disregarded from a strategical perspective, because it is thought to merely measure how the performance was, its true nature is fundamentally more far reaching.

First, in the process of PM, it is necessary to establish what performance is (i.e., the objectives of the alliance) and how these objectives are going to be reached (i.e., the underlying performance drivers). Here, potentially different goals and objectives of individual LSPs, but also different management approaches will meet and clash [17]. Second, PM needs to be changed over time as the internal context of cooperation changes (e.g., member companies entering or leaving the alliance), but also the external context of cooperation changes. Third, PM has a coordinating function across different functional units and helps to deal with complex and multiple institutional performance logics [53], which are likely to exist, by reducing the tension within and across the performance logics [53].

Fourth, through its different steps, it is an instrument to guide and control an organization [54, 55], in our case, a horizontal LSP cooperation, by prioritizing and by setting definite targets. This is not only reflected in the old saying "what gets measured gets done," but also in the corresponding research results, which emphasize that what actually is measured shapes what managers focus on and how they try to improve performance [15].

PM in an alliance has various attributes that makes it a likely source of conflicts. On a formal level, the questions about the *what* and *how* (i.e., *what* will be measured *how*) are origins of possible disagreements [18]. With respect to the *what*, disputes can arise when members discuss what the decisive aspects of the cooperation are (i.e., goals, objectives, and ambitions) and how they should be prioritized [56]. Subsequently, coordination is needed on target values for the corresponding performance indicators [17]. Yet, the multitude of opinions in a horizontal alliance may impede a consistent approach to PM resulting in the use of individual key performance indicators (KPIs) by the individual firms [57]. That, in turn, sows the seed for conflict as even using the same terminology for KPIs does not ensure that the alliance members actually measure and report the same things since the exact procedures to come up with values for the indicators may still vary [16, 58]. This potentially leads to situations in which members think that they are discussing the same things, but in reality talk about different things, not realizing that slight or even big differences between their understandings are present [17, 59]. A good example for this is service level, where even a seemingly straightforward indicator like OTIF (on-time-in-full) may be measured differently by each involved LSP, for example, by measuring time of arrival in different manners: One may measure arrival when the truck reaches the site of the consignee, while another may measure it when the truck docks to the unloading dock at the consignee, while a third may only measure once the goods are unloaded and scanned.

Concerning the *how* to measure, members need to find a compatible way of actually conducting PM. Here, without a coordinated approach to measurement each company may pursue individual PM processes [59, 60]. However, splitting up actually interdependent processes creates interface problems that can lead to conflicts [21] as addressed by the decoupling principle of [61], which states that interdependent activities should be carried out "under the same authority" (p. 70). One reason is that already small deviations in measurement processes can lead to a differing basis of results which potentially triggers conflict as in subsequent steps the alliance members discuss performance results that are actually not comparable.

Both aspects mentioned (the *what* and the *how*) easily lead to dissent among the alliance members, as ultimately, PM reflects the operation model, and thus, the priorities and goals as well as the processes of the horizontal alliance [57]. PM has a central role in steering the cooperation and in understanding its context [60]. Consequently, it is focal to all members and a field of conflict where differing opinions and interests clash.

2.4 Collaborative processes in PM for conflict prevention and reduction

Prior conflict literature has mainly focused on post-manifestation issues of conflict, covering topics such as which conflict management techniques exist and how conflicts can be resolved or minimized (e.g., [10, 62]). Yet, this neglects that alliance members can also act much earlier. Once conflict is present, it is rather difficult to control [63] and requires considerable effort to manage and resolve [10, 43, 62]. Therefore, attention should also be put on actively countering conflict already before and during its emergence by promoting preventive instead of only reactive means [11, 40].

Conflict stems from tension, manifests itself because of "the failure of alliance partners to coordinate" [37, p. 157], and is triggered when interaction takes place [19]. Therefore, efforts to limit the emergence of conflict via preventive measures need to focus on points of interaction. Here, from the general conflict avoidance perspective two collaborative processes have shown to offer substantial potential [19, 20]: joint action and information sharing. Joint action in PM is an extensive form of collaboration and refers to alliance members actively coordinating their PM activities [64]. It is consistent with Kozlowski and Bell's [65] call for more collaborative mechanisms in preventing conflict, compared to merely exchanging information. Information sharing with respect to PM as a less intense form of collaboration [20] refers to sharing important information regarding the process as well as the results of PM (i.e., values attained for the performance indicators).

The pre-existing notion in the literature is that how conflict emerges and manifests itself depends on the business context of the cooperation and that collaborative processes in PM are effective in reducing conflict. Based on this general understanding, we develop a specific theoretical model that outlines that this effect is contingent on the (a)symmetry of power within the cooperation, and that joint action is an effective approach in symmetrical power relations, while information sharing is effective in asymmetrical power relations. The corresponding conceptual model is displayed in Fig. 1.

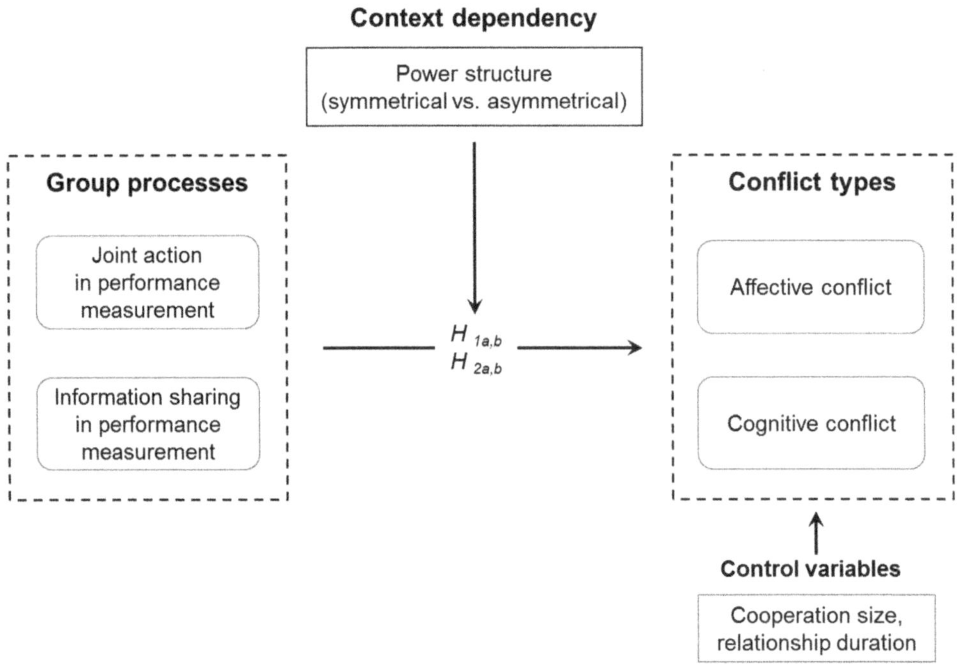

Fig. 1 Conceptual model

2.5 Context-dependent effectiveness of joint action in PM

Joint action can be understood as interpenetration of organizational boundaries by carrying out important activities (as in this case PM) in a cooperative or coordinated way and is a key element of relational governance [51]. It aims at involving all alliance members in the PM and for them to share their needs, concerns, and expectations [66] so that they effectively become "business partners" [51, 67]. The approach promotes an atmosphere of forbearance, mutual respect, and balanced reciprocity, and reduces ex-post transaction costs [5]. Joint action in PM reflects the argument of [22] that elements which "increase the strength of the ties between groups" (p. 522) help to prevent conflicts.

2.5.1 Symmetrical power alliances

In symmetrical power alliances, members are reciprocally dependent and must give and take to achieve individual and mutual benefit. From the RDT perspective, this implies that all of them have given up comparable levels of autonomy. Consequently, increasing the strength of ties via joint action and extending the coordination within the alliance does not impose additional constraints to them.

Following SCT, these alliances are shaped by expectations that are reflected in "norms based in informal social contracts," which "essentially frame their relationship" [26, p. 29]. One such key expectation in symmetrical power alliances is that members have equal say in terms of decisions made and processes conducted. The involvement in PM processes creates a platform to pursue this claim as it provides the opportunity to express possible concerns that can be resolved while collaborating with other members. Therefore, joint action in PM is well suited for symmetrical power alliances and reduces conflict in three ways:

First, being reciprocally dependent, the members are required to contribute to the success of the cooperation, but are also more willing to do so, implying that they will open up and establishing a mutual understanding about individual member's motives, opinions, and know-how [20, 21]. By this, misunderstandings and "mutual feelings of frustration" [68, p. 65] are avoided already early on. The members are enabled to put other members' behavior and actions into context and effectively transmit their own actions [69]. This creates the conditions to clear up possible task-related contradictions [20, 21].

Second, joint action leads to relational norms and mutual trust among members [70]. These act as a safeguard against partner misbehavior, improve coordination and reduce exchange hazards [71]. Here, exchanges in an alliance are embedded within a system of relational norms and social interactions which create mutual confidence that no party will exploit others' vulnerabilities even if there is such opportunity [72]. Further, member involvement and the corresponding collaboration increase commitment and ownership. As the companies feel that their voice and contribution to directly address points of concern at a rather early stage is valued, their motivation is increased, helping minimize the risk that dissent manifests into conflicts [22, 69].

Third, acting jointly in PM strengthens interpersonal relations [22, 73] that leads to a benevolent atmosphere and helps establishing pride for being a member of the alliance [74, 75]. Strong interpersonal ties and pride in the alliance will facilitate internal cohesion (i.e., an *esprit de corps*), a sense of belonging together, and maintaining their belief into the alliance even when exposed to internal or external challenges. Especially in settings of coopetition, this reduces the risk that the different members will be aggressive toward each other [76], which in turn, reduces potentials for affective conflicts.

2.5.2 Asymmetrical power alliances

In general, according to RDT, companies should choose the least constraining device to govern relations with their exchange partners that will secure resource access and maximize autonomy [45]. Here, asymmetrical power alliances assemble a heterogeneous set of LSPs in the sense that LSPs with a stronger dependence on the alliance, which have given up more autonomy, are combined with others with a weaker dependence that have given up less autonomy to secure their desired resources via the alliance. For the LSPs with more autonomy (and more power), joint action in PM, which implies stronger ties and more coordination, would result in added constraints and reduce both their autonomy and their power advantage over the other members. This would run counter to them striving to keep both their power advantage and their autonomy [25], and creates tension.

As they do not have the same reliance on the alliance, they are more likely to pursue their own agenda [48] and will not be interested in incorporating processes that allow weaker parties detailed insights into the stronger parties' management [48, 49]. Further, they tend to consider themselves to be in power, try "calling the shots" and in that sense dominate processes in the cooperation [50]—also PM processes. Applying joint action in PM in this setting would create tension from the RDT perspective between their striving for autonomy and how their autonomy would tend to be restraint by the joint action.

Applying joint action in PM in this setting would also create severe tension from the SCT perspective between the actual power situation and the expectations that would be created for the less powerful members via joint action. As outlined before, joint action entails close collaboration and mutuality, and promotes an atmosphere of forbearance, mutual respect, and balanced reciprocity. Further, it implies interacting on equal footing, where all members contribute to and influence the outcome to similar degrees. As a consequence, joint action in PM would create a social contract of mutual influence that is in contradiction to how the parties view and approach the alliance. Hence, strong collaboration via joint action in a context in which, due to the imbalanced distribution of power, the alliance members are not acting fully on par is not effective and creates tension. Instead, this is likely to be a source of conflict in itself.

Overall, aggregating the negative effects of joint action in PM in an asymmetrical setting and its positive effects in a symmetrical setting, it can be concluded that joint action in PM is effective in reducing conflict in symmetrical power settings and that it will be less effective the more asymmetrical the alliance is. This will be the case for both affective and cognitive conflicts because joint action addresses both levels of these conflicts: the task-related level of cognitive conflict as well as the personal level of affective conflict. Thus, we hypothesize:

$H_{1a,b}$ Joint action in PM is more effective in reducing a) affective and b) cognitive conflict in horizontal alliances with symmetrical power relation than in ones with asymmetrical power relation.

2.6 Context-dependent effectiveness of information sharing in PM

Information sharing with respect to PM refers to sharing important information regarding the process as well as the results of PM (i.e., values attained for the performance indicators). Such information sharing has been found to be vital [19, 20] as it facilitates transparency with regard to performance. Through "communication [...] task details, task progress, and reasoning for task decisions" [20, p. 384] are clarified. The alliance members are better able to understand other members' way of thinking, minimizing animosities and misinterpretations that could lead to both affective and cognitive conflicts. This helps reducing the level of uncertainty [77] and creating a common understanding [59], which improves coordination [78] and reduces the level of conflict [20]. Additionally, problems can proactively be prevented as more thorough decisions can be made [79]. This reduces the risk of errors, which, in turn, is associated with less conflict about task-related issues [20].

Yet, in comparison to joint action in PM, the extent and intensity of exchange is substantially smaller as information sharing focuses on informing alliance members about issues and results of PM and not integrating them into PM. It is less comprehensive in providing insights and a common understanding. Moreover, members merely receiving information from other members cannot rule out that the information provided is filtered or even falsified [80]. In the case of joint action, this situation is different as close collaboration increases the ability for verification and reduces risks of misinformation. Thus, information sharing is not as effective as joint action in establishing relational norms and trust and in serving as effective means for coordination and mitigating exchange hazards. Consequently, information sharing can be considered a "light" version of collaboration compared to joint action.

Thus, in alliances with symmetrical power relation, information sharing has only little or even nothing to add to the potentials created via joint action in PM, and thus, joint action will be substantially more effective in mitigating conflict in this context.

In contrast, when power is unevenly distributed, the less dependent members are less interested in giving up autonomy by establishing extensive collaborative processes as "constant coordination and mutual adjustment among group members is not necessary for the group to function successfully" [20, p. 390]. Because joint action in PM is not in the interest of the less dependent members, these alliances can instead only rely on information sharing in PM, which constitutes a less pronounced form of collaboration to mitigate conflict. From the SCT perspective, this approach entails social contracts that differ from those of joint action in the sense that it allows for one or a few parties to dominate decisions on what and how much information is exchanged. In that sense, information sharing is compatible with companies wanting to dominate, while still providing potential to prevent conflict from manifesting itself.

Consequently, information sharing in PM is an effective approach to reduce affective and cognitive conflicts in alliances with asymmetrical power relations, whereas in symmetrical power relations this is not the case:

$H_{2a,b}$ Information sharing in PM is more effective in reducing a) affective and b) cognitive conflict in horizontal alliances with asymmetrical power relation than in ones with symmetrical power relation.

3 Methodology

3.1 Sampling and data collection

For testing the hypotheses, primary data from LSPs were collected. We employed a key informant approach [81] and targeted senior managers of the LSPs to provide information regarding one horizontal LSP alliance they were active in (here relationships with a subcontracting nature where explicitly highlighted to be out of the scope of the survey). The executive management level was chosen as appropriate point of contact due to the rather small company size in the industry and in the sample (50 % of the surveyed LSPs have less than 100 employees—see Table 1) and the fact that the executive managers at LSPs are still involved even at the day-to-day level of cooperating with other LSPs in a horizontal alliance.

Table 1 Demographics of responding companies

Annual turnover (in €)	*N*	%
1–5 million	51	26
>5–25 million	57	30
>25–100 million	37	19
>100–500 million	22	11
>500 million–5 billion	9	5
>5 billion	2	1
Not specified	15	8
Total	193	100
Number of employees	***N***	**%**
1–50	64	33
51–100	31	16
101–500	44	23
501–1000	20	10
>1000	22	12
Not specified	12	6
Total	193	100

We derived the sample from two commercial databases that provide company data of German LSPs with annual revenue of more than €1 million. The executive managers received an email invitation with a personalized link to our web-based survey in German language. Overall, a response rate of 11.7 % was achieved. After discarding 18 responses due to incomplete data and four responses because they represented outliers as identified based on Mahalanobis distance, 193 usable responses remained for the analyses (see Table 1).

We followed established practices for testing non-response bias. First, answers of early and late responders were compared [82]. Second, a follow-up study was conducted with 18 randomly chosen non-respondents, who received a questionnaire with 12 items of the original survey [83]. Their responses were compared to the ones of the participants that completed the questionnaire in the first place. Neither method indicated significant differences in the data.

3.2 Measurement scales

For the constructs of the present hypothesized model, we used previously established measurement scales and adapted them to the current study. A qualitative pretest with ten logistics researchers and seven CEOs of LSPs was carried out to ensure face validity. In this process, few minor alterations were made iteratively to the construct measurements until no further changes were suggested. The measurement of the constructs relied on multi-item, 7-point Likert-type scales presented in the Appendix 1 and described in the following.

Joint action in PM was measured based on [64] and refers to the extent to which PM activities are integrated via coordination across alliance members. *Information sharing in PM* was captured based on the scale of [84] and refers to the extent to which the alliance members keep each other informed about critical and proprietary information concerning PM. *Cognitive conflict* and *affective conflict* were measured based on [33]. The construct for cognitive conflict measures the extent to which task oriented conflicts were experienced, whereas the construct for affective conflict identifies the extent to which emotional and personal incompatibilities or disputes were experienced.

To capture the power structure, we used two scales of [85]. The firms' dependence on the alliance was assessed by measuring to what extent the alliance was important to them. The alliance's dependency on the individual firm was measured by asking for the extent to which the respondent's company is important for the alliance. For the multi-group analysis, the sample was divided into two groups based on the power structure that was calculated as the absolute difference between the average score of the items of the first construct and the average score of the items of the second construct. Values up to 0.75 were considered as symmetrical power relation ($N = 72$) as the respondents' company depends on the cooperation to a similar degree to that the cooperation depends on it. Values above 0.75 were considered as asymmetrical power relation ($N = 121$). The rationale behind this cutoff was that if the majority of items differed by at least one between the two constructs, the relation would be considered asymmetrical. To ensure robustness, we also tested a cutoff of 1.0 with two groups of $N = 84$ and $N = 109$; the results of this model were consistent with those of the 0.75 cutoff that will be reported below in the Results section.

To validate the measurement scales, we conducted an exploratory factor analysis using SPSS, extracting four factors equivalent to our four focal constructs (see Appendix 2). Subsequently, we conducted a confirmatory factor analysis with AMOS to assess the reliability and validity of the measurement model, which revealed adequate fit ($\chi^2 = 173.0$ with $df = 81$; $\chi^2/df = 2.14$; CFI = 0.96; RMSEA = 0.077). All factor loadings were significant at $p < 0.001$, supporting convergent validity for the constructs. Item reliability exceeded 0.4 for all items. The lowest Cronbach's alpha and the lowest composite reliability for the latent constructs are 0.86 (see Appendix 1). Additionally, [86] procedure to test discriminant validity was used. The squared correlations between any pair of the four constructs used were lower than the average variance extracted of the corresponding constructs (see Appendix 3).

3.3 Control variables

We aimed to control for cooperation size [87] and relationship duration [88]. The first variable captures the number of companies engaged in the alliance, whereas the second variable captures the number of years the alliance exists. This is intended to account for any influence that the two variables may have on affective and cognitive conflicts. During normality tests, these two variables exhibited a relatively strong positive skew. Therefore, a square-root-transformation on the original data was performed and, subsequently, the square root of cooperation size and of relationship duration was used as control variables.

3.4 Results

The structural equations model was tested by conducting a multi-group analysis using AMOS to identify the moderating effects of the power structure. The fit indices showed adequate model fit for the hypothesized model ($\chi^2 = 323.0$ with $df = 210$; $\chi^2/df = 1.54$; CFI = 0.95; RMSEA = 0.05). Table 2 presents the results of the hypothesized relationships.

Hypotheses $H_{1a,b}$, which postulate that joint action in PM is more effective in reducing both forms of conflict in symmetrical power relations than in asymmetrical ones, are supported. In symmetrical power relations, joint action has a significant negative effect on affective conflict (−0.372; $p < 0.05$) and on cognitive conflict (−0.533; $p < 0.01$). In contrast, it has no significant effect in asymmetrical power relations, neither on affective conflict (0.178; n.s.), nor on cognitive conflict (−0.122; n.s.). The differences between symmetrical and asymmetrical power structures are significant for both paths.

Hypotheses $H2_{a,b}$ posit that information sharing in PM is more effective in reducing conflict in asymmetrical power relations than in symmetrical ones. Our results fully support hypothesis H_{2a}: Information sharing in PM has a significant negative effect on affective conflict in asymmetrical power relations (−0.455; $p < 0.01$), whereas the effect in a symmetrical power relation is nonsignificant (+0.078; n.s.). Furthermore, the structural paths are significantly different ($p < 0.01$). At the same time, the effect of information sharing in PM on cognitive conflict in

Table 2 Results of multi-group analysis with respect to power structure (unstandardized path coefficients)

	Power structure						Hypothesis
	Symmetrical		Asymmetrical		Difference		
	Estimate	p	Estimate	p	z-score	p	
Joint action → affective conflict (H1a)	**−0.372**	**0.045**	0.178	0.334	**2.091**	**0.018**	Support
Joint action → cognitive conflict (H1b)	**−0.533**	**0.003**	−0.122	0.525	**1.599**	**0.055**	Support
Information sharing → affective conflict (H2a)	0.078	0.620	**−0.455**	**0.002**	**−2.471**	**0.007**	Support
Information sharing → cognitive conflict (H2b)	0.082	0.554	−0.225	0.134	**−1.499**	**0.067**	Weak support
R^2 affective conflict	12.6 %		20.8 %				
R^2 cognitive conflict	29.6 %		12.2 %				

Bold values indicate significant relationships

symmetrical power relations is also negative, but above the significance threshold (−0.225; $p = 0.134$). But again, the moderation follows the hypothesized pattern (i.e., information sharing is more effective in asymmetrical power relations), and this difference is significant ($p = 0.067$), so that weak support for H_{2b} can be concluded.

Further, our results show that the explanatory power of the two collaborative processes is substantial for symmetrical power relations; the R^2 is 12.6 % for affective conflict and 29.6 % for cognitive conflict. In asymmetrical power relations, collaborative processes in PM account for a little lower but still substantial percentage of variance in conflict ($R^2 = 20.8$ % for affective conflict and $R^2 = 12.2$ % for cognitive conflict) (see Table 2). With regards to our control variables, we find that neither the square root of cooperation size, nor of relationship duration has a significant effect on affective and cognitive conflicts (all p values are above 0.4).

4 Discussion and implications

4.1 Theoretical implications

The results of this study present several important theoretical implications with respect to conflict in horizontal alliances. Even though conflict has been studied extensively in prior research (e.g., [10, 11, 33]), the focus, so far, was mainly on post-manifestation topics such as conflict management, neglecting that conflict could be prevented before emerging in the first place. The major drawback of the prior research that actually was concerned with conflict prevention is that it mostly builds on antecedents of conflict that are not really actionable, for example, the similarity of organizational climate [62] or goal uncertainty [38]. Exceptions are the studies of [20], who tested the effect of information sharing in student groups and its effect on conflict and performance, and [18] who presented a model of proactive approaches for handling conflicts such as selecting partner fit, partnership negotiation, and the build up of relational quality among members in international joint ventures. Here, our research provides one additional step by focussing on actionable collaborative processes (i.e., joint action and information sharing) in a concrete environment (i.e., PM) that can be applied to contribute to a reduction of the overall levels of affective and cognitive conflicts within a horizontal cooperation. This, at the same time, underscores that collaboration is an approach to improve the performance of supply chain relationships [89], and that PM is of key importance to horizontal alliances in that it not only is an area prone to conflict, but also one with the potential to mitigate conflict.

To further refine our assertions, we applied a context-specific view based on RDT and SCT focussing on autonomy of the involved LSPs and the social contracts embedded in the alliance. Here, our findings first underscore the prior assumption that how conflict emerges and manifests itself depends on the context [21, 22] and that the power structure within the cooperation is a relevant contextual variable [21, 24, 28].

Second, prior findings are expanded in showing that the effectiveness of collaboration is not equal for symmetrical and asymmetrical power relations. Overall, collaborative processes in PM are effective in preventing conflict in both setups. Yet, joint action is most effective in symmetrical power relations, while information sharing is most effective in asymmetrical power relations. Conducting a mutual approach such as joint action usually is based on alliance members that all give up autonomy in the alliance and are reciprocally dependent and hence interested in maintaining or even intensifying a close relationship [48]. Otherwise they would not be willing to both give and take in the relationship as well as provide insights into their motives, opinions, and know-how [48, 49]. Prior expectations were that information sharing is rather effective in mitigating the risk of conflict manifestation in asymmetrical power relations as in such a situation, the more powerful members are less interested in reducing their autonomy and altering social contracts by establishing processes that are involving in nature [20]. In this case, stronger members rather prefer information sharing where they can better steer which information is disclosed. Our results were able to support this. Thus, we show with our research that the suitability as well as the effectiveness of collaborative processes is very much dependent on the context (in this case, the power structure) in which they are applied.

4.2 Managerial implications

The results of our study are of importance to practice in that they provide guidance to managing horizontal alliances in particular. Our research provides managers with insights on how to prevent conflict potential to manifest itself in actual conflict.

The first important aspect is that managers need to shift their focus of attention from post-manifestation issues of conflict (i.e., conflict management techniques after the conflict has already emerged) to the question of how conflict can proactively be prevented. If conflict has already emerged, it is difficult to control [63] and necessitates considerable effort to manage and resolve it [10, 43, 62]. In this respect, the present research found that PM is an area where conflicts can emerge, because within this area different viewpoints and approaches meet and clash.

Building on this, the results show that PM, besides being a potential source of conflict, can also be utilized to remedy this issue and to reduce the overall level of conflict—both cognitive and affective—in an alliance. Keys to this are processes such as joint action and information sharing in PM, and such collaborative process can already be initiated at the start of the alliance via contractual provisions [90]. However, their effectiveness to proactively prevent the manifestation of conflict is highly dependent on the specific context of the cooperation and specifically the inherent power structure. In symmetrical power relations, where members are cooperating on equal footing, managers should focus on high participation and involvement during the PM. This does not further impede the autonomy of the alliance members and is consistent with the underlying social contracts of the cooperation. This situation is very different in asymmetrical power relation settings. Here, where dependency and autonomy across alliance members are unequally distributed, instead increased transparency through information sharing is effective while joint action of the alliance members is not effective in mitigating conflict as it reduces member autonomy and runs counter to the implicit social contracts of the alliance.

Thus, managers should be aware of the fact that, first, their focus should be on proactive rather than reactive measures and, second, that when taking measures these have to be chosen depending on the contextual situation they find themselves in.

5 Limitations and further research

In sum, this article provides sound results regarding conflict prevention in horizontal alliances in general and of LSPs in particular. However, the qualification of our conclusions necessitates an acknowledgment of limitations inherent to this study.

First, the study focus was on horizontal cooperation. Compared to vertical relationships, the mode of collaboration in horizontal alliances differs due to their often multilateral setup [3]. This complicates collaborative processes as more than two companies are to be included. Thus, due to the more straightforward arrangement of a buyer–supplier relationship in vertical settings, the effectiveness of the different collaborative processes in PM may be even more pronounced. In order to examine potential differences in the effectiveness, we encourage the replication of our study for vertical relationships. Second, this study used data from the logistics industry. This is an industry in which fear of competition is high while at the same time utilizing horizontal cooperation is very common. Therefore, we expect the conflict-reducing potential of collaborative process to be higher than in other industries, while we do not expect any differences in the underlying general mechanisms of conflict prevention. To investigate potential differences to other service industries, corresponding research is encouraged. Last, we limit our consideration to PM. We also suggest testing the role of joint action and information sharing in other areas of horizontal cooperation. Here, it could be promising to also view less central activities to view how conflict in peripheral areas may also infect collaboration in central activities.[2]

[2] We thank one anonymous reviewer for providing the idea for this promising future research.

Appendix 1

See Table 3.

Table 3 Constructs and questionnaire scale items

Measurement scales	Mean	SD
Joint action in PM[a] [64]		
Cronbach's alpha = 0.87; Composite reliability = 0.86; AVE = 0.61		
Joint action 1: Our performance measurement activities across the cooperation are well coordinated	4.46	1.86
Joint action 2: We systematically coordinate our performance measurement strategies with our cooperation partners	4.12	1.91
Joint action 3: We have processes to systematically transfer performance measurement knowledge across the cooperation partners	4.63	1.75
Joint action 4: Managers from different cooperation members meet periodically to examine how we can create synergies with respect to performance measurement across our cooperation	4.44	1.80
Information sharing in PM[a] [84]		
Cronbach's alpha = 0.92; Composite reliability = 0.91; AVE = 0.73		
Inf. sharing 1: We share our results of the cooperation performance measurement with our cooperation partners	4.31	2.16
Inf. sharing 2: Our cooperation partners share proprietary performance measurement results with us	4.10	2.06
Inf. sharing 3: We inform our cooperation partners in advance of changing needs concerning the cooperation performance measurement	4.35	2.11
Inf. sharing 4: In our cooperation it is common that partners are informed concerning occurrences and changes with respect to performance measurement	4.89	1.93
The cooperation partners keep us fully informed about performance measurement issues (e.g., goal deviations) (eliminated in scale refinement process)		
Affective conflict[a] [33]		
Cronbach's alpha = 0.86; composite reliability = 0.86; AVE = 0.68		
When taking joint decisions …		
Affec. conflict 1: … there is tension in the cooperation decision-making process	2.35	1.45
Affec. conflict 1: … we and our cooperation partners often come into conflict due to different personalities	2.41	1.40
Affec. conflict 3: … personal dislikes transform objective discussions into emotional conflicts.	2.13	1.32
… there often is disagreement in the cooperation concerning the results (eliminated in scale refinement process)		
Cognitive conflict[a] [33]		
Cronbach's alpha = 0.92; composite reliability = 0.91; AVE = 0.72		
The cooperation is characterized that we and our cooperation partners …		
Cogn. conflict 1: … have often disagreements over questions of content	2.57	1.37
Cogn. conflict 2: … have often to work through differences about the content of tasks	2.66	1.45
Cogn. conflict 3: … have often differences in opinions	2.81	1.41
Cogn. conflict 4: … have often different opinions concerning methods for problem resolution (new item)	2.97	1.44
Dependence of focal firm on alliance[a] [85]		
Cronbach's alpha = 0.90		
If our relationship was discontinued with this cooperation, we would have difficulty in keeping up our current business	4.32	2.25
This cooperation is crucial to our future performance	5.30	1.58
We are dependent on this cooperation	3.42	2.02
This cooperation is essential to round out our service offering	4.17	2.09
If our relationship was discontinued, we would have difficulty in replacing this cooperation	4.39	2.00
Dependence of alliance on focal firm[a] [85]		
Cronbach's alpha = 0.88		
If we discontinued being a member of this cooperation, this cooperation would have difficulty in keeping up business	4.43	2.08
We are important to this cooperation	5.39	1.55
We play a major role in contributing to the success of this cooperation	5.26	1.54
We are difficult to replace in this cooperation (new item)	4.34	2.02

Table 3 continued

Control variables	Mean	SD
Relationship duration [88]	11.44	8.96
Since how many years does this cooperation exist?		
Alliance size [87]		
How many companies are engaged in the cooperation?	*N*	%
2	35	18
3–10	60	31
11–20	23	12
21–50	19	10
51–100	29	15
>100	23	12
Not specified	4	2

[a] Corresponding items are measured on a 7-point Likert scale anchored by 1 = *strongly disagree* and 7 = *strongly agree*

Appendix 2

See Table 4.

Table 4 Exploratory factor analysis

Items[a]	Joint action	Inf. sharing	Affec. conflict	Cogn. conflict
Joint action 1	**0.825**	0.058	0.082	−0.006
Joint action 2	**0.819**	0.135	0.121	−0.003
Joint action 3	**0.862**	0.008	−0.102	0.076
Joint action 4	**0.849**	−0.110	−0.033	−0.033
Inf. sharing 1	−0.023	**0.961**	0.071	0.025
Inf. sharing 2	−0.046	**0.959**	0.028	−0.008
Inf. sharing 3	0.043	**0.852**	−0.074	−0.023
Inf. sharing 4	0.109	**0.723**	−0.112	−0.027
Affec. conflict 1	−0.120	0.082	**0.618**	0.326
Affec. conflict 2	0.056	0.020	**0.948**	−0.011
Affec. conflict 3	0.034	−0.086	**0.925**	−0.071
Cogn. conflict 1	0.037	−0.039	0.081	**0.833**
Cogn. conflict 2	−0.088	0.030	−0.001	**0.875**
Cogn. conflict 3	0.040	−0.045	−0.071	**0.963**
Cogn. conflict 4	0.049	0.017	−0.046	**0.933**
AVE	0.61	0.73	0.68	0.72
CR	0.86	0.91	0.86	0.91

Standardized factor loadings, average variance extracted, and composite reliabilities of the latent variables

Standardized factor loadings above 0.6 are denoted in bold

[a] Joint action = joint action in PM; inf. sharing = information sharing in PM; affec. conflict = affective conflict; cogn. conflict = cognitive conflict

Appendix 3

See Table 5.

Table 5 Comparison of AVE and squared correlations (Fornell-Larcker criterion)

	(1)	(2)	(3)	(4)
(1) Joint action in PM	0.61			
(2) Information sharing in PM	0.43	0.73		
(3) Affective conflict	0.07	0.06	0.68	
(4) Cognitive conflict	0.12	0.06	0.55	0.72

Values on the diagonal are estimates of average variance extracted (AVE), and values below the diagonal are the squared correlations between the constructs

References

1. Cruijssen F, Cools M, Dullaert W (2007) Horizontal cooperation in logistics: opportunities and impediments. Transp Res E Log 43(2):129–142
2. Pomponi F, Fratocchi L, Tafuri SR (2015) Trust development and horizontal collaboration in logistics: a theory based evolutionary framework. Supply Chain Manag 20(1):83–97
3. Schmoltzi C, Wallenburg CM (2011) Horizontal cooperations between logistics service providers: motives, structure, performance. Int J Phys Distrib 41(6):552–575
4. Carbone V, Stone MA (2005) Growth and relational strategies used by the European logistics service providers: rationale and outcomes. Transp Res E Log 41(6):495–510
5. Schmoltzi C, Wallenburg CM (2012) Operational governance in horizontal cooperations of logistics service providers: performance effects and the moderating role of cooperation complexity. J Supply Chain Manag 48(2):53–74
6. Bengtsson M, Kock S (2014) Coopetition—Quo vadis? Past accomplishments and future challenges. Ind Mark Manag 43(2):180–188
7. Wild N, Zhou L (2011) Ethical procurement strategies for international aid non-government organisations. Supply Chain Manag 16(2):110–127
8. Christoffersen J (2013) A review of antecedents of international strategic alliance performance: synthesized evidence and new directions for core constructs. Int J Manag Rev 15(1):66–85
9. Parayitam S, Dooley RS (2011) Is too much cognitive conflict in strategic decision-making teams too bad? Int J Confl Manag 22(4):342–357
10. Parry ME, Song M, Spekman RE (2008) Task conflict, integrative potential and conflict management strategies in joint ventures. IEEE Trans Eng Manag 55(2):201–218
11. Rahim MA (2002) Toward a theory of managing organizational conflict. Int J Confl Manag 13(3):206–235
12. Das TK, Teng BS (2003) Partner analysis and alliance performance. Scan J Manag 19(3):279–308
13. Janowicz-Panjaitan M, Krishnan R (2009) Measures for dealing with competence and integrity violations of interorganizational trust at the corporate and operating levels of organizational hierarchy. J Manag Stud 46(2):245–268
14. Bititci US, Carrie AS, McDevitt L (1997) Integrated performance measurement systems: a development guide. Int J Oper Prod Manag 17(5):522–534
15. Melnyk SA, Bititci U, Platts K, Tobias J, Andersen B (2014) Is performance measurement and management fit for the future? Manag Acc Res 25(2):173–186
16. Forslund H (2012) Performance management in supply chains: logistics service providers' perspective. Int J Phys Distrib 42(3):296–311
17. Forslund H, Jonsson P (2007) Dyadic integration of the performance management process: a delivery service case study. Int J Phys Distrib 37(7):546–567
18. Le Nguyen H (2011) Partnership strategies: pro-active and preactive approach in conflict management in international joint ventures. Int J Bus Manag 6(9):38–45
19. Korsgaard MA, Jeong SS, Mahony DM, Pitariu AH (2008) A multilevel view of intragroup conflict. J Manag 34(6):1222–1252
20. Moye NA, Langfred CW (2004) Information sharing and group conflict: going beyond decision making to understand the effects of information sharing on performance. Int J Confl Manag 15(4):381–410
21. Vaaland T, Håkansson H (2003) Exploring interorganizational conflict in complex projects. Ind Mark Manag 32(2):127–138
22. Wall JA, Callister RR (1995) Conflict and its management. J Manag 21(3):515–558
23. Nyaga GN, Lynch DF, Marshall D, Ambrose E (2013) Power asymmetry, adaptation and collaboration in dyadic relationships involving a powerful partner. J Supply Chain Manag 49(3):42–65
24. Tjosvold D, Cho YH, Park HH, Liu C, Liu WC, Sasaki S (2001) Interdependence and managing conflict with sub-contractors in the construction industry in East Asia. Asia Pac J Manag 18(3):295–313
25. Pfeffer J, Salancik GR (1978) The external control of organizations: a resource dependence perspective. Harper & Row, New York
26. Dunfee TW, Smith NC, Ross WT (1999) Contracts and social marketing ethics. J Mark 63(3):14–32
27. Wallenburg CM, Schäffler T (2014) The interplay of relational governance and formal control in horizontal alliances: a social contract perspective. J of Supply Chain Manag 50(2):41–58
28. Dant RP, Schul PL (1992) Conflict resolution processes in contractual channels of distribution. J Mark 56(1):38–54
29. Lin X, Germain R (1998) Sustaining satisfactory joint venture relationships: the role of conflict resolution strategy. J Int Bus Stud 29(1):179–196
30. Stern LW, Reve T (1980) Distribution channels as political economies: a framework for comparative analysis. J Mark 44(3):52–64
31. Das TK, Teng BS (2000) Instabilities of strategic alliances: an internal tensions perspective. Organ Sci 11(1):77–101
32. Wilhelm MM (2011) Managing coopetition through horizontal supply chain relations: linking dyadic and network levels of analysis. J Oper Manag 29(7–8):663–676
33. Amason AC (1996) Distinguishing the effects of functional and dysfunctional conflict on strategic decision making: resolving a paradox for top management teams. Acad Manag J 39(1):123–148
34. De Dreu CKW, Weingart LR (2003) Task versus relationship conflict, team effectiveness and team member satisfaction: a meta-analysis. J Appl Psychol 88(4):741–749
35. Jehn KA (1997) A qualitative analysis of conflict types and dimensions in organizational groups. Adm Sci Q 42:530–557
36. Jehn KA, Greer L, Levine S, Szulanski G (2008) The effects of conflict types, dimensions, and emergent states on group outcomes. Group Decis Negot 17(6):465–495
37. Das TK, Kumar R (2010) Interpretive schemes in cross-national alliances: managing conflicts and discrepancies. Cross Cult Manag Int J 17(2):154–169

38. Mooney AC, Holahan PJ, Amason AC (2007) Don't take it personally: exploring cognitive conflict as a mediator of affective conflict. J Manag Stud 44(5):733–758
39. Deutsch M (1969) Conflicts: productive and destructive. J Soc Issues 25(1):7–42
40. Andrade L, Plowman DA, Duchon D (2008) Getting past conflict resolution: a complexity view of conflict. Emerg Complex Organ 10(1):41–48
41. Jehn KA (1997) Affective and cognitive conflict in work groups: increasing performance through value-based intragroup conflict. In: De Dreu CKW, Van de Vliert E (eds) Using conflict in organizations. Sage, Thousand Oaks, pp 87–100
42. Hall RH (1972) Organization, structure and process. Prentice Hall, Englewood Cliffs
43. Wallenburg CM, Raue JS (2011) Conflict and its governance in horizontal cooperations of logistics service providers. Int J Phys Distrib 41(4):385–400
44. Claycomb C, Frankwick GL (2004) A contingency perspective of communication, conflict resolution and buyer search effort in buyer-supplier relationships. J Supply Chain Manag 40(1):18–34
45. Davis G, Cobb JA (2010) Chapter 2 resource dependence theory: past and future. In: Schoonhoven CB, Dobbin F (eds) Stanford's organization: theory renaissance 1970–2000 (Research in the sociology of organizations, vol 28). Emerald Group Publishing Limited, pp 21–42
46. Heide JB, Wathne KH, Rokkan AI (2007) Interfirm monitoring, social contracts, and relationship outcomes. J Mark Res 44(3):425–433
47. Edwards JC, Karau SJ (2007) Psychological contract or social contract? development of the employment contracts scale. J Lead Organ Stud 13(3):67–78
48. Lusch RF, Brown JR (1996) Interdependency, contracting and relational behavior in marketing channels. J Mark 60(4):19–38
49. Frazier GL (1999) Organizing and managing channels of distribution. J Acad Mark Sci 27(2):226–240
50. Jarratt D, Morrison M (2003) Dependence and the application of power and control in major business relationships: a study of manufacturing and service firms in the business-to-business sector. J Strateg Mark 11(4):235–253
51. Heide JB, John G (1990) Alliances in industrial purchasing: the determinants of joint action in buyer–supplier relationships. J Mark Res 27(1):24–36
52. Aldrich H (2007) Organizations and environments. Stanford University Press, Redwood City
53. Carlsson-Wall M, Kraus K, Messner M (2016) Performance measurement systems and the enactment of different institutional logics: insights from a football organization. Manag Acc Res. doi:10.1016/j.mar.2016.01.006
54. Neely A, Gregory M, Platts K (1995) Performance measurement system design: a literature review and research agenda. Int J Oper Prod Manag 15(4):80–116
55. Wouters M, Wilderom C (2008) Developing performance-measurement systems as enabling formalization: a longitudinal field study of a logistics department. Acc Org Soc 33(4–5):488–516
56. Rey-Marston M, Neely A (2010) Beyond words: testing alignment among inter-organizational performance measures. Meas Bus Excell 14(1):19–27
57. Lohman C, Fortuin L, Wouters M (2004) Designing a performance measurement system: a case study. Eur J Oper Res 156(2):267–286
58. Lai KH, Ngai EWT, Cheng TCE (2002) Measures for evaluating supply chain performance in transport logistics. Transp Res E Log 38(6):439–456
59. Simatupang TM, Wright AC, Sridharan R (2002) The knowledge of coordination for supply chain integration. Bus Process Manag J 8(3):289–308
60. Holmberg S (2000) A systems perspective on supply chain measurements. Int J Phys Distrib 30(10):847–868
61. Stinchcombe A, Heimer CA (1985) Organizational theory and project management. Norwegian University Press, Oslo
62. Fey CF, Beamish PW (2000) Joint venture conflict: the case of russian international joint ventures. Int Bus Rev 9(2):139–162
63. Amason AC, Thompson KR, Hochwarter WA, Harrison AW (1995) Conflict: an important dimension in successful management teams. Organ Dyn 24(2):20–35
64. Bonner J, Kim D, Cavusgil S (2005) Self-perceived strategic network identity and its effects on market performance in alliance relationships. J Bus Res 58(10):1371–1380
65. Kozlowski SWJ, Bell BS (2003) Work groups and teams in organizations. In: Borman WC, Ilgen DR, Klimoski RJ (eds) Handbook of psychology: industrial and organizational psychology. Wiley-Blackwell, New York, pp 412–469
66. Nyaga GN, Whipple JM, Lynch DF (2010) Examining supply chain relationships: do buyer and supplier perspectives on collaborative relationships differ? J Oper Manag 28(2):101–114
67. Kim K (1999) On determinants of joint action in industrial distributor–supplier relationships: beyond economic efficiency. Int J Res Mark 16(3):217–236
68. Etgar M (1979) Sources and types of intrachannel conflict. J Retail 55(1):61–75
69. Song M, Xie J, Dyer B (2000) Antecedents and consequences of marketing managers' conflict-handling behaviors. J Mark 64(1):50–66
70. Bertelli AM, Smith CR (2010) Relational contracting and network management. J Publ Adm Res Theor 20:i21–i40
71. Lai F, Tian Y, Huo B (2012) Relational governance and opportunism in logistics outsourcing relationships: empirical evidence from China. Int J Prod Res 50(9):2501–2514
72. Kale P, Singh H, Perlmutter H (2000) Learning and protection of proprietary assets in strategic alliances: building relational capital. Strateg Manag J 21(3):217–237
73. Tjosvold MM (1989) A manager's view of applying conflict knowledge. In: Tjosvold D, Johnson DW (eds) Productive conflict management. Team Media, Minneapolis, pp 216–230
74. Menon A, Bharadwaj SG, Howell R (1996) The quality and effectiveness of marketing strategy: effects of functional and dysfunctional conflict in intraorganizational relationships. J Acad Mark Sci 24(4):299–313
75. Rose GM, Shoham A, Neill S, Ruvio A (2007) Manufacturer perceptions of the consequences of task and emotional conflict within domestic channels of distribution. J Bus Res 60(4):296–304
76. Barclay W (1991) Interdepartmental conflict in organizational buying: the impact of the organizational context. J Mark Res 28(2):145–159
77. Kwon IWG, Suh T (2004) Factors affecting the level of trust and commitment in supply chain relationships. J Supply Chain Manag 40(2):4–14
78. Lee HL, So KC, Tang CS (2000) The value of information sharing in a two-level supply chain. Manag Sci 46(5):626–643
79. Stank T, Crum M, Arango M (1999) Benefits of interfirm coordination in food industry supply chains. J Bus Log 20(2):21–41
80. Li S, Lin B (2006) Accessing information sharing and information quality in supply chain management. Decis Support Syst 42(3):1641–1656
81. Phillips L, Bagozzi RP (1986) On measuring organizational properties of distribution channels: methodological issues in the use of key informants. Res Mark 8(1):313–369
82. Wagner SM, Kemmerling R (2010) Handling nonresponse in logistics research. J Bus Log 31(2):357–382
83. Lambert DM, Harrington TC (1990) Measuring nonresponse bias in customer service mail surveys. J Bus Log 11(2):5–25

84. Mohr J, Spekman RE (1994) Characteristics of partnership success: partnership attributes, communication behavior and conflict resolution techniques. Strateg Manag J 15(2):135–152
85. Ganesan S (1994) Determinants of long-term orientation in buyer–seller relationships. J Mark 58(2):1–19
86. Fornell C, Larcker DF (1981) Evaluating structural equation models with unobservable variables and measurement error. J Mark Res 18(1):39–50
87. Rubin de Celis JC, Lipinski J (2007) R&D alliances and the effect of experience on innovation: a focus on the semiconductor industry. J Lead Organ Stud 14(1):26–37
88. Schreiner M, Kale P, Corsten D (2009) What really is alliance management capability and how does it impact alliance outcomes and success? Strateg Manag J 30(13):1395–1419
89. Adams FG, Richey RG, Autry CW, Morgan TR, Gabler CB (2014) Supply chain collaboration, integration, and relational technology: how complex operant resources increase performance outcomes. J Bus Log 35(4):299–317
90. Lumineau F, Henderson JE (2012) The influence of relational experience and contractual governance on the negotiation strategy in buyer–supplier disputes. J Oper Manag 30(5):382–395

5

Development of hardware system using temperature and vibration maintenance models integration concepts for conventional machines monitoring: a case study

Michael Kanisuru Adeyeri[1] · Khumbulani Mpofu[1] · Buliaminu Kareem[2]

Abstract This article describes the integration of temperature and vibration models for maintenance monitoring of conventional machinery parts in which their optimal and best functionalities are affected by abnormal changes in temperature and vibration values thereby resulting in machine failures, machines breakdown, poor quality of products, inability to meeting customers' demand, poor inventory control and just to mention a few. The work entails the use of temperature and vibration sensors as monitoring probes programmed in microcontroller using C language. The developed hardware consists of vibration sensor of ADXL345, temperature sensor of AD594/595 of type K thermocouple, microcontroller, graphic liquid crystal display, real time clock, etc. The hardware is divided into two: one is based at the workstation (majorly meant to monitor machines behaviour) and the other at the base station (meant to receive transmission of machines information sent from the workstation), working cooperatively for effective functionalities. The resulting hardware built was calibrated, tested using model verification and validated through principles pivoted on least square and regression analysis approach using data read from the gear boxes of extruding and cutting machines used for polyethylene bag production. The results got therein confirmed related correlation existing between time, vibration and temperature, which are reflections of effective formulation of the developed concept.

Keywords Maintenance model · Agent hardware system · Conventional machines · Machine conditions monitoring

✉ Michael Kanisuru Adeyeri
adeyerimichaeltut@gmail.com; adeyerimk@tut.ac.za

Khumbulani Mpofu
mpofuk@tut.ac.za

Buliaminu Kareem
bkareem@futa.edu.ng

1 Department of Industrial Engineering, Tshwane University of Technology, Pretoria, South Africa

2 Department of Mechanical Engineering, The Federal University of Technology, Akure, Nigeria

Introduction

The manufacturing world is fast evolving, revolving and transforming to E-manufacturing especially in the developed world where technological advancement is changing rapidly every second which calls for the introduction and use of sophisticated, knowledge-based and highly intelligent machines for industrial uses. The maintenance of these machines which are purely done on e-maintenance platform as corroborated by Iung et al. (2009) are assisted by the use of inbuilt mechanism through artificial intelligence, expert system, neural network, agents and multi agents software that tends to automatically effect changes, carries out repairs and suggest possible means to avert the intending failures and total breakdown. With this dramatic change, consideration is no more given to the developing and underdeveloped countries as well as the less privileged (those who are into production and manufacturing activities at small scale level) who due to high cost could not adopt the use of these hybrid machines. This category of manufacturers has no option than to continue using the conventional machines at their reach and maintain them using the customary maintenance culture.

It is worth noting at this juncture to define what conventional machines are. According to Adeyeri et al. (2012), conventional machines are "machines which are operated manually. These machines are controlled by cams, gears, levers, or screws. Examples of these machines are Lathe, grinding machine, flaking machine, extruding machine and just to mention a few. They indeed needed special attention to safe guard or vouch safe for their functionality and optimal performance as compared to the non-conventional machines which are controlled automatically by integrated computer". In view of this, there should be a platform that bridges the gap between the rich manufacturing industries and the less privileged in the maintenance world or a platform that provides a face lift for the customary approach for maintenance practices. The customary maintenance technique is breakdown maintenance (which is also called unplanned maintenance, or run-to-failure maintenance), takes place only at breakdowns. Therefore this article gives an attempted concept of using an embedded approach in embedding sensors (agent hardware) for monitoring machinery conditions from the perspective view of vibration and temperature effects on machines performance.

The rest of this paper is structured as follows. "Literature review" section offers a brief review of the research in the area of maintenance and sensors description. "Methodology" section outlines the methods involved in developing the hardware components. The validation and verification of the development system are enclosed in "Results and discussion" section. Finally, last section summarizes conclusions of this work and outline guidelines for possible future work.

Literature review

Many researchers have written extensively on maintenance from various dimensions of types, framework, concepts and simulation to modelling theory. The works of Jardine et al. (1999), Koc and Lee (2001), Tao et al. (2003), Lee and Scott (2006), Lu et al. (2007), Mahantesh et al. (2008), Derigent et al. (2009) and Ahmad et al. (2011) were focused on maintenance types be it condition based, preventive and opportunistic. Some work results on maintenance plan, policies, strategies, review and framework had been showcased by Ucar and Qiu (2005), Tsang (2002), Albino et al. (1992), Yuan and Chaing (2000), Hausladen and Bechheim (2004), Dufuaa et al. (2001), Levrat et al. (2008), Iung et al. (2009), Muller et al. (2008) and Peng et al. 2010).

Some research outputs of Albino et al. (1992), Marquez and Herguedes (2002), Marquez et al. (2003), Zineb and Chadi (2001) and Ashraf (2004) are on maintenance models development for evaluation and optimal throughput. Results output through the application of simulation integration, artificial intelligence, neural network, genetic algorithm and knowledge-based expert into solving maintenance problems are already published by Dufuaa et al. (2001), Zineb and Chadi (2001), Andijani and Dufuaa (2002), Greasly (2005), Oladokun et al. (2006), Mahantesh et al. (2008), Voisin et al. (2010), Babaei et al. (2011) and Jasper et al. (2011).

Of recent times, in early 2000, the e-maintenance paradigm emerged. Technological development revealed that e-maintenance platform utilizes internet networking, intranetting and Extra-netting based on web technology, sensors application, wireless communications and mobile accessories (Iung and Marquez (2006) and Iung et al. (2009)). With this emerging technology, the concept of this work is therefore hinged on the application of sensors technology in assisting in the monitoring of machines behaviour from the view of vibration and temperature effects.

Lee et al. (2006) discussed that multiple degradation indicators built on sensor signals are the most potent tools needed for the real-time condition monitoring as potential failures could be attributed to many correlated degradation processes. It is in view of this that the present concept of hardware development is initiated. It is on the principle of embedded system.

Embedded systems are electronic systems that are intelligent enough to operate on their own without receiving operating commands from an external source. They are standalone systems which carry out their functions automatically with little supervision and operation. Examples of these systems are Global System for Mobile communication (GSM) handsets, calculators, microwave ovens, digital scrolling display, cameras, mouse, security alarm system, etc. At the centre of any embedded system is microcontroller. The remaining parts of this section are for the literature support of the core components for the proposed hardware (microcontroller, vibration and temperature sensors).

Microcontrollers

A microcontroller is a single chip, self-contained computer which incorporates all the basic components of a personal computer on a much smaller scale. Functionally, it is a programmable single chip which controls a process or a system. Microcontrollers are typically used as embedded controllers where they control part of a larger system such as an appliance, automobile, scientific instrument or a computer peripheral. Physically, a microcontroller is an integrated circuit with pins along each side which are used for power, ground, oscillator, input/output ports, interrupt request signals, reset and control.

There are various manufacturers of microcontrollers like Microchip (PIC series), Atmel (AVR series), Actel,

Motorola (MPC series), Maxim integrated products, Texas instruments, Sharp (ARM series), Panasonic, etc. The microcontroller unit can be programmed in any language ranging from Assembly language to Basic, C, Pascal, as long as a compiler exists to convert the code to its machine equivalent (Muhammad et al. 2011).

Vibration and temperature sensors

The ADXL345 is a small, thin, low power, three-axis accelerometer with high resolution (13-bit) measurement at up to ±16 g. Digital output data are formatted as 16-bit twos complement and is accessible through either a SPI (3- or 4-wire) or I2C digital interface. The ADXL345 is well suited for mobile device applications. It measures the static acceleration of gravity in tilt-sensing applications, as well as dynamic acceleration resulting from motion or shock. Its high resolution (4 mg/LSB) enables measurement of inclination changes <1.0° (OTW 2009). The ACCEL Board is the vibration sensor board made by Microelectronika, it is built around ADXL345, and the means of communication with the microcontroller is through the Serial Peripheral Interface (SPI). The board can measure vibration in three axes, with a resolution of 16 g (Atmel 2010).

Temperature measurement can be accomplished using several types of sensing mechanisms. Temperature measurement systems generally consist of a sensor, a transmitter, an external power supply (for some types of systems), and the wiring that connects these components. The temperature measurement sensors most commonly used in engineering applications are thermocouples, Resistance Temperature Detectors (RTDs), and infrared (IR) thermometers.

For simplicity, reliability, and relatively low cost, thermocouples are widely used. They are self-powered, eliminating the need for a separate power supply to the sensor. Thermocouples are durable when they are appropriately chosen for a given application. Thermocouples also can be used in high-temperature applications, such as incinerators (Grieb 1992).

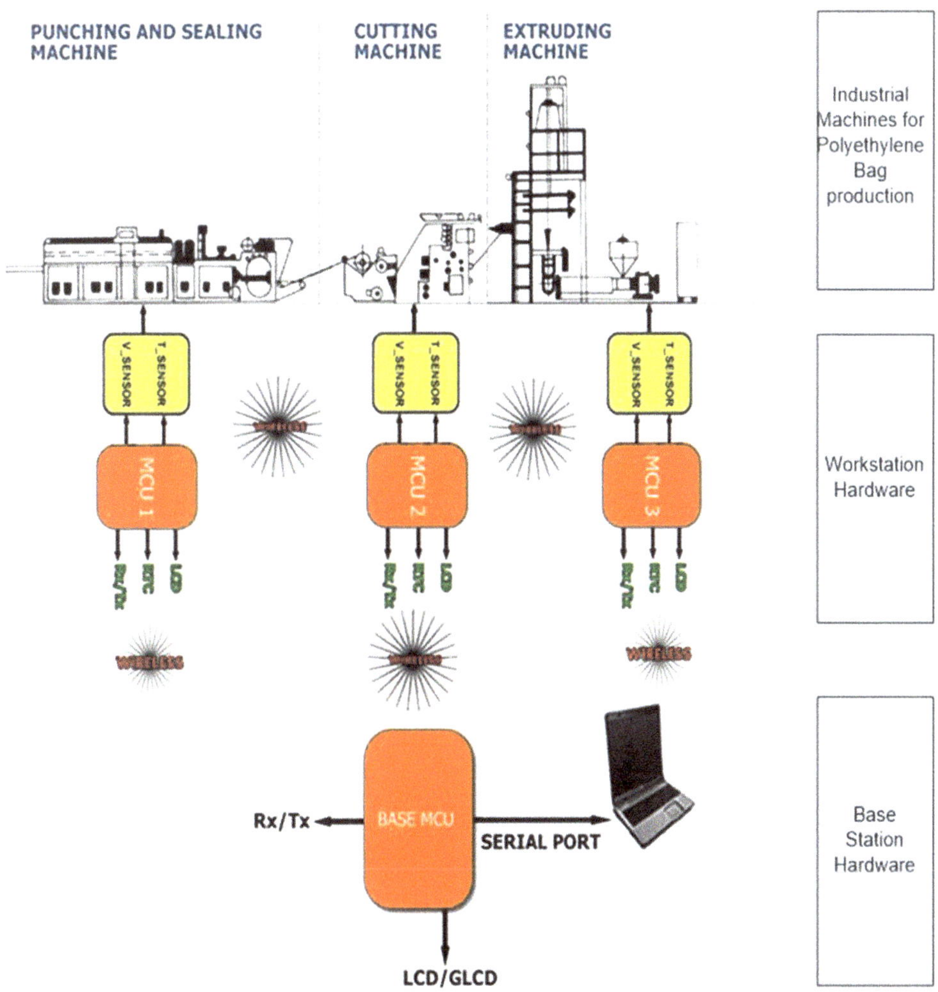

Fig. 1 Hardware architecture of the designed concept

Methodology

The description of the methods behind this concept of maintenance monitoring shall be viewed under monitoring concept formulation, model formulation for vibration and temperature, algorithm formulation, materials needed for the hardware developed and its circuitry and calibration of the developed hardware.

Hardware development for maintenance monitoring concept

The block diagram representing the hardware concept and design for the condition monitoring is as shown in Fig. 1. As depicted, the embedded design consists of a base station embedded system connected directly to a PC and remote machine monitoring systems known as the substation or work station connected directly to the machines which can be produced in the number of machines in the industrial plant of interest.

It is worth noting that only the workstation hardware would be discussed fully in this research article.

Materials for hardware development

The tools and equipment needed for the hardware design and its implementation are as listed:

1. Computer hardware: (Vibration Transducer (ADXL345, AD595, type K thermocouple), AVR Microcontroller (Atmega16; Atmega32); Real Time clock (RTC); Liquid Crystal Display (LCD), MJ MRF24J40MA (2.4 GHz zigbee) transceiver, transformer, capacitor, resistors, MiKroC board platform, and Vero board.
2. Computer software: Microsoft visual studio, C# and C language
3. Auxiliary tools: soldering lead, soldering iron, and USB.

Model formulation

Maintenance based model is formulated from the view of vibrational and temperature effect on machinery.

Temperature-based maintenance model

Adeyeri et al. (2012) assumed that if T_i is the initial temperature value, and T_o defined as the measured temperature before the predicted value of temperature at next planned time of measurement or reading, then the temperature deteriorating factor, U_T is expressed as

$$U_T = \frac{T_i - T_o}{T^c_m} \tag{1}$$

where T^c_m is critical temperature limit level.

Therefore, $T_i^{t_a}$, which is the predicted value of temperature at next planned time of measurement or reading taken, would be

$$T_i^{t_a} = \left[T_i^o + T_i^o U_T\right]_b^{t_n} \tag{2}$$

Simplifying Eq. (2) gives

$$T_i^{t_a} = T_i^o[1 + U_T]_b^{t_n} \tag{3}$$

where t_n is the periodic time numbering of readings and b is a function of speed, environmental condition and product demand.

Therefore, if $T_i^{t_a} \geq T^c_m$, then maintenance is required, otherwise do not.

Vibration-based maintenance model

Dynamic components of machinery do give rise to one form of vibration or another. This vibration generated by its dynamic components is a potent parameter for condition monitoring. Condition monitoring based on vibration measurement and analysis can be carried out either on-load or off-load.

When dealing with Vibration-Based Maintenance (VBM), the condition of significant parts cannot be assessed effectively, i.e. with high certainty, without considering both probabilistic and deterministic aspects of the degradation process. Modelling the time for maintenance action and predicting the value of the vibration level when damage of a significant component is detected are examples of the probabilistic part. However, issues related to machine function, failure analysis and diagnostics are examples of the deterministic part.

Alsyouf (2004) formulated a sub model for predicting vibration level during the next period and until the next measuring moment as stated in Eq. (4). The equation formulated assumed that V be the dependent variable representing the predicted value of the vibration level and that V is a function of three independent variables (V_i, Z, t) and three parameters (a, b, c), for $i = 1, 2 \ldots n$, and i is the number of measuring opportunities.

$$V_{i+1} = \left[V_i + ae^{-(b_i t_{i+1} z_i^{c_i})}\right]_k. \tag{4}$$

where V_{i+1} is the predicted value of the vibration level at the next planned measuring time. t_{i+1} the elapsed time since the damage is initiated and its development is detected. V_i the current vibration level value and Z_i the deterioration factor, i.e. the function of the current and anticipated future load and previous deterioration rate.

Fig. 2 Flowchart of condition based maintenance monitoring

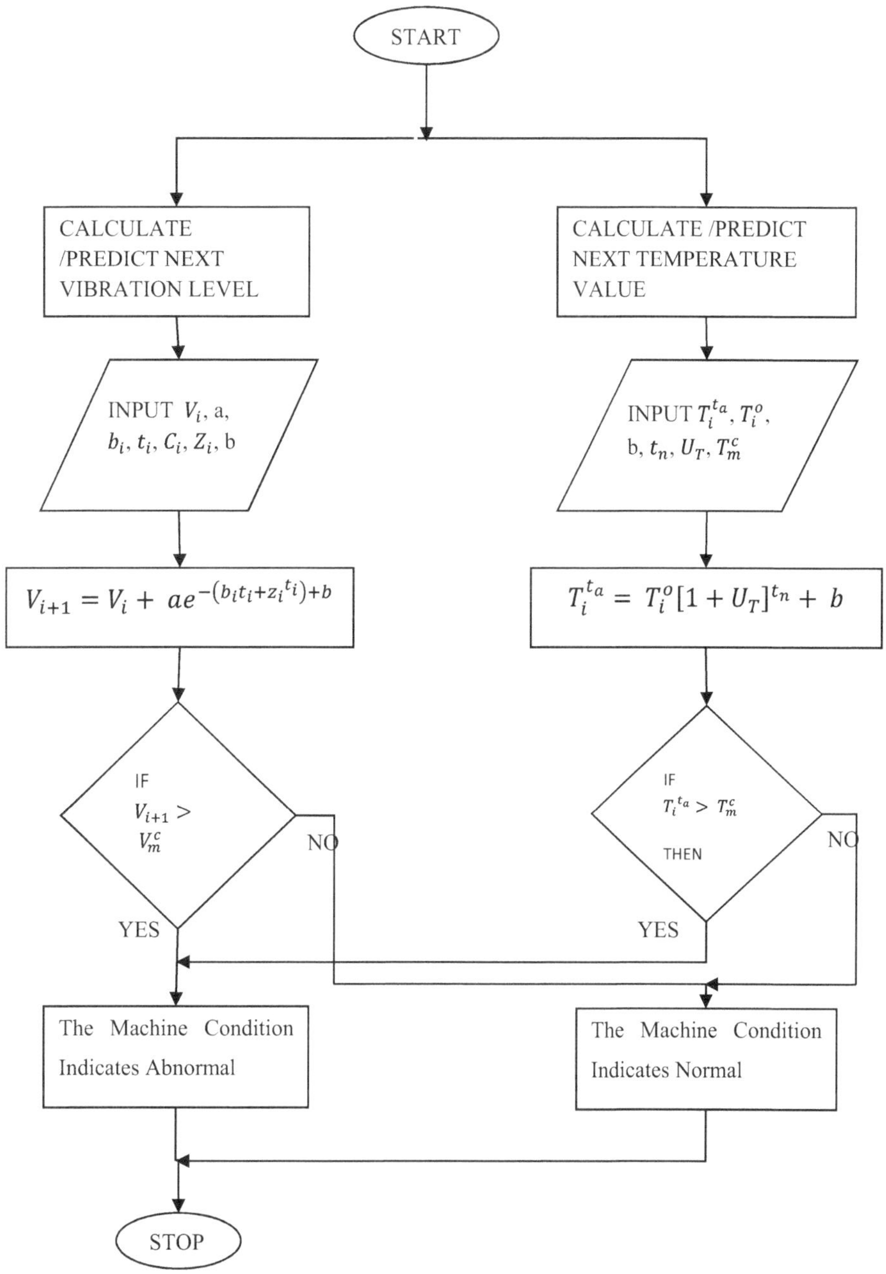

$$Z_i = \left| \frac{V_i - V_O}{V_C} \right| \tag{5}$$

where V_c is the critical vibration level which is to be supplied by the manufacturer, V_o is the measured vibration before V_{i+1}, a the gradient (slope) by which the value of the vibration level varied since it started to deviate from its normal state (V_o) due to initiation of damage until detecting it at V_p, b_i and c_i are on-linear model's constants, K is a function of loading, speed of machine, environmental condition, and E_i the model error, which is assumed to be identical, independent and normally distributed with zero mean and constant variance, N (0, σ).

When $V_{i+1} \geq V_c$ then maintenance is required, otherwise do not.

Mechanical vibrations a characterised in severity by one and all of these three basic parameters, viz a viz: displacement, velocity or acceleration. Velocity which is best suited for intermediate frequency range has found to give the best indication of severity over the wildest range of frequencies and hence has the wildest application in condition monitoring.

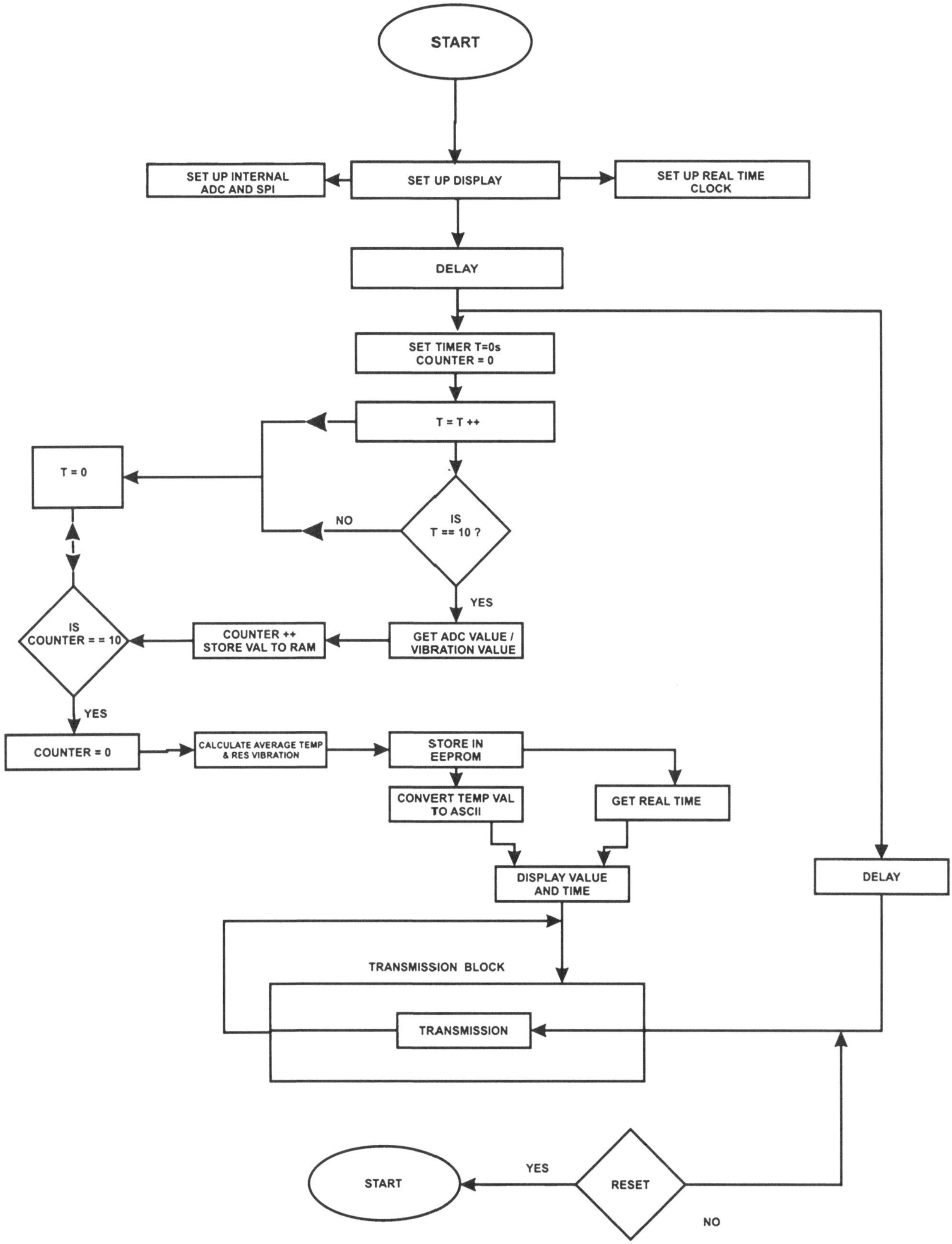

Fig. 3 Flowchart for workstation module

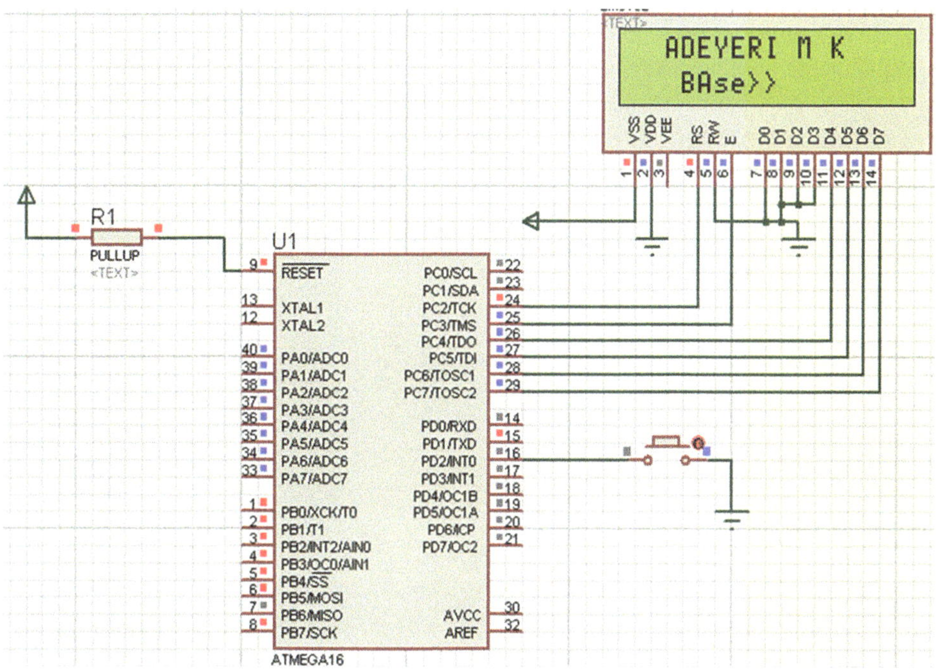

Fig. 4 LCD customization for text display

Model flowchart

The flowchart for implementing the vibration and temperature modeis as shown in Fig. 2.

The flowchart used in coding the temperature sensor and the vibration sensor so as to give room for the computer machine interface under the C language is as shown in Fig. 3.

Development of hardware circuitry and interface to the microcontrollers

As depicted in the block diagram of Fig. 1, at the centre of each of the monitoring hardware, there is a microcontroller interfaced to the sensors and other peripherals to enhance human interaction with the system.

According to the pictorial representation, the following hardware was interfaced to the monitoring system:

1. Liquid crystal display(LCD),
2. Real time clock (RTC), and
3. Transceiver.

These are as discussed as regards their operations.

Base station circuitry

Broadly speaking, the base station circuitry is divided into:

1. Serial port circuitry
   The serial port circuitry handles the communication connection of the computer machines hardware interface of the base station with the computer system.
2. Wireless transceiver circuitry
   Another important functionality added to this project is the ability of each monitoring system to be able to communicate with the base station system. In order to establish this, a wireless device was used. The wireless device used for this project is MRF24J40 Zigbee wireless transceiver. An internal transmit/receive (TR) switch combines the transmitter and receiver circuits into differential RFP and RFN pins. These pins are connected to impedance matching circuitry and antenna. An external power amplifier (PA) and could be controlled via the GPIO pins.
3. The display (LCD) circuitry.
   The schematic of Fig. 4 shows the circuitry displayed when it was being simulated on PROTEUS Professional for circuit designs (NIC 2011).

Workstation and base station circuitry design

The workstation circuitry is divided into: temperature sensor circuitry, vibration sensor circuitry, real time clock circuitry, display circuitry, and wireless transceiver circuitry.

The workstation and base station circuitry is as shown in Fig. 5, while Fig. 6a–c shows how the workstation hardware was built.

The development of the hardware was carried out in the following sequence.

1. The schematics design of the hardware was done with the aid of look-up tables, data sheets of the various components being interfaced on Proteus ISIS software.

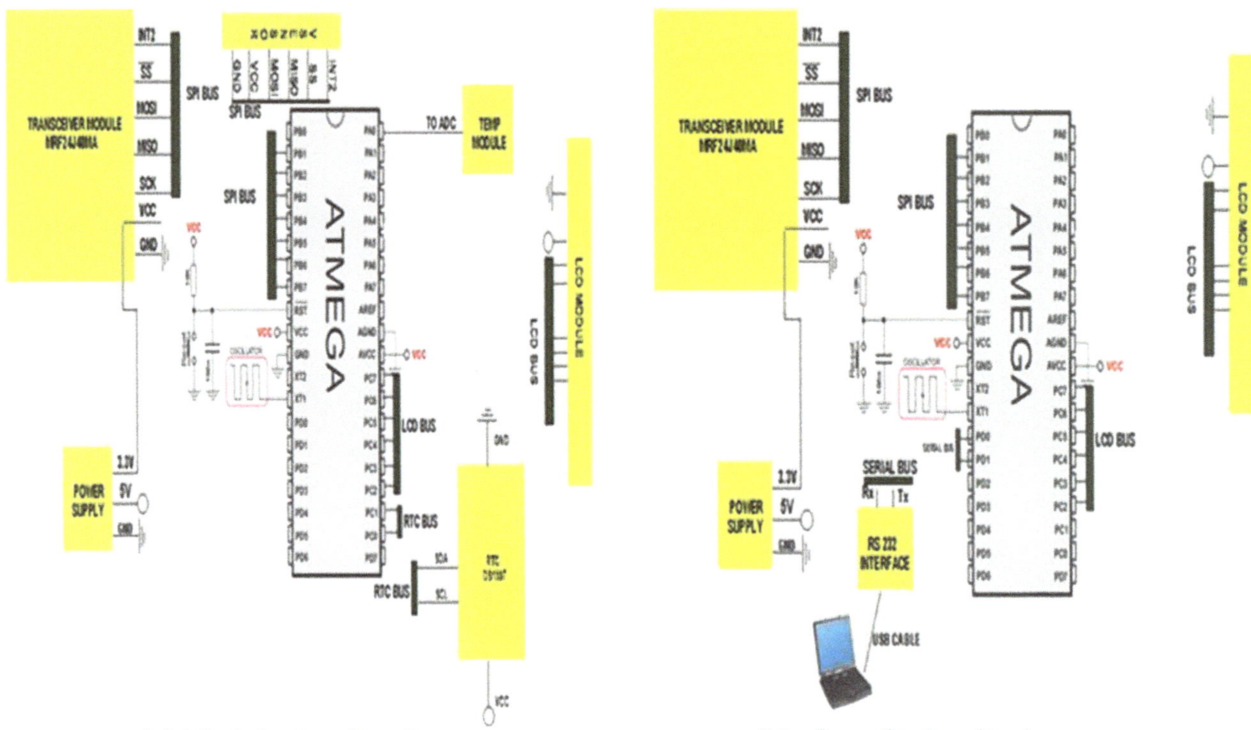

(a) Work Station Circuitry (b) Base Station Cicuitry

Fig. 5 Circuit diagram for the developed work station and base station

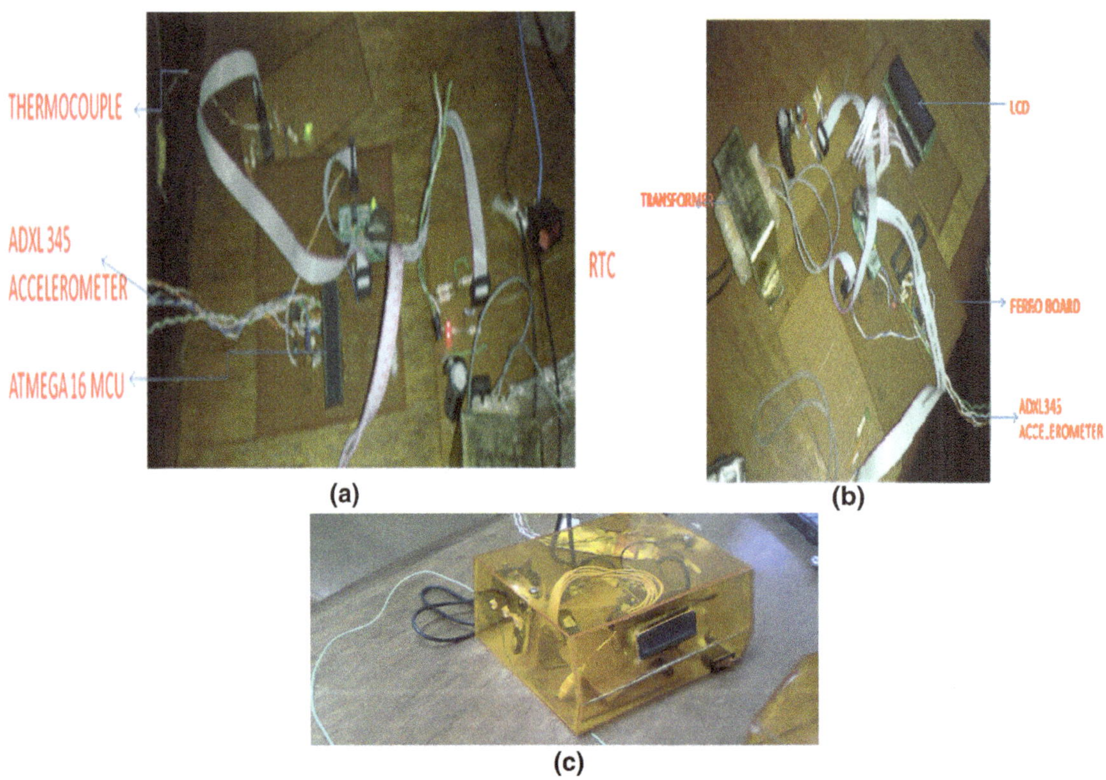

Fig. 6 Workstation hardware components on board at point of testing and the assembled hardware

2. Each component was placed in position on the Vero board and held in connection using soldering leads and a hot soldering iron. This shows the implementation of the designed circuitry for the data acquisition system with some of the components soldered on the Vero board.
3. The embedded programming of ATMEGA32 microcontroller was done by interfacing it with a PC using a programming codes and the MikroC IDE. Program codes written in the C language were converted to their hexadecimal equivalent and written to the EEPROM memory of the microcontroller. The completed device is shown in Fig. 6. All components used were soldered on Vero board and a casing was built for them to give a protection of the device's circuitry.

Hardware circuitry and interface to the microcontrollers of the base station

As depicted in the block diagram of Fig. 5b, the monitoring systems are meant to sense the required conditioning parameters such as temperature and vibration, store the data and each transmits the stored data to the base station over wireless on request by the base station. The base station on the other hand, on receiving the data, transmits the data to the PC via serial port to a standalone windows software application developed to analyse the data and make necessary decisions to enhance maintenance of the machines and productivity. To enhance interactivity with this base station by the users, the following design provision is used:

1. Serial port communication between PC and Base station device, and
2. Graphical liquid crystal display (GLCD) interface.

Power supply

The entire system was powered by a central power supply, which is capable of providing 5, 3.3 and 3.0 V. The design is composed of a voltage transformer, which steps down the 220 V supply to about 16 V. The different sections of the system that need power supply get them from this central supply. The microcontroller for instance needs an absolute 5.0 V supply, the RTC needs 3.0 Volts and the ACCEL board needs 3.3 V and so on. These distributions were made through the effective use of ICU connectors of two by five and two by four jumpers.

Calibration of hardware

The calibration of the temperature thermocouple type K AD595 and the vibration ADXL345 sensors were calibrated based on their specification from their manufacturers (MickroElectronika).

Temperature calibration

To produce a temperature proportional output of 10 mV/°C, and provide an accurate reference junction over the rated operating temperature range, the AD595 is again trimmed at the factory to match the transfer characteristics of Type K thermocouple at +25 °C. At this calibration temperature, the Seebeck coefficient, the rate of change of thermal voltage with respect to temperature at a given temperature is 0.44 μV/°C for a Type K. This corresponds to a gain of 247.3 for the AD595 to realize a 10 mV/°C out-put. Although the device is trimmed for a 250 mV output at +25 °C, an input offset error is induced in the output amplifier resulting in offsets of 11 μV for the AD595 (Joe 2011). To determine the actual output voltage from the AD595, the voltage output, V_o is therefore expressed as

$$V_o = (V_k + 11\ \mu V) \times 247.3 \tag{6}$$

where V_k is the type K voltage. And the corresponding temperature, T °C would be

$$T^{\circ}C = \frac{V_o}{10\ mV/^{\circ}C} \tag{7}$$

$$\text{At } 25^{\circ}C, \quad V_o = 250\ mV \tag{8}$$

$$\text{and at } X^{\circ}C,\ V_o \text{ is} \quad V_o = V_x\ mV \tag{9}$$

where V_x is the voltage read out at corresponding temperature of X °C.

From Eqs. (7) and (9), it implies that change in temperature would be

$$\Delta T = (X - 25)^{\circ}C \tag{10}$$

and change in voltage be

$$\Delta V = (V_x - 250)\ mV \tag{11}$$

With reference from Eq. (7), change in temperature is therefore expressed as

$$\Delta T\ ^{\circ}C = \frac{\Delta V_o}{10\ mV/^{\circ}C} \tag{12}$$

Therefore,

$$X - 25 = \frac{\Delta V}{10\ mV/^{\circ}C} \tag{13}$$

Further simplification of Eq. (13) gives X to be

$$X = \left(\frac{\Delta V}{10\ mV/^{\circ}C} + 25\right) \tag{14}$$

when $\Delta V = V_o - 0.25$

Table 1 Vibration and temperature readings from extruder

Time (min)	V (mm/s)	T (°C)	Time (min)	V*(mm/s)	T (°C)	Time (min)	V (mm/s)	T (°C)
1	1.000	35.50	51	0.980	45.80	101	1.420	50.90
2	1.000	35.52	52	1.100	45.90	102	1.440	51.00
3	1.000	35.54	53	1.100	46.00	103	1.460	51.00
4	1.010	35.56	54	1.100	46.10	104	1.480	51.00
5	1.010	35.58	55	1.100	46.20	105	1.500	51.00
6	1.010	35.60	56	1.100	46.30	106	1.520	51.00
7	1.010	35.62	57	1.100	46.40	107	1.540	51.00
8	1.010	36.02	58	1.100	46.50	108	1.560	51.00
9	1.020	36.42	59	1.100	46.60	109	1.550	51.00
10	1.020	36.82	60	1.100	46.70	110	1.540	51.00
11	1.020	37.22	61	1.100	46.80	111	1.530	51.00
12	1.020	37.62	62	1.100	46.90	112	1.520	51.00
13	1.020	38.02	63	1.100	47.00	113	1.510	51.00
14	1.030	38.42	64	1.100	47.10	114	1.500	51.00
15	1.030	38.82	65	1.100	47.20	115	1.490	51.00
16	1.030	39.22	66	1.100	47.30	116	1.480	51.00
17	1.020	39.62	67	1.100	47.40	117	1.470	51.00
18	1.020	40.02	68	1.100	47.50	118	1.460	51.00
19	1.020	40.42	69	1.100	47.60	119	1.450	51.00
20	1.010	40.82	70	1.100	47.70	120	1.440	51.00
21	1.010	41.22	71	1.100	47.80	121	1.430	51.00
22	1.010	41.62	72	1.100	47.90	122	1.420	51.00
23	1.000	42.02	73	1.100	48.00	123	1.410	51.00
24	1.000	42.42	74	1.100	48.10	124	1.400	51.00
25	1.000	42.82	75	1.100	48.20	125	1.390	51.00
26	1.000	43.22	76	1.100	48.30	126	1.380	51.00
27	0.990	43.62	77	1.100	48.40	127	1.370	51.00
28	0.990	44.02	78	1.100	48.50	128	1.360	51.00
29	0.990	44.50	79	1.100	48.60	129	1.350	51.00
30	0.980	44.50	80	1.100	48.70	130	1.340	51.00
31	0.980	44.50	81	1.100	48.80	131	1.330	51.00
32	0.980	44.50	82	1.100	48.90	132	1.320	51.00
33	0.970	44.50	83	1.100	49.00	133	1.310	51.00
34	0.970	44.50	84	1.100	49.10	134	1.300	51.00
35	0.970	44.50	85	1.100	49.20	135	1.290	51.00
36	0.970	44.50	86	1.100	49.30	136	1.280	51.00
37	0.960	44.50	87	1.120	49.40	137	1.270	51.00
38	0.960	44.50	88	1.140	49.50	138	1.260	51.00
39	0.960	44.50	89	1.160	49.60	139	1.250	51.00
40	0.950	44.60	90	1.180	49.70	140	1.240	51.00
41	0.960	44.70	91	1.200	49.80	141	1.260	51.00
42	0.960	44.80	92	1.220	49.90	142	1.280	51.00
43	0.960	44.90	93	1.240	50.00	143	1.300	51.00
44	0.960	45.00	94	1.260	50.10	144	1.320	51.00
45	0.960	45.10	95	1.280	50.20	145	1.340	51.00
46	0.970	45.20	96	1.300	50.30	146	1.360	51.00
47	0.970	45.30	97	1.320	50.40	147	1.380	51.00
48	0.970	45.40	98	1.340	50.50	148	1.400	51.00
49	0.970	45.50	99	1.360	50.60	149	1.420	51.00
50	0.970	45.60	100	1.380	50.70	150	1.440	51.00

V vibration, *T* temperature

Table 2 Correlation of time, vibration and temperature data

	Time of readings (min)	Vibration value (mm/s)	Temp (°C)
Time of readings (min)	1		
Vibration value (mm/s)	0.840634	1	
Temp (°C)	0.917332	0.746613	1

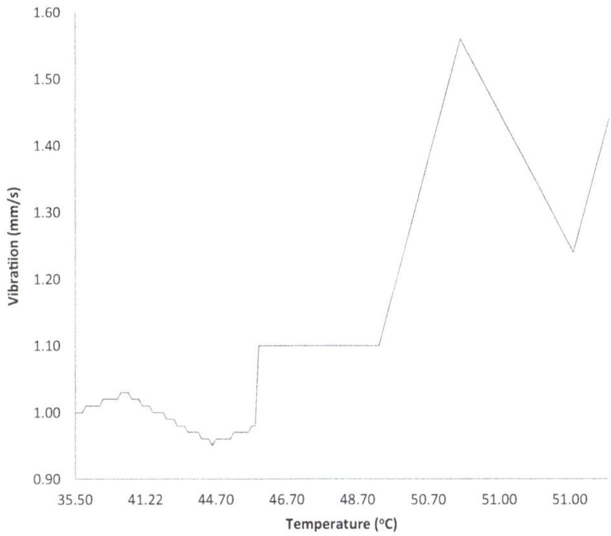

Fig. 7 Graph of vibration against temperature

$$\Delta T = X - 25$$

and "X °C" which is the present read out value would be

$$X = \Delta T + 25. \quad (15)$$

Vibration calibration

Accel SPI board of ADXL345 measures vibration in x, y and z axes. Adopting the principle of OTW (2009), the effective calibrated vibration V_b was got from the expression of (Eq. 16). As

$$V_b = \sqrt{x^2 + y^2 + z^2} \quad (16)$$

The hardware was used in taking readings on their embedment into machines to see if the calibration is perfectly done. The readings obtained and the regression

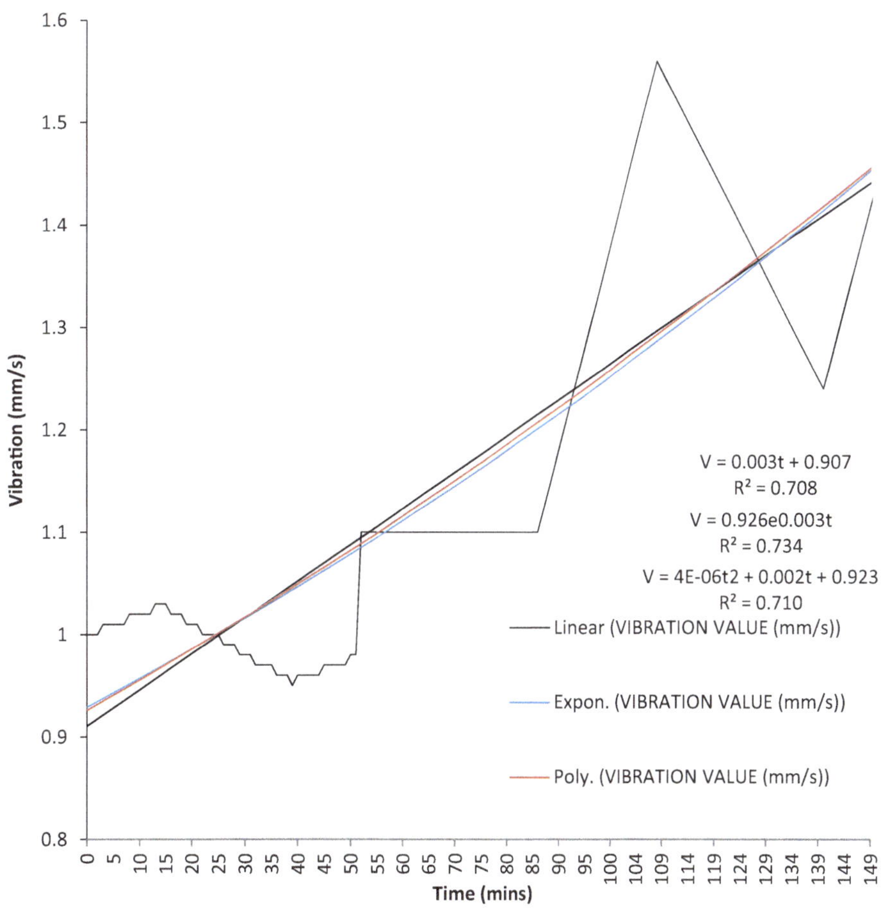

Fig. 8 Curve fitting on vibration data for vibration model calibration

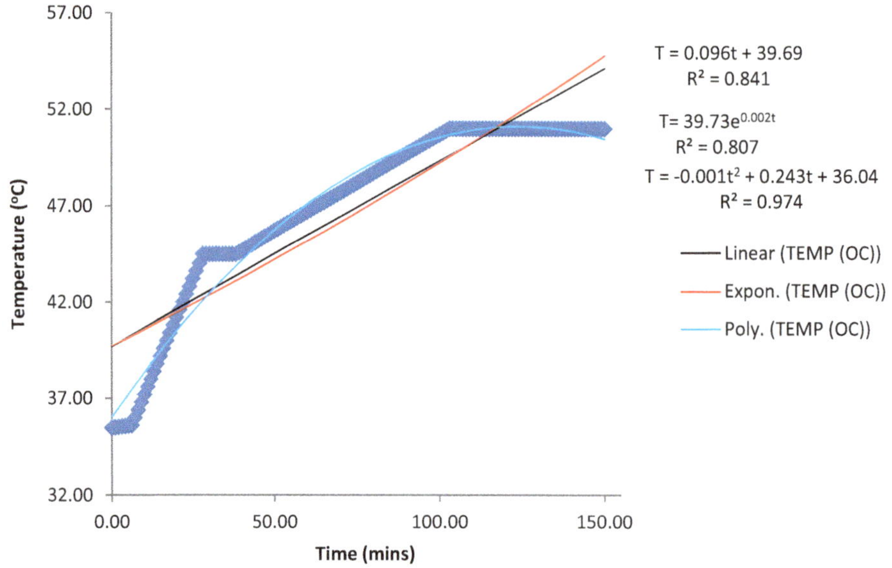

Fig. 9 Temperature data curve fitting for model calibration

Table 3 Observed and predicted vibration data from gear box of extruder

Time (min)	OV (mm/s)	PV (mm/s)	Time (min)	OV (mm/s)	PV (mm/s)	Time (min)	OV (mm/s)	PV (mm/s)
1	1.010	0.930	51	1.000	1.080	101	1.300	1.260
2	1.010	0.930	52	1.000	1.080	102	1.300	1.260
3	1.010	0.930	53	1.000	1.090	103	1.300	1.260
4	1.010	0.940	54	1.100	1.090	104	1.300	1.270
5	1.010	0.940	55	1.100	1.090	105	1.300	1.270
6	1.010	0.940	56	1.100	1.100	106	1.300	1.280
7	1.010	0.940	57	1.100	1.100	107	1.300	1.280
8	1.000	0.950	58	1.100	1.100	108	1.220	1.280
9	1.000	0.950	59	1.100	1.110	109	1.220	1.290
10	1.000	0.950	60	1.100	1.110	110	1.220	1.290
11	1.000	0.960	61	1.100	1.110	111	1.220	1.290
12	1.000	0.960	62	1.100	1.120	112	1.220	1.300
13	1.000	0.960	63	1.100	1.120	113	1.220	1.300
14	1.000	0.960	64	1.100	1.120	114	1.220	1.310
15	1.020	0.970	65	1.100	1.130	115	1.220	1.310
16	1.020	0.970	66	1.100	1.130	116	1.290	1.310
17	1.020	0.970	67	1.120	1.130	117	1.290	1.320
18	1.020	0.980	68	1.140	1.140	118	1.290	1.320
19	1.020	0.980	69	1.160	1.140	119	1.290	1.330
20	1.020	0.980	70	1.180	1.140	120	1.290	1.330
21	1.020	0.990	71	1.200	1.150	121	1.240	1.330
22	1.020	0.990	72	1.120	1.150	122	1.240	1.340
23	1.080	0.990	73	1.120	1.160	123	1.240	1.340
24	1.080	0.990	74	1.120	1.160	124	1.240	1.350
25	1.080	1.000	75	1.120	1.160	125	1.240	1.350
26	1.080	1.000	76	1.120	1.170	126	1.240	1.350
27	1.080	1.000	77	1.120	1.170	127	1.240	1.360
28	1.080	1.010	78	1.120	1.170	128	1.280	1.360
29	1.080	1.010	79	1.120	1.180	129	1.280	1.370
30	1.080	1.010	80	1.120	1.180	130	1.280	1.370

Table 3 continued

Time (min)	OV (mm/s)	PV (mm/s)	Time (min)	OV (mm/s)	PV (mm/s)	Time (min)	OV (mm/s)	PV (mm/s)
31	1.080	1.020	81	1.120	1.180	131	1.280	1.370
32	1.080	1.020	82	1.120	1.190	132	1.280	1.380
33	1.080	1.020	83	1.120	1.190	133	1.280	1.380
34	1.080	1.020	84	1.120	1.190	134	1.280	1.390
35	1.080	1.030	85	1.120	1.200	135	1.280	1.390
36	1.080	1.030	86	1.120	1.200	136	1.340	1.400
37	1.080	1.030	87	1.200	1.200	137	1.340	1.400
38	1.000	1.040	88	1.200	1.210	138	1.340	1.400
39	1.000	1.040	89	1.200	1.210	139	1.340	1.410
40	1.000	1.040	90	1.200	1.220	140	1.340	1.410
41	1.000	1.050	91	1.200	1.220	141	1.340	1.420
42	1.000	1.050	92	1.200	1.220	142	1.340	1.420
43	1.000	1.050	93	1.200	1.230	143	1.340	1.430
44	1.000	1.060	94	1.200	1.230	144	1.340	1.430
45	1.000	1.060	95	1.200	1.230	145	1.340	1.430
46	1.000	1.060	96	1.200	1.240	146	1.340	1.440
47	1.000	1.070	97	1.200	1.240	147	1.260	1.440
48	1.000	1.070	98	1.300	1.250	148	1.260	1.450
49	1.000	1.070	99	1.300	1.250	149	1.260	1.450
50	1.000	1.070	100	1.300	1.250	150	1.260	1.460

OV observed vibration values, *PV* predicted vibration values

Table 4 Observed and predicted temperature data from electric motor of cutting machine

Time (min)	OT (°C)	PT (°C)	Time (min)	OT (°C)	PT (°C)	Time (min)	OT (°C)	PT (°C)
1	36.00	36.53	52	42.00	47.82	102	48.00	53.86
2	36.00	36.80	53	43.00	47.99	103	48.00	53.93
3	36.00	37.07	54	43.00	48.15	104	48.00	53.99
4	36.00	37.34	55	43.00	48.32	105	48.00	54.06
5	36.00	37.61	56	43.00	48.48	106	48.00	54.12
6	36.00	37.87	57	43.00	48.65	107	48.00	54.19
7	38.00	38.13	58	43.00	48.80	108	48.00	54.25
8	38.00	38.39	59	43.00	48.96	109	49.00	54.30
9	38.00	38.65	60	43.00	49.12	110	49.00	54.36
10	38.00	38.90	61	43.00	49.27	111	49.00	54.41
11	38.00	39.15	62	43.00	49.42	112	49.00	54.47
12	38.00	39.40	63	43.00	49.57	113	49.00	54.52
13	38.00	39.65	64	44.00	49.72	114	49.00	54.56
14	39.00	39.90	65	44.00	49.87	115	49.00	54.61
15	39.00	40.15	66	44.00	50.01	116	49.00	54.66
16	39.00	40.39	67	44.00	50.15	117	49.00	54.70
17	39.00	40.63	68	44.00	50.29	118	49.00	54.74
18	39.00	40.87	69	44.00	50.43	119	49.00	54.78
19	39.00	41.11	70	44.00	50.56	120	49.00	54.81
20	39.00	41.34	71	44.00	50.70	121	50.00	54.85
21	39.00	41.58	72	44.00	50.83	122	50.00	54.88
22	39.00	41.81	73	44.00	50.96	123	50.00	54.91
23	39.00	42.04	74	44.00	51.09	124	50.00	54.94

Table 4 continued

Time (min)	OT (°C)	PT (°C)	Time (min)	OT (°C)	PT (°C)	Time (min)	OT (°C)	PT (°C)
24	40.00	42.26	75	44.00	51.21	125	50.00	54.96
25	40.00	42.49	76	44.00	51.34	126	50.00	54.99
26	40.00	42.71	77	45.00	51.46	127	50.00	55.01
27	40.00	42.93	78	45.00	51.58	128	50.00	55.03
28	40.00	43.15	79	45.00	51.70	129	50.00	55.05
29	40.00	43.37	80	45.00	51.81	130	50.00	55.06
30	40.00	43.58	81	45.00	51.93	131	50.00	55.08
31	40.00	43.80	82	45.00	52.04	132	50.00	55.09
32	40.00	44.01	83	46.80	52.15	133	50.00	55.10
33	41.00	44.22	84	46.80	52.26	134	50.00	55.11
34	41.00	44.43	85	46.80	52.36	135	50.00	55.12
35	41.00	44.63	86	46.80	52.46	136	50.00	55.12
36	41.00	44.83	87	46.80	52.57	137	50.00	55.12
37	41.00	45.04	88	46.80	52.67	138	52.00	55.12
38	41.00	45.24	89	46.80	52.76	139	52.00	55.12
39	41.00	45.43	90	46.80	52.86	140	52.00	55.12
40	41.00	45.63	91	46.80	52.95	141	52.00	55.11
41	41.00	45.82	92	47.00	53.05	142	52.00	55.10
42	42.00	46.01	93	47.00	53.14	143	52.00	55.10
43	42.00	46.20	94	47.00	53.22	144	52.00	55.08
44	42.00	46.39	95	47.00	53.31	145	52.00	55.07
45	42.00	46.58	96	47.00	53.39	146	52.00	55.05
46	42.00	46.76	97	47.00	53.48	147	52.00	55.04
47	42.00	46.94	98	47.00	53.56	148	52.00	55.02
48	42.00	47.12	99	47.00	53.63	149	52.00	55.00
49	42.00	47.30	100	48.00	53.71	150	52.00	54.97
50	42.00	47.47						

OT observed temperature values, *PT* predicted temperature values

analyses of curve fittings on them for verification of the calibration were fully discussed in subsequent section.

Results and discussion

A polyethylene bag production industry was used as case study for the implementation of the developed hardware. The key machines utilized are extruding machine, cutting, sealing and punching machines were the main machines in which the built hardware were used on.

Data verification and analysis

The temperature and vibration models developed were verified using some of the data collected from the machines by carrying out curve fitting on them.

Table 1 gives the extracted data of machines' vibration and temperature. A correlation test was performed to know if the parameters considered were related. The correlation result displayed in Table 2 shows that vibration tends to increase as time increases since the value of R^2 value is 0.841 which is almost 1. Also temperature is correlated with time since the R^2 value is 0.917.

Consequently, the R^2 value of 0.7466 arrived at gave the indication that there exit a correlation relationship between vibration and temperature. Plotting the values of these factors against each other, a nonlinear relationship was got as depicted on Fig. 7. However, the relationship is non-linear, but there is indication that as vibration increases with time, temperature also increases.

The vibration verification was done using the developed hardware system to retrieve data from the polyethylene bag production machines. Some data were extracted from the retrieved data as seen in Table 2. The verification principle adopted is the curve fitting/Goodness of fit approach. The data in Table 2 was plotted and the result is as shown in Fig. 8. From this figure, three curves fitting were done, namely; linear, exponential and polynomial, while Eqs. (17), (18) and (19), respectively, described their

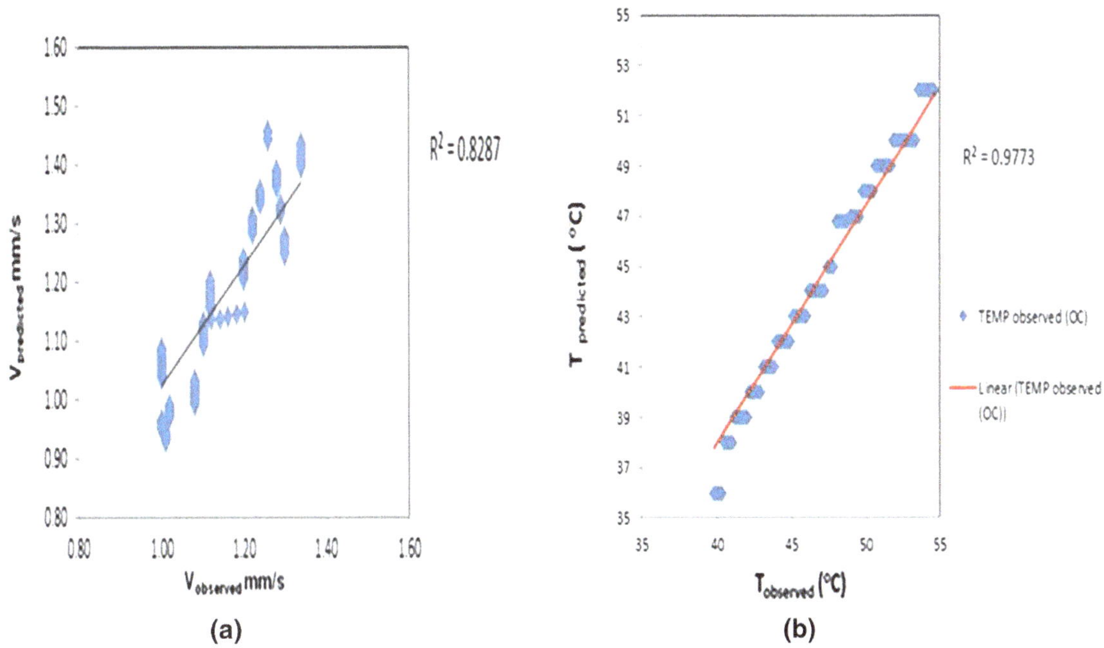

Fig. 10 **a** Graph of vibration model validation, **b** temperature model validation

mathematics formulation and their corresponding R^2 values.

$$V = 0.003t + 0.91 \quad (17)$$

and its $R^2 = 0.706$

$$V = 0.928\ell^{0.003t} \quad (18)$$

and its $R^2 = 0.734$

$$V = 0.000004t^2 + 0.002t + 0.925 \quad (19)$$

and its $R^2 = 0.708$

The exponential line or curve of best fit was picked based on the value of R^2, which is closest to value one (1) as compared to the other curves. This is a confirmation of the formulated vibration model of Eq. (4).

Similarly, the temperature calibration was done using the developed hardware system for both the substation and base-station to retrieve data from the polyethylene bag production machines. The temperature data in Table 2 was plotted and the result is as shown in Fig. 8. From this figure, three curves fitting were done, namely; linear, exponential and polynomial. The equations describing the various curves and their corresponding R^2 values are as stated in Eqs. (20), (21) and (22).

$$T = 0.096t + 39.69 \quad (20)$$

and its $R^2 = 0.841$

$$T = 39.73\ell^{0.002t} \quad (21)$$

and its $R^2 = 0.807$

$$T = -0.000t^2 + 0.243t + 36.04 \quad (22)$$

and its $R^2 = 0.974$

Table 5 Anova analysis of the observed and predicted vibration data of gear box of extruder

Groups	Count	Sum	Average	Variance
Summary				
Time (min)	150	11325	75.5	1887.5
OV (mm/s)	150	172.44	1.1496	0.013026685
PV (mm/s)	150	175.89	1.1726	0.023730779

Source of variation	SS	*df*	MS	*F*	*p* value	*F* crit
ANOVA						
Between groups	552627.245	2	276313.6	439.1654172	3.2E − 106	3.015899
Within groups	281242.9769	447	629.1789			
Total	833870.2219	449				

Table 6 Anova analysis of the observed and predicted temperature data of electric motor of cutting machine

Groups	Count		Sum	Average		Variance
Summary						
Time (min)	148		11,173	75.49324324		1904.673423
OT (°C)	148		6631.2	44.80540541		20.99697739
PT (°C)	148		7310.05	49.39222973		31.99060928
Source of variation	SS	*df*	MS	*F*	*p* value	*F* crit
ANOVA						
Between groups	81,106.23258	2	40,553.11629	62.14525816	1.67E−24	3.016175
Within groups	287,776.1685	441	652.55367			
Total	368,882.4011	443				

The linear line or curve of best fit was picked because the value of R^2 of the linear curve is closer to one than that of the exponential curve. Also the equation of polynomial curve of second order is attesting to the fact that it should be linear as the coefficient of t^2 is −0.00. And this supported the formulated temperature model of Eq. (3) (Fig. 9).

Model validation

The model equations developed were used to predict values for machines vibration and temperature. The observed readings from the hardware and the predicted values of the developed equations for vibration and temperature are as tabulated in Tables 3 and 4, respectively.

As depicted in Fig. 10a, the predicted vibration data were plotted against the observed values to determine the vibration model validity.

The *R* value of 0.909 (which is almost 91 %) got from the plotted data as contained in Fig. 10a showed that the vibration model calibrated on the sensor is accurate to define the vibration behavior of the machine since it is closer to the value of one (1).

Also, Fig. 10b displayed the one-to-one plotting of the data in Table 4. The predicted temperature values were plotted against the observed values so as to determine the temperature model validity.

Consequently, using Tables 3 and 4 for the ANOVA analysis of the observed and predicted data, Tables 5 and 6 resulted therein.

The *R* value of 0.988 (which is almost 99 %) got from the plotted data as contained in Fig. 10b showed that the temperature model calibration is accurate to define the temperature characteristics of the machine since it is closer to the value of one (1). Also, the ANOVA analysis of the vibration and temperature data observed and predicted shows that the respective *p* values of 3.2×10^{-106} and 1.67×10^{-24} is an indication that the calibration of the hardware systems is highly significant.

Conclusion

This paper suggests a novel mechanism for machine condition monitoring of conventional machines using the developed hardware. The condition monitoring of production machines as means of monitoring machines' condition was achieved through the hardware developed for workstation and base station. The work-station hardware through the ADXL345 vibration sensor and type K thermocouple among other components embedded in it monitored the vibration and temperature of the machines, respectively. The behavioral patterns of the machines communicated to the base station hardware through a wireless transceiver were received from the computer system connected to the base station through universal serial bus port.

The paper established that there is correlation between time, vibration and temperature. As vibration increases with time, temperature also increases. Also, it has provided a platform of real time monitoring for conventional machines.

Due to lack of appropriate test equipment, it is recommended that further work should be done on work station hardware development that would accommodate the detection of fluid lubrication wear that occurs during operation and severe fatigue wear resulting from rolling contact fatigue in dry lubrication. Future work can be carried out by extending this approach to a system with multiple deterioration measures, and with multiple subsystems or multiple deterioration failure modes.

Acknowledgments TN Limited is highly appreciated for providing support in using their machines for testing the developed hardware system. Also, the Innovative Manufacturing and Design Solution Research Group, Department of Industrial Engineering, Tshwane University of Technology is held in high esteem.

References

Adeyeri MK, Kareem B, Aderoba AA, Adewale OS (2012) Temperature based embedded programming algorithm for conventional machines condition monitoring proceedings of the fourth international conference on future computing technology, international academy, research and industrial association (IARIA), Nice-France, pp 51–57

Ahmad R, Kamaruddin S, Azid I, Almanar I (2011) Maintenance management decision model for preventive maintenance strategy on production equipment. J Ind Eng Int 7(13):22–34

Albino V, Carella G, Okogbaa O (1992) Maintenance polices in just-in-time manufacturing lines. Int J Prod Res 30:369–382

Alsyouf I (2004) Cost effective maintenance for competitive advantages. Thesis for the degree of doctor of philosophy (terotechnology), School of Industrial Engineering, Växjö University, Sweden

Andijani A, Duffuaa S (2002) Critical evaluation of simulation studies in maintenance systems. Prod Plan Control 13(4):336–341

Ashraf WL (2004) A decision analysis model for maintenance policy selection using a CMMS. J Qual Maint Eng 10(3):191–202

Atmel (2010) 8-bit Microcontroller with 16 K bytes in-system programmable flash ATmega16. Atmel Incorporation, San Jose

Babaei H, Shahanaghi K, Bakhsha A (2011) A new approach for constraining failure probability of a critical deteriorating system Yard crane scheduling in port container terminals using genetic algorithm. J Ind Eng Int 7(12):52–59

Derigent W, Thomas E, Levrat E, Iung B (2009) Opportunistic maintenance based on fuzzy modelling of component proximity. CIRP Ann Manuf Technol 58:29–32

Duffuaa SO, Ben-Daya M, Al-Sultan KS, Andijani AA (2001) A generic conceptual simulation model for maintenance systems. J Qual Maint Eng 7(3):34–40

Greasly A (2005) Using system dynamics in a discrete-event simulation study of a manufacturing plant. Int J Oper Prod Manag 25(6):534–548

Grieb B (1992) Temperature measurement in process control. Advances in instrumentation and control, proceedings of the annual meeting of the instrument society of America, vol 47, Instrument Society of America, Research Triangle Park, North Carolina

Hausladen I, Bechheim C (2004) E-maintenance platform as a basis for business process integration. In: Proceedings of INDIN04, international conference on industrial informatics, pp 46–51

Iung B, Marquez CA (2006) Special issue on e-Maintenance. Comput Ind 57(6):473–606

Iung B, Levrat E, Marquez AC, Erbe H (2009) Conceptual framework for e-Maintenance: illustration by e-Maintenance technologies and platforms. Annu Rev Control 33(2009):220–229

Jardine AKS, Joseph T, Benjevic D (1999) Optimising condition-based maintenance decisions for equipment subject to vibration monitoring. J Qual Maint Eng 5(3):192–202. (http://www.emerald-library.com). Retrieved: June, 2012

Jasper V, Warse K, Hans W (2011) Managing condition-based maintenance technology, a multiple case study in the process industry. J Qual Maint Engi 17 (1): 40–62. (http://www.emerald-library.com. June, 2012)

Joe M (2011) Application note on thermocouple signal conditioning using the AD594/AD595. One Technology Way Norwood, USA. (http://www.analog.com. Accessed 26 Nov 2011)

Koc M, Lee J (2001) A system framework for next-generation e-Maintenance system. 2nd International symposium on environmentally conscious design and inverse manufacturing

Lee J, Scott LW (2006) Zero-breakdown machines and systems: productivity needs for next-generation maintenance. Eng Asset Manag 2006:31–43

Lee J, Ni J, Djurdjanovic D, Qiu H, Liao H (2006) Intelligent prognostic tools and e-maintenance. Comput Ind 57:476–489

Levrat E, Iung B, Marquez CA (2008) e-Maintenance: review and conceptual framework. Prod Plan Control 19(4):408–429

Lu S, Tu YC, Lu H (2007) Predictive condition-based maintenance for continuously deteriorating systems. Qual Reliab Eng Int 23:71–81. doi:10.1002/qre.823

Mahantesh N, Ramachandra A, Satosh Kumar AN (2008) Artificial intelligence-based condition monitoring for plant maintenance. Assem Autom 28(2):143–150

Marquez AC, Herguedas SA (2002) Models for maintenance optimization: a study for repairable systems and finite time periods. Reliab Eng Syst Saf 75(3):367–377

Marquez CA, Gupta JND, Herguedas SA (2003) Maintenance policies for a production system with constrained production rate and buffer capacity. Int J Prod Res 41(9):1909–1926

Muhammad AM, Sarmad N, Sepehr N (2011) The AVR microcontroller and embedded systems using assembly and C. Prentice Hall, London

Muller A, Suhner MC, Iung B (2008) Proactive maintenance for industrial system operation based on a formalised prognosis process. Reliab Eng Syst Saf 93(2):234–253

NIC-National Instruments Corporation (2011) Proteus software, National Instruments Corporation, Mopac Expwy Austin TX 78759-3504. http://www.ni.com/academic/circuits. Accessed 08 May 2011

Oladokun VO, Charles-Owaba OE, Nwaouzru CS (2006) An application of artificial neural network to maintenance management. J Ind Eng Int 2(3):19–26

OTW-One Technology Way (2009) "ADXL345 Data sheet", analog devices, One Technology Way, USA. http://www.analog.com. Accessed 08 Oct 2011

Peng Y, Dong M, Zuo MJ (2010) Current status of machine prognostics in condition-based maintenance—a review. Int J Adv Manuf Technol 50:293–313

Tao B, Ding H, Xion YL (2003) IP sensor and its distributed networking application in e-Maintenance. In: Proceedings of the 2003 IEEE international conference on systems, man and cybernetics, vol 4, pp 3858–3863

Tsang A (2002) Strategic dimensions of maintenance management. JQME 8(1):7–39

Uca M, Qiu RG (2005) e-Maintenance in support of e-automated manufacturing systems. J Chin Inst Ind Eng 22(1):1–10

Voisin A, Levrat E, Cocheteux P, Iung B (2010) Generic prognosis model for proactive maintenance decision support: application to pre-industrial e-maintenance test. J Intell Manuf 21:177–193

Yuan J, Chaing J (2000) Optimal maintenance policy for a production system subject to aging and shocks. J Qual Maint Eng 6(3):200–216

Zineb S, Chadi S (2001) Maintenance integration in manufacturing systems: from the modeling tool to evaluation. Int J Flex Manuf Syst 13(3):267–285

6

Analysis of implementation of Tradable Green Certificates system in a competitive electricity market: a game theory approach

Meysam Ghaffari[1] · Ashkan Hafezalkotob[1] · Ahmad Makui[2]

Abstract This paper investigates three models to implement Tradable Green Certificates (TGC) system with aid of game theory approach. In particular, the competition between thermal and renewable power plants is formulated in three models: namely cooperative, Nash and Stackelberg game models. The price of TGC is assumed to be determined by the legislative body (government) which is fixed. Numerical examples presented in this paper include sensitivity analysis of some key parameters and comparison of the results of different models. In all three game models, the parameters that influence pricing of the TGC based on the optimal amounts are obtained. The numerical examples demonstrate that in all models: there is a reverse relation between the price of electricity and the TGC price, as well as a direct relation between the price of electricity and the share of green electricity in total electricity generation. It is found that Stackelberg model is an appropriate structure to implement the TGC system. In this model, the supply of electricity and the production of green electricity are at the highest level, while the price of electricity is at the lowest levels. In addition, payoff of the thermal power plant is at the highest levels in the Nash model. Hence this model can be an applicatory structure for implementation of the TGC system in developing countries, where the number of thermal power plants is significantly greater than the number of renewable power plants.

Keywords Green electricity · Tradable Green Certificate · Game theory · Renewable energy

✉ Ashkan Hafezalkotob
a_hafez@azad.ac.ir; Hafezalkotob@iust.ac.ir

[1] Department of Industrial Engineering, Islamic Azad University, South Tehran Branch, Entezari Aly, Oskoui St, Choobi Brg, 1151863411 Tehran, Iran

[2] Department of Industrial Engineering, Iran University of Science and Technology, Hengam St, Resalat Sq, 1684613114 Tehran, Iran

Introduction

In the energy sector, climate change and energy security are significant factors affecting policies, regulations and investment (REN21 2012; Bazilian et al. 2011). With respect to growing concerns about climate changes, many countries have pursued policies to develop clean energy and set mandatory targets for renewable-source and low-carbon emission. For instance, European Union (EU) proposes a goal of 20 % share of renewable energy sources in the Union's total energy consumption by 2020 (Zhou 2012).

In global primary energy, the share of renewable energy could increase from the current 17 to 30 or 75 %, and in some nations exceed even 90 %, until 2050 (Johansson et al. 2012). Renewable energy (RE) considerably influences over many areas such as: strengthening economic growth to promote industrial development and employment, contribute to the transition toward a low carbon development growth for reduction of the greenhouse gas emissions, enhancement of technology diversification and hedging against fuel price volatility to increase supply adequacy, and facilitating the access to electricity to promote rural development and social welfare (Azuela and Barroso 2012; Fargione et al. 2008).

Electricity industry is one of the most important sources of pollution and RE plays a key role in the electricity generation. Most nations have pursued some policies to support the electricity generation from the renewable energy sources as one of the ways to curb global warming. In this regard, two of the most common practices are feed-in tariff and TGC systems (Tamás et al. 2010).

Some studies tried to answer the question of how does the feed-in tariff could affect selection of the energy resources. For example, Mahmoudi et al. (2014) proposed a computational framework for helping the government to determine the optimal taxes and subsidies for each individual electric power plant in a competitive electricity market, regarding the emitted pollutants of the power plants.

Taxes and subsidies on some technologies may help the government to achieve sustainable development objectives. The existing literature on TGC proposes that when the substantial investments in RE are already in place and the technologies are at a mature stage, switching to implementation of a TGC system is an appropriate alternative (Ciarreta et al. 2014).

A TGC system is introduced as renewable portfolio standards (RPS) or renewable obligations (RO) recognized in the RE Sector where the producers, retailers, consumers and distributors are required to accept obligation of a certain share in the production or consumption of electricity from renewable sources (Aune et al. 2012). The main objective of the TGC system is increasing the share of credit for green electricity generation from renewable sources, with the minimum possible cost for the government (Vogstad 2005).

Tamás et al. (2010) showed TGC system more efficient from feed-in tariff. RPS laws or TGC system use in 25 countries at national level and 54 states/provinces in the United stated, Canada, and India (REN21 2014). The Renewable Obligation was introduced in the UK in 2002 to support generation of green electricity. The RO target started at 3 % for the first period 2002–2003, increased annually by 1 % until it reach to 15.4 % in 2015–2016 (Zhou 2012). The UK increased the level of support for offshore wind producers under its green certificate scheme to 0.26 USD/KWh. At the beginning of 2012, the Norwegian–Swedish TGC market lunched to develop renewable capacity to produce 26/4TWh up to 2020. Romania implemented new law aimed at limiting the capacity expansion, growth of new players and more interesting for investors of TGC market (REN21 2014).

In this paper, competition of the power plants is modeled in the electricity market and the TGC system under producers' obligations. Therefore, some models are developed for two situations: competitive (Nash and Stackelberg equilibriums) and cooperative situations. Furthermore, adopting a numerical example, the impact of minimum share of electricity supply from RE sources and price of certificates on total supply and price of electricity, moreover, payoff and production of the power plants.

The reminder of the paper is organized as follows. "Literature review" section briefly discusses the related literature. "Prerequisites and assumptions" section describes the prerequisites and assumptions. "Model formulation" section provides the formulations of power plants problems. "Game theory models" section presents three game theory models for implementation of TGC system. "Numerical examples and sensitivity analysis" section discusses a numerical examples along with a set of sensitivity analyses. "Conclusion" section provides the conclusions and several directions for future research.

Literature review

TGC are financial assets provided for green electricity producers for the amount of green electricity measured and fed into the electricity grid. The TGC may be considered as a market-oriented environmental subsidy (Vogstad 2005; Boots 2003). In other words, the renewable power plants that generate electricity from RE (green electricity), benefit from a double source of income, from the sale of both physical electricity and green certificates (Farinosi et al. 2012).

A system of TGC is both an economic mechanism that supports RE production and a regulatory instrument available for public authorities to reach a specified goal for RE production. The market for TGC consists of supply and demand for certificates (Nielsen and Jeppesen 2003). Demand is created by a politically determined target for the share of electricity production or consumption from RE. Based on the policies of each country, any point on the electricity supply chain can be required to obligation of the set of targets. As shown in Fig. 1, the obligation can be placed at: supply, transmission, distribution and consumption electricity (Mitchell and Anderson 2000).

The TGC are generated by producers of green electricity. A certificate is issued for a certain amount of the green electricity generated. The size of certificate can be 1 MW/h or higher units of the green electricity produced. The certificates can be sold by the renewable power plant separately from the physical electricity. Every entity in the electricity supply chain like producers (except the green electricity producers), distributors, retailers, importers and

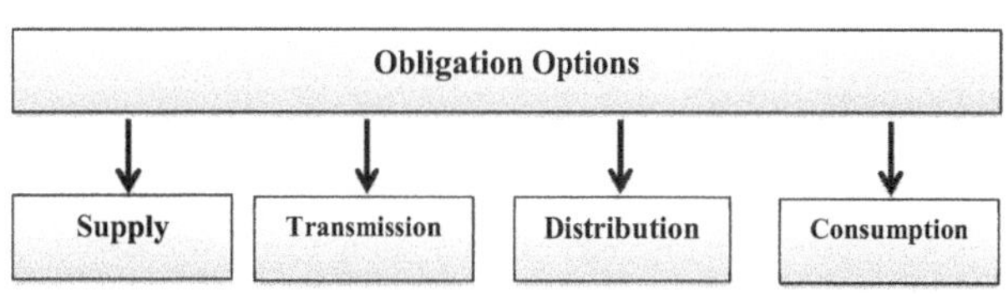

Fig. 1 Obligation option

consumers can be obliged to purchase a certain portion of the certificates from a renewable power plant. Financial market for the certificates may be created from interaction between the green electricity producer (as the TGC supplier) and the obligated entity (as the TGC demand). For instance, because the customer's obligation is considered in Denmark (Nielsen and Jeppesen 2003), interaction between customers and green electricity producers creates a market for TGC. In this approach, the consumers are obliged to consume a minimum quota of the green electricity, by purchasing the related certificates.

In designing of the TGC system, there are four mechanisms to organize the demand for certificates (Schaeffer et al. 2000):

1. An obligation on an entity in the electricity supply chain, to purchase a certain number of certificates within a certain period,
2. Setting a fixed price at which the certificates can be sold to a certain actor,
3. A tendering process aiming at buying the certificates,
4. Voluntary demand.

There are a few formal analyses of the TGC system (Tamás et al. 2010). Amundsena and Mortensen (2001) investigated the electricity and TGC markets in the case of Denmark assuming a perfect competition. They showed that an increase in the mandatory quota of the green electricity decreases the total supply and increases the electricity price. In the same case and method, Jensen and Skytte (2003) demonstrated that there is a linear relationship between the electricity price and the certificate price. They showed that the linear coefficient depends on the mandatory quota of green electricity by assuming a perfect competition on the certificates market and monopolistic competition on the electricity market. In a case study of Italy, Lorenzoni (2003) explained a formal implementation of the TGC system in 2002 and showed possible trends of the quantity and price of the certificates in the coming years. Verbruggen (2004) described details of the TGC system in some regions of Belgium and analyzed the established TGC system in Flemish region.

Ford et al. (2007) used system dynamics method to anticipate the price of certificate in a market TGC, to promote generation of the electricity from wind energy. They concluded that the certificate price climbs rapidly in the early years after a market opens. After a few years, it would lead to this fact that the electricity generated from the wind energy exceeds the requirement. Zhou and Tamas (2010) investigated the influences of integrating the production of green and thermal electricity on performance of the TGC system. They assumed that both the electricity and the certificate markets are imperfect. They showed that total supply of the electricity is greater under integration than when in disintegration; whereas, the price of TGC in an integrated market is higher than that of the disintegrated market.

Colcelli (2012) by quality method discussed the problem of legal nature of TGC in Italy and concluded that TGC be regarded as good. Currier (2013) examined a Cournot electricity oligopoly operated under TGC system with producer obligation. He calculated parametric optimal percentage requirement using Bound branches algorithm to sure maximum social welfare. Fagiani et al. (2013) by system dynamic approach analyzed the performance of feed-in tariff and TGC markets. They simulated electricity market a period which cover 39 years from 2012 to 2050 in case of Spain and showed Tariffs could obtain better efficiency but also low effectiveness or over-investment, moreover, TGC performances benefit from higher social discount rates. Ciarreta et al. (2014) analyzed implementation of TGC system in Spain. They modeled interaction between the electricity pool and TGC market and analyzed this, through solving a sequential game. They studied the retailer regulation design that would give lead to a decreasing TGC demand and simulated the impact of same regulation on the TGC price.

Currier and Sun (2014) investigated performance of TGC system in electricity market under alternative market structure. They demonstrated that an oligopolistic market structure may create more welfare than a competitive market structure. Fagiani and Hakvoort (2014) analyzed the impact of regulatory changes on TGC price volatility in Swedish market and a bigger Swedish/Norwegian market. By econometrics approach, they showed regulatory change harms TGC market and bigger Swedish/Norwegian market has not resulted in lower volatility yet.

Most researchers investigated the electricity market and the TGC market with economic analysis and systems dynamic methods. Moreover, most of the previous researches have concentrated on implementation of the TGC systems in a specific country. To the best of authors' knowledge, there is no research in this context which adopts the game theory approach. Analysis based on game theory approach helps to policy makers for market structure design for electricity and TGC market. Some studies consider to market structure in the case of imperfect and perfect competition generally by simple economic method. In this paper, we aim model market structure of electricity and TGC markets in case of imperfect competition Cournot oligopoly and monopoly under fixed TGC price policy.

Prerequisites and assumptions

For simplicity of this research, we concentrate on interaction of two power plants: green and thermal electricity producers. These power plants compete together in the electricity markets under the TGC system. Under the TGC system, a thermal electricity producer is obliged to acquire a minimum number of green certificates. This number corresponds to a percentage (quota) of the yearly thermal electricity generated.

It is assumed that the minimum quota and price of the certificates are set by the lawgiver. This means that the price of certificates is fixed and not determined by the market equilibrium of supply and demand.

Assumptions

The proposed models in this paper are based on the following assumptions:

1. Power plants have no limitation on consumption of the resources.
2. The price of certificates is only at fixed prices in the long term similar to the former feed-in tariffs.
3. The electricity price is set under a national supply and demand mechanism (in the local market).
4. There are no limitations on the supply and demand for the electricity as well as the certificates.
5. There are no excess supply and demand in the markets of electricity and certificates.

Notations

Before describing the payoff functions for the companies, the indices, parameters and decision variables are explained below:

Parameters

- α the minimum mandatory quota (percentage) of green electricity, $0 \leq \alpha \leq 1$;
- π_R the profit function of renewable power plant;
- π_T the profit function of thermal power plant;
- π the total payoff of centralized power plant, $(\pi = \pi_R + \pi_T)$;
- C_T the cost function of thermal power plant;
- C_R the cost function of renewable power plant;
- P_c the price of green certificates ($/MWh), $P_c > 0$;
- γ the cap price of electricity, $\gamma > 0$;
- β the price elasticity of electricity supply; $\beta > 0$.

Decision variables

- P_e the wholesale price of electricity ($/MWh), $P_e > 0$;
- q_T the quantity production of electricity from non-renewable energy sources (MW), $q_T \geq 0$;
- q_R the production of electricity from renewable energy sources (MW), $q_R \geq 0$;
- Q the total supply of electricity (MW), $Q \geq 0 (Q = q_T + q_R)$.

Model formulation

Producer of renewable power

We adopted profit functions proposed by Currier and Sun (2014), and considering relation between wholesale price and end-user price of electricity explained by Amundsen and Bergman (2012). In their model, producer of green electricity can sell both electricity generated on the electricity market as well as certificates on separate market. The cost of renewable power plant is function of electricity generated from renewable sources. Therefore, profit maximization problem for renewable power plant can be formulated as follows:

$$\text{Max} \quad \pi_R = [P_e + P_c]q_R - C_R(q_R)$$

S.t:

$$q_R \geq 0 \tag{1}$$

This means that a renewable producer can receive P_c for each unit in addition to the electricity price. Cost of the renewable producer is dependent only on the actual amount of electricity production. Under the TGC system, a renewable producer would receive per unit "subsidy" P_c.

Producer of thermal power

A producer of the thermal power can sell the electricity generated in the electricity market. It is obligated to supply a certain proportion of the green electricity from total electricity supplied on the grid. It can fulfill their obligation by either supplying the green electricity or by purchasing the TGC.

The cost of a thermal power plant is a function of the electricity generated from the non-renewable sources. Therefore, the profit maximization problem for the producer of thermal power is as follows:

$$\text{Max} \quad \pi_T = [P_e - P_c\alpha]q_T - C_T(q_T)$$

S.t:

$$q_T \geq 0 \tag{2}$$

Thermal producer can receive P_e for each unit of electricity. Cost of the thermal power is dependent only on the actual amount of the electricity production. It is obligated to payment for buying the TGC from the renewable producer, to compensate for the unfulfilled requirement. Therefore, the thermal producer under the TGC system virtually pays a per unit "tax" αP_c as in Eq. (2). In our model, only a thermal power plant is obligated to hold a number of the TGC equal to α times its production.

Cournot model

According to the Cournot model, the price is a function of the production quantity. Kreps and Scheinkman (1983) discussed that if the producers first determine their capacity, and only later are allowed to set a price, the outcome will be the Cournot equilibrium.

Thus, it can be assumed that the electricity price is a function of the total electricity generated by the renewable and non-renewable sources.

$$P_e = \gamma - \beta Q = \gamma - \beta(q_R + q_T) \tag{3}$$

where γ the cap is the price of electricity and β is the price elasticity of the electricity supply. Meanwhile, $Q = (q_R + q_T)$ is the total electricity supply. It is assumed that $\beta > 0$.

Cost function

It is assumed that the cost function of the power plants is a quadratic function. The cost functions for the renewable and thermal power plant can be described as follows:

$$C_R(q_R) = a_R q_R^2 + b_R q_R + c_R \tag{4}$$

$$C_T(q_T) = a_T q_T^2 + b_T q_T + c_T \tag{5}$$

In Eqs. (4) and (5), it is assumed that $a_R, b_R, a_T, b_T > 0$, and the marginal production costs are increasing. Jensen and Skytte (2003) used the same model for the cost function of the power plants.

Profit maximization problem for power plants

Substituting Eqs. (3), (4) and (5) into Eqs. (1) and (2), the problems of power plants can be described as follows.

The profit maximization problem for the producer of green electricity is given below:

$$\text{Max} \quad \pi_R = [\gamma - \beta(q_R + q_T)]q_R + P_c q_R - a_R q_R^2 - b_R q_R - c_R$$

S.t:

$$q_R \geq 0; \quad a_R, b_R > 0 \tag{6}$$

The profit maximization problem for the producer of thermal power is given below:

$$\text{Max} \quad \pi_T = [\gamma - \beta(q_R + q_T)]q_T - P_c q_T - a_T q_T^2 - b_T q_T - c_T$$

S.t:

$$q_T \geq 0; \quad a_T, b_T > 0 \tag{7}$$

Game theory models

Nash equilibrium

Nash equilibrium (NE) solution is one of the fundamental solution concepts in the game theory. NE solution is where the strategy of each player is the best response against strategies of the rivals. Because of deviation from NE would lead to reduction of player's profit, none of the players has motivation to reject this strategy. The NE of the game is defined as follows (Krause et al. 2006):

In a game of n players, the strategy profile $P^* = (P_1^*, \ldots\ldots, P_n^*)$ is a NE if for all I $i = \{1, \ldots\ldots, n\}$ there is:

$$U_i = (P_1^*, \ldots\ldots, P_n^*) \geq (P_1^*, \ldots\ldots, P_{i-1}^*, P_i, P_{i+1}^*, \ldots\ldots . P_n^*) \tag{8}$$

where U_i is the utility function of the ith player.

Several algorithms have been developed for computing of NE. The interested reader may refer to Krause et al. (2004) and Porter et al. (2008). In this study, an NE approach is used for the Cournot game to calculate the price equilibrium of the electricity in a competitive market under a green certificate system.

Based on the NE, q_T^* and q_R^* will be obtained from Eqs. (6) and (7) first, then with substitution of q_T^* and q_R^* into π_R and π_T, respectively, the maximum profit of the power plant will be obtained as π_T^*, π_R^*.

Proposition 1 *If the profit function of the power plants is concave, the optimal amounts of production for the green and thermal power plants in the Nash model are*

$$q_{R[N]}^* = \frac{AP_c + B\beta + 2a_T C}{2D + 2\beta^2} \tag{9}$$

$$q_{T[N]}^* = -\frac{P_c(2\alpha a_R + 2\beta\alpha + \beta) + \beta(-b_R + 2b_T - \gamma) + 2a_R(b_T - \gamma)}{2D + 2\beta^2} \tag{10}$$

where $A = \alpha\beta + 2a_T + 2\beta$, $B = -2b_R + b_T + \gamma$, $C = -b_R + \gamma$, $D = 2a_R a_T + 2a_R\beta + 2a_T\beta + \beta^2$. *[N] Denotes the optimum amounts in the Nash model.*

Proof of all the propositions are given in "Appendix". Substituting Eqs. (9) and (10) into Eqs. (6) and (7), an optimal payoff of the renewable and thermal power plants is obtained in the Nash game model as follows:

$$\pi^*_{R[N]} = (AP_c + B\beta + 2a_T C)\left[\gamma - \beta\left(\frac{P_c(2\alpha a_R + 2\beta\alpha + \beta) + \beta(-b_R + 2b_T - \gamma) + 2a_R(b_T - \gamma) + AP_c + B\beta + 2a_T C}{(2D + 2\beta^2)^2}\right) - \frac{a_R(AP_c + B\beta + 2a_T C)}{(2D + 2\beta^2)^2} - \frac{b_R + P_c}{2D + 2\beta^2}\right] - c_R \tag{11}$$

$$\pi^*_{T[N]} = \left(\frac{P_c(2\alpha a_R + 2\beta\alpha + \beta) + \beta(-b_R + 2b_T - \gamma) + 2a_R(b_T - \gamma)}{2D + 2\beta^2}\right)\left[\gamma - \frac{a_t}{2D + 2\beta^2} + P_c\alpha - b_T - \beta\left(\frac{-P_c(2\alpha a_R + 2\beta\alpha + \beta) - \beta(-b_R + 2b_T - \gamma) - 2a_R(b_T - \gamma) + AP_c + B\beta + 2a_T C}{2D + 2\beta^2}\right)\right] - c_T \tag{12}$$

Since the TGC price is determined by the government and it is fixed to help the government for pricing the TGC, the parameters that influence the price of TGC is found based on the optimal amounts. Substituting $q^*_{R[N]}$ and $q^*_{T[N]}$ into Eq. (3) gives:

$$P^*_{c[N]} = \frac{-P^*_e(2\beta^2 + 2D) - \beta^2(B + b_R - 2b_T - \gamma) - \beta(2Ca_T - 2a_R b_T + 2a_R\gamma) + 2D\gamma}{(A\beta + 2\alpha a_R + 2\alpha\beta + \beta)} \tag{13}$$

Note that $P^*_c = P_c(P^*_e)$. Equation (13) shows that there is a linear relationship between the electricity price and the TGC price in the Nash game model. The linear coefficient is negative and depends on the minimum quota of the green electricity (α).

Cooperative game

In this section, a cooperative game approach is applied to the problem of thermal–green power plants with respect to the TGC system. Using this approach, the thermal and renewable power plants work together to determine Q and P_e. It is possible to examine whether the thermal power plant allocates a portion of its capacity to produce the green electricity to get more profit considering a situation in which it competes with renewable power plants or not? To calculate the optimal amounts under a cooperative situation, the new model will be obtained from summation of Eqs. (6) and (7).

$$\begin{aligned} \text{Max} \quad \pi = {} & [\gamma - \beta(q_R + q_T)]q_R + P_c q_R - a_R q_R^2 \\ & - b_R q_R - c_R + [\gamma - \beta(q_R + q_T)]q_T \\ & - P_c q_T - a_T q_T^2 - b_T q_T - c_T \end{aligned}$$

S.t :

$$q_R, q_T \geq 0; \quad a_R, a_T, b_R, b_T > 0 \tag{14}$$

Hessian matrix of π in Eq. (14) is: $H = \begin{bmatrix} -2\beta - 2a_R & -2\beta \\ -2\beta & -2\beta - 2a_T \end{bmatrix}$; the utility function π is a concave function on (q_R, q_T) if and only if the Hessian matrix H is negative definite.

Proposition 2 *If* $\det(H) > 0$, *the optimal amounts for production of the green and thermal power plants in the cooperative game model will be:*

$$q^*_{R[C]} = \frac{P_c(A - a_T - \beta) + \beta(B + b_R - \gamma) + a_T C}{D - \beta^2} \tag{15}$$

$$q^*_{T[C]} = \frac{P_c(A - 2a_T - \beta + \alpha a_R) + \beta(B + b_R - \gamma) + a_T C}{D - \beta^2} \tag{16}$$

where [C] *denotes the optimum amounts in the cooperative game model.*

Inserting Eqs. (15) and (16) into (6) and (7), the optimal payoff of the renewable and thermal power plants in the Nash game model are found as follows:

$$\pi^*_{R[C]} = \left(\frac{P_c(A - a_T - \beta) + \beta(B + b_R - \gamma) + a_T C}{D - \beta^2}\right) \left[P_c + C - \beta\left(\frac{P_c(\alpha a_R + 2A - 2\beta - 3a_T) + 2\beta(b_R + B + \gamma) + 2a_T C}{D - \beta^2}\right) - \frac{a_R(P_c(A - a_T - \beta) + \beta(B + b_R - \gamma) + a_T C)}{D - \beta^2}\right] - c_R \tag{17}$$

$$\pi^*_{T[C]} = \left(\frac{P_c(A - 2a_T - \beta + \alpha a_R) + \beta(B + b_R - \gamma) + a_T C}{D - \beta^2}\right) \left[\gamma - \beta\left(\frac{P_c(\alpha a_R + 2A - 2\beta - 3a_T) + 2\beta(b_R + B + \gamma) + 2a_T C}{D - \beta^2}\right) - \frac{a_T(P_c(A - 2a_T - \beta + \alpha a_R) + \beta(B + b_R - \gamma) + a_T C)}{D - \beta^2} - b_T - P_c\alpha\right] - c_T \tag{18}$$

Similar to the previous section, P_c is calculated by substituting $q^*_{R[C]}$ and $q^*_{T[C]}$ into Eq. (3) as follows:

$$P^*_{c[C]} = \frac{P^*_e(\beta^2 - D) - 2B\beta^2 - 2Ca_T\beta - 2\beta^2 b_R + \beta^2\gamma + D\gamma}{2\beta(\alpha a_R + A - 2a_T - \beta)} \tag{19}$$

Equation (19) indicates that there is a linear relationship between the electricity price and the TGC price in a cooperative game model. The linear coefficient depends on the minimum quota of the green electricity (i.e., α) but the positive or negative linear coefficient depends on other parametric values.

Non-cooperative Stackelberg games

This section considers the relationship between thermal and renewable power plants using a non-cooperative structure. The interaction between these power plants will be regarded as a Stackelberg game, where one of the participants, i.e., the leader, has the initiative and can enforce its strategy on the its rival, i.e., the follower. The leader makes the first move and the follower reacts by playing the best move according to the available information. The objective of the leader is to design its move in such a way as to maximize its profit after considering all the rational moves that can be advised by the follower.

The renewable producer–Stackelberg model takes the renewable power plant as the leader and the thermal power plant as the follower. In this section, the renewable producers first generates electricity and sells it to the distributors then the thermal producers with knowledge about the issued certificates and the remaining market share will produce and sell its electricity to the market. Considering that aim of the TGC system is supporting than increasing the share of the electricity produced from RE, the thermal producer–Stackelberg model is not investigated as a thermal producer leader.

Proposition 3 *The optimal amounts of production of green and thermal power plants in Stackelberg game model are:*

$$q^*_{R[S]} = \frac{AP_c + B\beta + 2a_T C}{2D} \tag{20}$$

$$q^*_{T[S]} = -\frac{P_c(A\beta + 2D\alpha) + \beta^2 B + Ca_T\beta + 2Db_T - 2D\gamma}{4D(\beta + a_T)} \tag{21}$$

where [S] *refers to the optimum amounts in Stackelberg model.*

Inserting Eqs. (20) and (21) into (6) and (7), the optimal payoff of the renewable and thermal power plants in Stackelberg model is obtained as follows:

$$\pi^*_{R[S]} = (AP_c + B\beta + Ca_T)\left[\left(\frac{P_c}{2D}\right) - \frac{a_R + 2Db_R}{4D^2} + \left(\frac{\gamma - \beta}{2D}\right)\left(\frac{(-2D(\alpha P_c + b_T - \gamma) + (AP_c + B\beta + Ca_T)\beta + 2(\beta + a_T)(AP_c + B\beta + Ca_T)}{8D^2(\beta + a_T)}\right)\right] - c_R \tag{22}$$

$$\pi^*_{T[S]} = \left(P_c\alpha + \frac{(AP_c + B\beta + Ca_T)\beta}{2D} + b_T - \gamma\right)\left[\gamma - \left(\frac{(\beta + a_T)(2D(P_c\alpha + b_T + \gamma) + \beta(AP_c + B\beta + Ca_T))}{8D(\beta + a_T)^2}\right)\right] - c_T \tag{23}$$

$$P^*_{c[S]} = \frac{-4DP^*_e(\beta^2 + D) - \beta^2(2Ba_T + B\beta + Ca_T) - \beta(2Ca_T^2 - 2Db_T - 2D\gamma) + 4Da_T\gamma}{\beta(2Aa_T + A\beta - 2D\alpha)} \tag{24}$$

Substituting $q^*_{R[S]}$ and $q^*_{T[S]}$ into Eq. (3) gives:

Same as the other models, in Stackelberg game model, there is a linear relationship between the electricity price and the TGC price. The linear coefficient is dependent on the mandatory quota of the green electricity (), but the positive or negative linear coefficient depends on some other parameter values.

Numerical examples and sensitivity analysis

In this section, a number of numerical examples are presented with the aim of illustrating some significant features of the models established in the previous sections. A sensitivity analysis of the main parameters of these models will also be performed. Note that Examples (1–2) illustrate the renewable producer–Stackelberg, Nash equilibrium and cooperative game models, respectively.

Example 1 The changes of $q^*_R, q^*_T, q^*_T, Q^*, P^*_e$ with respect to the changes of α are investigated. Let the parameters be set as below:

$\gamma = 150;\quad \beta = 0.4;\quad a_R = 0.6;\quad a_T = 0.4;\quad b_R = 11;\quad b_T = 8;\quad c_T,\quad c_R = 101;\quad P^*_c = 18.$

Table 1 lists the results of this example in three game models. Some important results in the table are also graphically displayed in Figs. 2, 3 and 4.

The results of Example 1 show that in every three models, by increasing α, all q^*_R, π^*_R and P^*_e increase, however, q^*_T, π^*_T, Q^* and π^* decrease. The value of Q^* in the Stackelberg model will be greater than that of the Nash and cooperative models (see Fig. 2). In other words, electricity supply in Stackelberg model is set in the highest level and this can lead to social welfare improvement.

The renewable power plant acquires the maximum payoff in the cooperative model where the payoff of the thermal power plant is minimum. Moreover, the thermal power plant acquires the maximum payoff in Nash model, but the payoff of the renewable power plant is minimum. So, if electricity market structure follows cooperative scenario, thermal power plant will be eliminated from market quickly.

As can be seen from Fig. 4, total payoff of the both power plants in the cooperative model is the highest and the lowest in the Stackelberg model. Additionally, Fig. 3 illustrates that P^*_e will be at the highest level in the cooperative model and at the lowest level in the Stackelberg model. This means end-users' welfare in Stackelberg scenario can be more than other scenarios. As can be observed in Table 1: π^*_R has the lowest value in the Nash game model and the highest value in the cooperative model, while π^*_T is minimum in the cooperative model and maximum in the Nash model. q^*_R shows the lowest value in the Nash model and the highest value in the Stackelberg model, while q^*_T has the lowest value in the cooperative model and the highest value in the Nash model. Therefore,

Table 1 Sensitivity analysis of game theory models with respect to α

α	0.05	0.1	0.15	0.2	0.25
Nash game model					
q_R^*	125.80	126.35	126.91	127.46	128.02
q_T^*	103.16	101.88	100.61	99.33	98.06
Q^*	228.96	228.24	227.52	226.80	226.08
π_R^*	7179	7243	7308	7373	7437.67
π_T^*	4581	4466	4353	4241	4129.97
π^*	11,760	11,710	11,661	11,613	11,567.64
P_e^*	58.42	58.70	58.99	59.28	59.57
Stackelberg game model					
q_R^*	166.91	167.65	168.38	169.12	169.85
q_T^*	84.47	83.11	81.76	80.40	79.04
Q^*	251.38	250.76	250.14	249.52	248.9
π_R^*	7649	7717	7786	7855	7924.55
π_T^*	3039	2939	2840	2743	2648.11
π^*	10,688	10,656	10,626	10,599	10,572.66
P_e^*	49.45	49.70	49.94	50.19	50.44
Cooperative game model					
q_R^*	149.06	153.30	157.55	161.79	166.04
q_T^*	24.83	19.95	15.07	10.19	5.31
Q^*	173.89	173.25	172.62	171.98	171.34
π_R^*	11,599.94	11,933.20	12,266.45	12,599.71	12,932.96
π_T^*	1651.10	1297.69	948.68	604.06	263.83
π^*	13,251.05	13,230.89	13,215.13	13,203.76	13,196.79
P_e^*	80.44	80.70	80.95	81.21	81.46

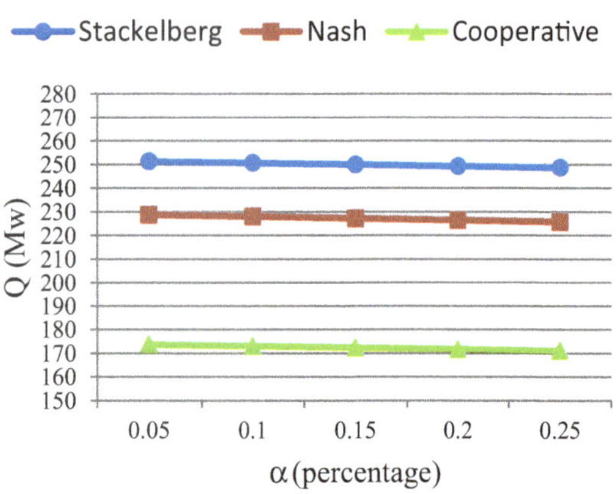

Fig. 2 Changes of total electricity supply versus α

the cost of pollution in the Nash model will be more than other scenarios.

Example 2 In this example, the changes of $q_R^*, q_T^*, \pi_R^*, \pi_T^*, \pi^*, Q^*$ and P_e^* are investigated with the changes of P_c. Let the parameters be set as below:

$$\gamma = 150; \quad \beta = 0.4; \quad a_R = 0.22; \quad a_T = 0.04;$$
$$b_R = 16, \quad b_T = 8; \quad c_T, c_R = 101; \quad \alpha = 0.1.$$

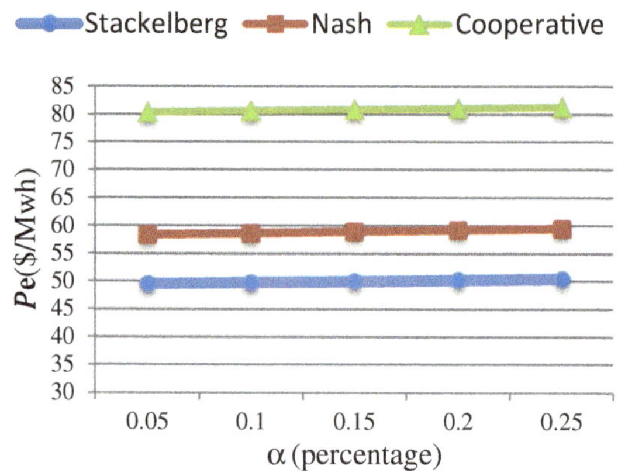

Fig. 3 Changes of price of electricity versus α

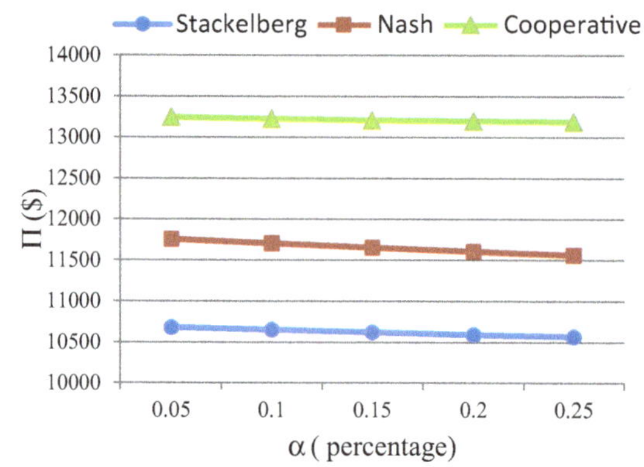

Fig. 4 Changes of total payoff versus α

Table 2 summarizes the results of this example in three models. Some important results of Table 2 are also graphically illustrated in Figs. 5, 6 and 7. As reported in the table, the results of Example 2 show that in each of the three models, increasing P_c^* will reduce P_e^* and q_T^* while increase Q^* and q_R^*. In other words, with TGC increasing the electricity price decreases and electricity supply increases at the same time. So it is expected that implementation of TGC system leads welfare improvement in all scenarios. As expected before from Eqs. (1) and (2), P_c has a direct relation with π_R^* and an inverse relation with π_T^*. In every three models, π^* and π_R^* increases by increasing P_c but π_T^* decreases.

In the Nash and Stackelberg game models P_e^* decreases and Q^* increases faster than the cooperative game model with respect to increasing P_c^*. Results of Examples 1-2 imply that P_e^* has the lowest value in the Stackelberg model and the highest value in the cooperative model (Figs. 3, 5). Q^* has the highest value in the Stackelberg model and the lowest value in the cooperative model (see Figs. 2, 6). Moreover, π^* has the lowest value in the Stackelberg

Table 2 Sensitivity analysis of game theory models with respect to α

P_c	10	30	50	70
Nash game model				
q_R^*	75.52	95.27	115.03	134.79
q_T^*	125.90	114.65	103.39	92.14
Q^*	201	210	218	227
π_R^*	3435	5527	8103	11,164
π_T^*	6874	5682	4603	3634
π^*	10,308	11,209	12,706	14,798
P_c^*	69.43	66.03	62.63	59.23
Stackelberg game model				
q_R^*	91.18	115.04	138.90	162.76
q_T^*	118.78	105.66	92.55	79.43
Q^*	210	221	231	242
π_R^*	3542	5698	8353	11,507
π_T^*	6107	4811	3667	2675
π^*	9649	10,510	12,020	14,182
P_c^*	66.01	61.72	57.42	53.13
Cooperative game model				
q_R^*	30.85	73	116	159
q_T^*	132.18	91	50	9
Q^*	163.03	165	166	168
π_R^*	2120	5918	10,567	16,067
π_T^*	9218	6239	3342	527
π^*	11,338	12,157	13,909	16,594
P_c^*	84.79	84.15	83.5	82.9

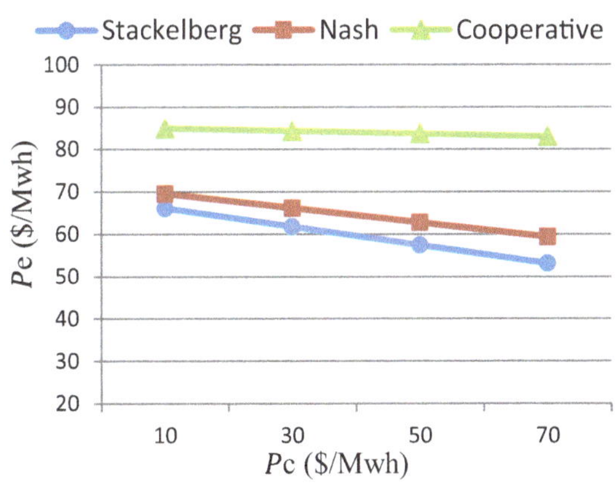

Fig. 5 Changes of electricity price versus P_c

model and the highest value in the cooperative model (Figs. 4, 7).

Example 3 In this example, the changes of α versus P_c in three Nash, Stackelberg and cooperative game models [i.e., Eqs. (13), (19), (24)] are evaluated. Let:

$$\gamma = 150;\quad \beta = 0.4;\quad P_e = 50;\quad a_R = 0.6;\quad a_T = 0.04;\quad b_R = 11,\quad b_T = 8;\quad c_R = 80,\quad c_T = 20.$$

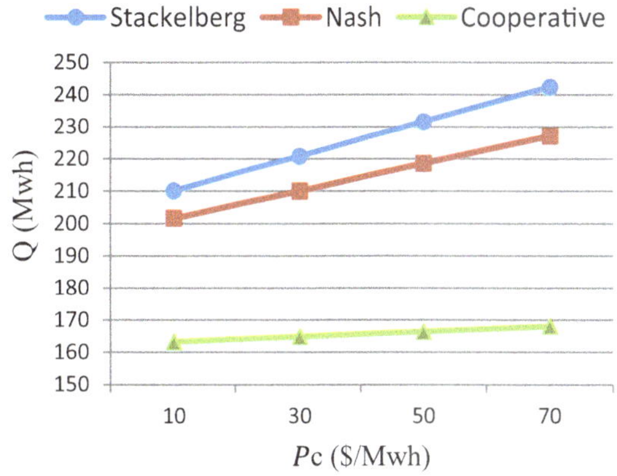

Fig. 6 Changes of supply of electricity versus P_c

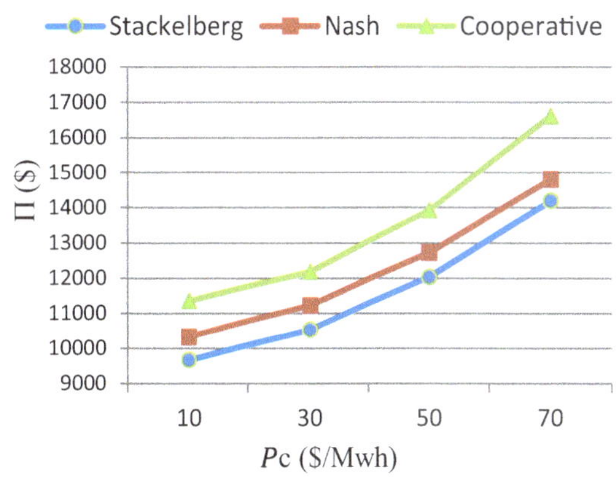

Fig. 7 Changes of total payoff versus P_c

Figure 8 depicts the results of this example in the models. The obtained results show that by increasing, certificate price P_c^* increases in the cooperative and Stackelberg game models while P_c^* decreases in the Nash game model. In the cooperative game model, P_c^* is the highest level in comparison with the other models.

The results of this example can be stated as follows: in the countries which their electricity market structures follow the Nash model, when the green electricity share increases, certificates price decreases and this leads to reduction of renewable power plants profit. This may signify that the TGC system has no appropriate incentives to produce green power sufficiently. Because in this case, renewable producer earned low profit from TGC sale.

The results of this paper can be useful for both public and private investors in the green electricity generation and other electricity producers. Therefore, the policy makers of government may adopt these models to design an implementation structure of the TGC system and to determine the objectives for generation of the green electricity.

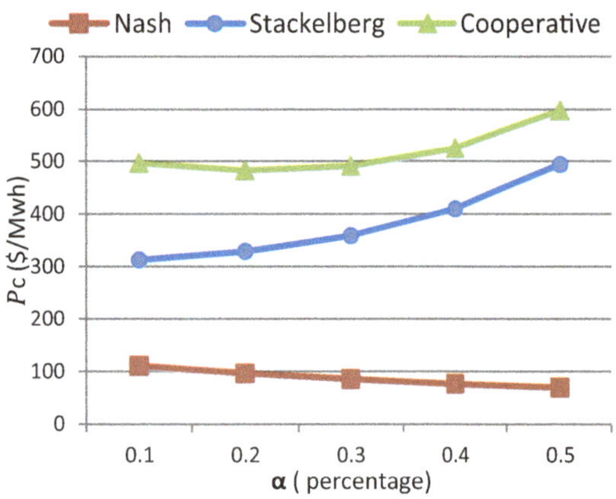

Fig. 8 Changes of P_c versus α

Pricing of the TGC is a challenging problem for the government, the parameters which are effective on the TGC price were shown in various game models. Finally, we summarize the numerical example results of game theory models for TGC system. Table 3 draws a comparison among optimal values of the models.

Conclusion

This paper considers the problem of interaction between the thermal and renewable power plants under TGC system conditions. We proposed three game theory models for TGC system, namely: cooperative, Nash and renewable-producer–Stackelberg models. These models were analyzed to implement the TGC system under the producer's obligation, assuming fixed prices for the certificates. Through a comprehensive sensitivity analysis, the effect of some main parameters of the model on the thermal and the renewable's decisions were evaluated. We showed that there is a reverse relation between price of the electricity and price of the certificates. In addition, price of electricity has a direct relation with the minimum quota. We found that the electricity supply in the cooperative game is at the lowest level, while the price of electricity is at the highest level. In the Stackelberg model, the price of electricity is at the lowest level and the supply of electricity and the production of green electricity are greater than the other models. In the Nash model, the payoff of the thermal power plant is at the maximum level and the payoff of the renewable power plant is at the minimum level.

There are several directions for the future research. First, producer's obligation option in the TGC system is considered, while the other obligation in the TGC system is both challenging and interesting. Second, time constraints were not considered for validation of the certificates. Using time variables in modeling of the TGC system can yield useful results. Third, this paper considers national trade in the electricity market and the TGC system. It is found that modeling the international trade in both of the markets with the game theory approach is interesting. Finally, applying other game theory's models to analyze implementation of the TGC system can be considered. For example, modeling of the TGC system in the incomplete information mode by Bayesian models is both interesting and challenging.

Appendix

Proof for Proposition 1 If the second order driven for Eq. (6) is negative, the profit function of green producer will be concave. The first-order condition for Eq. (6) is:

$$\frac{\partial \pi_R}{\partial q_R} = (P_c + \gamma) - (\beta q_T + 2\beta q_R + 2a_R q_R + b_R) = 0. \tag{25}$$

Equation (26) is negative if $(P_c + \gamma) < (\beta q_T + 2\beta q_R + 2a_R q_R + b_R)$. The second-order condition for Eq. (6) is as follows:

$$\frac{\partial^2 \pi_R}{\partial^2 q_R} = -(+2\beta + 2a_R). \tag{26}$$

Since the amounts of β and a_R are positive, the second-order condition is negative $\left(\frac{\partial^2 \pi_R}{\partial^2 q_R} < 0\right)$; therefore, the profit function of the green producer is concave.Similarly, if the second order driven for Eq. (7) is negative, the profit function of the thermal producer will be concave. The first-order condition for Eq. (7) is as follows:

$$\frac{\partial \pi_T}{\partial q_T} = \gamma - (2\beta q_T + \beta q_R + P_c\alpha + 2a_T q_T + b_T) = 0. \tag{27}$$

Table 3 Comparison of three game theory models

Profit of power plants	Amount of generated electricity	Price of electricity and TGC
$\pi^*_{R[C]} > \pi^*_{R[S]} > \pi^*_{R[N]}$	$q^*_{R[S]} > q^*_{R[C]} > q^*_{R[N]}$	$P^*_{e[C]} > P^*_{e[N]} > P^*_{e[S]}$
$\pi^*_{T[N]} > \pi^*_{T[S]} > \pi^*_{T[C]}$	$q^*_{T[N]} > q^*_{T[S]} > q^*_{T[C]}$	$P^*_{c[C]} > P^*_{c[S]} > P^*_{c[N]}$
$\pi^*_{[C]} > \pi^*_{[N]} > \pi^*_{[S]}$	$Q^*_{[S]} > Q^*_{[N]} > Q^*_{[C]}$	

Equation (27) is negative if $\gamma < (2\beta q_T + 2\beta q_R + \alpha P_c + 2a_T q_T + b_T)$. The second-order condition for Eq. (7) yields:

$$\frac{\partial^2 \pi_T}{\partial^2 q_T} = -(2\beta + 2a_T). \tag{28}$$

Since the amounts of β and a_T are positive, the second-order condition is negative $\left(\frac{\partial^2 \pi_R}{\partial^2 q_R} < 0\right)$; hence, the profit function of the thermal producer will be concave. From solving Eqs. (28) and (26), it follows that the optimal production of power plants are:

$$q^*_{R[N]} = \frac{AP_c + B\beta + 2a_t C}{2D + 2\beta^2}, \quad q^*_{T[N]} = -\frac{P_c(2\alpha a_R + 2\beta\alpha + \beta) + \beta(-b_R + 2b_T - \gamma) + 2a_R(b_T - \gamma)}{2D + 2\beta^2}. \quad \square$$

Proof for Proposition 2 The first-order condition for profit function of the power plants in Eq. (18) yields:

$$\frac{\partial \pi_R}{\partial q_R} = P_c + \gamma - 2\beta(q_T + q_R) - 2a_R q_R - b_R = 0, \tag{29}$$

$$\frac{\partial \pi_T}{\partial q_T} = \gamma - 2\beta(q_T + q_R) - \alpha P_c - 2a_R q_T - b_T = 0. \tag{30}$$

Solving Eqs. (29) and (30), we have:

$$q^*_{R[C]} = \frac{P_c(A - a_T - \beta) + \beta(B + b_R - \gamma) + a_T C}{D - \beta^2}.$$

$$q^*_{T[C]} = \frac{P_c(A - 2a_T - \beta + \alpha a_R) + \beta(B + b_R - \gamma) + a_T C}{D - \beta^2}. \quad \square$$

Proof for Proposition 3 To solve the model, q_T is first obtained as a function of q_R, then the order condition is first examined for a profit function of the thermal power plant of Eq. (30); the best response strategy of thermal power plant is computed as follows:

$$q_T = \frac{\alpha P_c + q_R + b_T - \gamma}{2(\beta + a_T)}. \tag{31}$$

Inserting Eq. (31) into Eq. (7) gives:

$$\pi_R = P_c q_R + \left[\gamma - \beta - \left(-\frac{\alpha P_c + \beta q_R + b_T - \gamma}{2(\beta + a_T)} + q_R\right)\right] q_R - a_R q_R^2 - b_R q_R - c_R. \tag{32}$$

The first-order condition for Eq. (32) yields:

$$\frac{\partial \pi_R}{\partial q_R} = P_c - \beta\left(-\frac{\beta}{2(\beta + a_T)} + 1\right) q_R + \gamma - \beta\left(-\frac{\alpha P_c + \beta q_R + b_T - \gamma}{2(\beta + a_T)} + q_R\right) - 2a_R q_R - b_R = 0. \tag{33}$$

The profit function of the renewable power plant is concave if the second-order condition for Eq. (33) is negative. The second-order condition for the renewable power plant gives:

$$\frac{\partial^2 \pi_R}{\partial^2 q_R} = -\frac{2a_R a_T + 2a_R \beta + \beta^2}{\beta + a_T}. \tag{34}$$

Regarding the assumption and parameter values, Eq. (34) is negative. Therefore, the profit function of the renewable power plant in this section is found to be concave. From Eq. (33), it follows that the optimal green electricity production is:

$$q^*_{R[S]} = \frac{AP_c + B\beta + 2a_T C}{2D}.$$

Inserting $q^*_{R[S]}$ into Eq. (31), the optimal black electricity production is:

$$q^*_{T[S]} = -\frac{P_c(A\beta + 2D\alpha) + \beta^2 B + C a_T \beta + 2D b_T - 2D\gamma}{4D(\beta + a_T)}. \quad \square$$

Compliance with ethical standards

Conflict of interest The authors declare that they have no competing interests.

References

Amundsen E, Bergman L (2012) Green certificates and market power on the Nordic power market. Energy J 33(2):101–117

Amundsena E, Mortensen J (2001) The Danish green certificate system: some simple analytical results. Energy Econ 23(5):489–509

Aune F, Dalen H, Hagem C (2012) Implementing the EU renewable target through green certificate markets. Energy Econ 34(4):992–1000

Azuela GE, Barroso LA (2012) Design and performance of policy instruments to promote the development of renewable energy: emerging experience in selected developing countries. World Bank Publications

Bazilian M, Hobbs BF, Blyth W, MacGill I, Howells M (2011) Interactions between energy security and climate change: a focus on developing countries. Energy Policy 39(6):3750–3756

Boots M (2003) Green certificates and carbon trading in the Netherlands. Energy Policy 31(1):43–50

Ciarreta A, Paz Espinosa M, Pizarro-Irizar C (2014) Switching from feed-in tariffs to a Tradable Green Certificate market. Interrelat Between Financ Energy 54:261–280

Colcelli V (2012) The problem of the legal nature of green certificates in the Italian legal system. Energy Policy 40:301–306

Currier K (2013) A regulatory adjustment process for the determination of the optimal percentage requirement in an electricity market with Tradable Green Certificates. Energy Policy 62:1053–1057

Currier K, Sun Y (2014) Market power and welfare in electricity markets employing Tradable Green Certificate systems. Int Adv Econ Res 20(2):129–138

Fagiani R, Hakvoort R (2014) The role of regulatory uncertainty in certificate markets: a case study of the Swedish/Norwegian market. Energy Policy 65:608–618

Fagiani R, Barquín J, Hakvoort R (2013) Risk-based assessment of the cost-efficiency and the effectivity of renewable energy support schemes: certificate markets versus feed-in tariffs. Energy Policy 55:648–661

Fargione J, Hill J, Tilman D, Polasky S, Hawthorne P (2008) Land clearing and the biofuel carbon debt. Science 319(5867):1235–1238

Farinosi F, Carrera L, Mysiak J, Breil M, Testella F (2012) Tradable certificates for renewable energy: the Italian experience with hydropower. In: 2012 9th international conference on the European Energy Market (EEM). IEEE, pp 1–7

Ford A, Vogstad K, Flynn H (2007) Simulating price patterns for Tradable Green Certificates to promote electricity generation from wind. Energy Policy 35(1):91–111

Jensen S, Skytte K (2003) Interactions between the power and green certificate markets. Energy Policy 30(5):425–435

Johansson TB, Nakicenovic N, Patwardhan A, Gomez-Echeverri L (2012) Global energy assessment: toward a sustainable future. Cambridge University Press, Cambridge

Krause T, Andersson G, Ernst D, Vdovina-Beck E, Cherkaoui R, Germond A (2004) Nash equilibria and reinforcement learning for active decision maker modelling in power markets. In: 6th IAEE European conference: modelling in energy economics and policy

Krause T, Beck EV, Cherkaoui R, Germond A, Andersson G, Ernst D (2006) A comparison of Nash equilibria analysis and agent-based modelling for power markets. Int J Electr Power Energy Syst 28(9):599–607

Kreps DM, Scheinkman JA (1983) Quantity precommitment and Bertrand competition yield Cournot outcomes. Bell J Econ 14(2):326–337

Lorenzoni A (2003) The Italian green certificates market between uncertainty and opportunities. Energy Policy 31(1):33–42

Mahmoudi R, Hafezalkotob A, Makui A (2014) Source selection problem of competitive power plants under government intervention: a game theory approach. J Ind Eng Int 10(3):1–15

Mitchell C, Anderson T (2000) The implication of tradable green certificate for UK. Int J Ambient Energy 21(3):161–168

Nielsen L, Jeppesen T (2003) Tradable Green Certificates in selected European countries—overview and assessment. Energy Policy 31(1):3–14

Porter R, Nudelman E, Shoham Y (2008) Simple search methods for finding a Nash equilibrium. Games Econ Behav 63(2):642–662

REN21 (2012) Renewables 2012 global status report. REN21 Secretariat, Paris

REN21 (2014) Renewables 2014 global status report. REN21 Secretariat, Paris

Schaeffer GJ, Boots MG, Mitchell C, Timpe C, Cames M, Anderson T (2000) Options for design of tradable green certificate systems. Energy Research Centre of the Netherlands, Petten

Tamás MM, Shrestha SB, Zhou H (2010) Feed-in tariff and Tradable Green Certificate in oligopoly. Energy Policy 38(8):4040–4047

Verbruggen A (2004) Tradable Green Certificates in Flanders (Belgium). Energy Policy 32(2):165–176

Vogstad K (2005) Combining system dynamics and experimental economics to analyse the design of Tradable Green Certificates. In: Proceedings of the 38th annual Hawaii international conference on system sciences, 2005 (HICSS'05). IEEE, pp 58a–58a

Zhou H (2012) Impacts of renewables obligation with recycling of the buy-out fund. Energy Policy 46:284–291

Zhou H, Tamas MM (2010) Impacts of integration of production of black and green energy. Energy Econ 32(1):220–226

7

Adaptive call admission control and resource allocation in multi server wireless/cellular network

Madhu Jain[1] · Ragini Mittal[2]

Abstract The ever increasing demand of the subscribers has put pressure on the capacity of wireless networks around the world. To utilize the scare resources, in the present paper we propose an optimal allocation scheme for an integrated wireless/cellular model with handoff priority and handoff guarantee services. The suggested algorithm optimally allocates the resources in each cell and dynamically adjust threshold to control the admission. To give the priority to handoff calls over the new calls, the provision of guard channels and subrating scheme is taken into consideration. The handoff voice call may balk and renege from the system while waiting in the buffer. An iterative algorithm is implemented to generate the arrival rate of the handoff calls in each cell. Various performance indices are established in term of steady state probabilities. The sensitivity analysis has also been carried out to examine the tractability of algorithms and to explore the effects of system descriptors on the performance indices.

Keywords Cellular network · Admission control · Handoff priority · Handoff guarantee · Guard channel · Subrating · Balking · Reneging · Blocking

✉ Ragini Mittal
m234ragini@gmail.com

Madhu Jain
madhufma@iitr.ernet.in; drmadhujain@yahoo.co.in

1 Department of Mathematics, IIT Roorkee, Roorkee 247667, India

2 Department of Mathematics, St. John's College, Agra 282002, India

Introduction

With the advancement of technology, the cellular mobile user population has been growing at a rapid pace. New generation wireless/cellular networks are designed to support adaptive multimedia service by controlling individual ongoing flow of calls. The admission control is employed to decide how a connection or call should be admitted into the network. A new connection or call is said to be blocked if its requests for resources cannot be met by the network. Call blocking probability is an important quality of service (QoS) parameter in cellular networks. One of the key QoS issues in wireless/cellular network is how to control the handoff dropping. Handoff is the situation in which a call in one cell attempts to migrate into a neighboring cell and if the level of resources required by the call cannot be satisfied by the new (neighboring) cell, the handoff is denied and call is dropped. In real time networks, the dropping of a call in progress is considered to have more negative impact from users' perspective than the blocking of a newly requested call. Thus, the optimal strategies for prioritizing handoff calls verses new calls are to be needed to maintain the bandwidth (resource) reserves for the future handoff. Jain and Rakhee (2003) developed an integrated traffic model using guard channel scheme in which priority is given to handoff calls over the new calls. The problem of resource reservation and admission control was studied by Rashad and Kantardzic (2006) for wireless mobile network. Jain et al. (2008) developed a handoff priority-based channel allocation scheme for the cellular radio network based on the channel sub-rating with balking, reneging and buffer. A handoff scheme with wireless access points (WAP) was proposed by Shet et al. (2010). This scheme uses WAP as the ad-hoc routing station, to connect to the base stations. Jain and Mittal (2011) analyzed the performance of a double layer cellular network using new call bounding scheme and directed retry scheme. A study on minimization of handoff failure in 2G/3G cellular network was presented by Kumar et al. (2012). Jain et al. (2014a, b) proposed a call admission control and resource allocation of OFDM wireless networks with buffer and subrating. Again, Jain and Mittal (2015) suggested a call admission control for soft handoff coverage in CDMA cellular system with balking and reneging.

In wireless/cellular networks, to ensure that ongoing calls are not dropped while the owner mobile stations room among cells, the handoff calls should be admitted with guarantee. Moreover, some calls such as emergency rescue or business transactions cannot be dropped before service completion. These applications will require the handoff guaranteed service. For guaranteed and handover routing in low earth orbit constellation, Galtier (2001) suggested the geographical reservation scheme. Huang et al. (2004) proposed a traffic model consisting of three service classes: handoff priority, handoff guarantee and best effort. To reduce the blocking probability of handoff voice calls in progress, a few researchers have proposed a subrating scheme. According to this scheme the reserve channels are splitted into two half rate channels to serve more than one call. Jain and Rakhee (2001) suggested the subrating channel assignment for PCS networks for integrated voice/data cellular traffic model. Jain (2003) and Jain et al. (2003a, b) proposed prioritized handover scheme with subrating. Again, Jain (2005) developed a queuing model with cutoff priority scheme, in which the subrating of guard channels was provided to serve more handover voice calls. Islam and Murshed (2007) investigated an advance resource reservation and call admission control scheme for cellular networks to achieve efficient results in case of network congestion. Wu et al. (2009) developed analytic models based on 1-D Markov process in microcell and 2-D Markov process in macrocell, in which a call admission control scheme based on the channel sub-rating is used. A mathematical model to estimate the dropping probabilities of cellular wireless networks by queuing handoff instead of reserving guard channels was proposed by Samanta et al. (2010). Hashemin and Fatemi Ghomi (2012) developed a mathematical model for the resource allocation in stochastic networks using multi-objective decision making algorithm. A heuristic method for consumable resource allocation in multi-class dynamic PERT networks was considered by Yaghoubi et al. (2013). Jain et al. (2015) presented an ANN model for multi channel infinite buffer queue under N-policy.

Buffering of the calls is one of the methods to reduce the blocking of the incoming calls. In the buffering process if the arriving call finds all the channels in the target cell occupied, it may be queued and if any channel is released it is assigned to the next call waiting in the queue (buffer). Lin (2003) has studied a dynamic resource allocation policy for GPRS with buffering mechanisms. Balking and reneging are two important aspects of customer's behaviors. When an incoming call finds all the channels busy with other calls, then the call may not join the queue (buffer); this state is considered as balking. In case of reneging, after joining the queue for some time, the call leaves the system due to impatience or some other reasons. Haghighi and Mishev (2006) discussed a general parallel finite buffer multiserver priority queuing system with balking and reneging. A channel allocation scheme was suggested by Tang and Li (2007) to evaluate the performance of the mobile cellular network in which a victim buffer is employed to non-real time calls. The impact of customer's balking and impatience behavior was examined by Artalejo and Pla (2009) on Markovian multiserver model for telecommunication system. Sharma and Purohit (2011) proposed two queueing models (1) priority handoff queueing model with reserve channels and (2) handoff queueing model with channel subrating for wireless cellular network. Jain et al. (2013) and Jain et al. (2014a, b) suggested handoff prioritized call admission control schemes for the cellular radio system having integrated traffic with impatience behavior. An $M^X/G/1$ retrial queueing system with k-phases of heterogeneous service under different vacation policies and impatient calls was proposed by Mittal and Jain (2015).

Now-a-days, there is a speedy growth of mobile users. As the available frequency spectrum is limited, it must be efficiently utilized. The main issue in the cellular networks is to decide the number of frequency channels that should be assigned to each cell so that a pre-defined level of grade of service (GoS) can be achieved. Jain et al. (2003a, b) suggested the channel allocation scheme for cellular network to optimally allocate the channel to each cell. Choi et al. (2006) developed a QoS aware selective feedback model and a method for optimal resource allocation. A dynamic channel allocation scheme with efficient channel reservation for handoff calls was proposed by Krishna and Iyengar (2008). Chowdhury et al. (2009) suggested a channel allocation algorithm that assigns optimally minimum channels in a distributed manner. A noble integer-programming problem was formulated by Wu et al. (2011) to optimize the channel allocation and power control for incoming calls. An optimal channel assignment scheme with the objective of maximizing bandwidth with fairness consideration to equalize the bandwidth assignment of flows was investigated by Jayarin and Ravi (2012). Kia et al. (2013) presented a multi-objective mixed-integer nonlinear programming mathematical model to design a group layout of a cellular manufacturing system in a dynamic environment. For multi rate wireless networks, optimization models and optimization algorithms were suggested by Ning et al. (2015).

In this investigation, we develop a cellular traffic model with integrated traffic and handoff guaranteed service. To give the priority to handoff calls over the new calls, guard channel scheme is used along with sub-rating scheme and buffer. The iterative algorithm is used to generate the arrival rate of handoff calls. The optimal allocation algorithm is suggested to allocate the optimal number of guard channels and ordinary channel in each cell of cellular cluster having K cells. The rest of the paper is organized as follows. Section 2 deals with traffic model along with the assumptions and notations. Various performance indices are established in Sect. 3. Optimal allocation algorithm and iterative algorithm are given in Sect. 4. Sensitivity analysis is carried out in Sect. 5. Finally, conclusion is drawn in Sect. 6.

The traffic model

To develop the traffic model, we consider a wireless/cellular network, consisting of a cluster of K hexagonal microcells of uniform size. The jth cell has c_j ($j = 1, 2,\ldots,K$) channels to serve the incoming calls which are classified into four types (1) new voice calls (2) new data calls (3) handoff voice calls, and (4) handoff data calls. The network provides two types of services, the handoff guaranteed service and handoff prioritized service. However, it is assumed that the handoff guaranteed service can traverse 'L' cells at most. The handoff prioritized service is provided at low price in comparison to handoff guaranteed service. In handoff guaranteed service, the handoff dropping is not allowed whereas a low handoff dropping can be tolerated in case of handoff prioritized scheme. To give the priority to handoff calls, r_j channels out of c_j channels in each cell are reserved and are allowed to be occupied by handoff attempts only. The rest of channels, i.e., $s_j = c_j - r_j$ serve all types of calls including handoff guaranteed calls while providing the service the handoff guaranteed calls are preference in the cells (see Fig. 1). The sub-rating of reserve channels is considered in order to accommodate more handoff voice calls. To reduce the blocking of handoff voice calls, there is provision of a finite buffer wherein handoff voice calls can wait. The handoff voice calls may balk or renege from the system. All the calls arrive in Poisson fashion and the call holding time and cell residence times are assumed to be exponentially distributed. As the arrival rate of new and handoff calls is interdependent, we suggest an iterative algorithm to generate the arrival rate of the handoff calls. The number of channels to be allocated in each cell of a cluster which minimize the overall blocking probability of calls, we suggest an optimal channel allocation algorithm which helps to assign optimal number of unreserved channels or ordinary channels (s_j) and guard channel (r_j) to each cell in the cluster.

The following notations are used for formulating the traffic model:

TC	Number of channels allocated to the cluster
N	Buffer size for handoff voice calls
$1/\mu$	Mean call holding time
$1/\eta$	Mean cell residence time of each port
N	Reneging rate
B	Joining probability of handoff voice calls
$\lambda_{j,nv}$ ($\lambda_{j,nd}$)	Arrival rates for new voice (new data) calls in the jth cell ($j = 1, 2,\ldots,K$)
$\lambda_{j,hv}$ ($\lambda_{j,hd}$)	Arrival rates for handoff voice (handoff data) calls in the jth cell ($j = 1, 2,\ldots,K$)
$\Lambda_{j,n}$	Arrival rate of new calls in the jth cell ($j = 1, 2,\ldots,K$) cell such that $\Lambda_{j,n} = \lambda_{j,nv} + \lambda_{j,nd}$
$\Lambda_{j,h}$	Arrival rate of handoff calls in the jth cell ($j = 1, 2,\ldots,K$) cell such that $\Lambda_{j,h} = \lambda_{j,hv} + \lambda_{j,hd}$
Λ_j	Integrated arrival rate of calls in the jth cell ($j = 1, 2,\ldots,K$); $\Lambda_j = \Lambda_{j,n} + \Lambda_{j,h}$
$P_{j,i}$	Steady state probability that there are i calls in the jth cell ($j = 1, 2,\ldots,K$)
$B_{j,n}$	Blocking probability of new calls in the jth cell ($j = 1, 2,\ldots,K$)
$B_{j,hd}$	Blocking probability of handoff data calls in the jth cell ($j = 1, 2,\ldots,K$)
$B_{j,hv}$	Blocking probability of handoff voice calls in the jth cell ($j = 1, 2,\ldots,K$)
$B_{j,h}$	Blocking probability of handoff calls in the jth cell ($j = 1, 2,\ldots,K$)
$D_{j,G}$	Dropping probability of handoff guaranteed calls in the jth cell ($j = 1, 2,\ldots,K$)
B_j	Overall blocking probability of calls in the jth cell ($j = 1, 2,\ldots,K$)

It is to be noted that we consider the performance model of an individual cell so that for the sake of simplicity, we

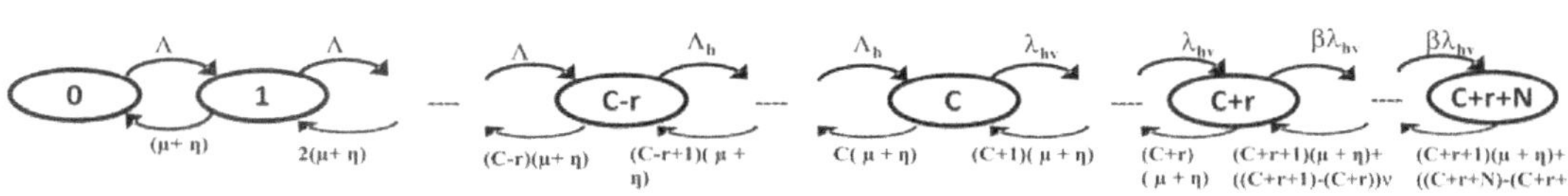

Fig. 1 State transition diagram

have dropped the suffix j. In addition, denote $s = c - r$ (Fig. 1).

The state dependent arrival and service rate are given as

$$\Lambda_i = \begin{cases} \Lambda, & 0 \le i \le s \\ \Lambda_h, & s+1 \le i < c \\ \lambda_{hv}, & c \le i < c+r \\ \beta\lambda_{hv}, & c+r \le i \le c+r+N \end{cases} \quad (1)$$

and

$$\mu_i = \begin{cases} i(\mu+\eta), & 1 \le i \le c+r-1 \\ (c+r)(\mu+\eta) + (i-(c+r))v, & c+r \le i \le c+r+N \end{cases} \quad (2)$$

The steady state probabilities for the suggested model are obtained using product type results (c.f. Gross and Harris 2003). Thus,

$$P_i = \begin{cases} \dfrac{\Lambda^i}{i!(\mu+\eta)^i} P_0, & 0 \le i \le s \\ \dfrac{\Lambda^s \Lambda_h^{i-s}}{i!(\mu+\eta)^i} P_0, & s+1 \le i \le c \\ \dfrac{\Lambda^s \Lambda_h^r \lambda_{hv}^{i-c}}{i!(\mu+\eta)^i} P_0, & c+1 \le i \le c+r \\ \dfrac{\Lambda^s \Lambda_h^r \lambda_{hv}^r (\beta\lambda_{hv})^{i-(c+r)}}{(c+r)!(\mu+\eta)^{c+r} \prod_{j=1}^{i-(c+r)} [(c+r)(\mu+\eta)+jv]} P_0, & c+r+1 \le i \le c+r+N \end{cases} \quad (3)$$

where P_0 can be calculated using the normalizing condition

$$\sum_{i=0}^{c+r+N} P_i = 1 \quad (4)$$

Performance measures

Using steady state probabilities, we can establish various performance indices as follows:

- The blocking probability of new calls is given by

$$B_n = \sum_{i=s+1}^{c+r+N} P_i \quad (5)$$

- The blocking probability of handoff data and handoff voice calls is obtained as

$$B_{hd} = \sum_{i=c+1}^{c+r+N} P_i \quad \text{and} \quad B_{hv} = P_{c+r+N} \quad (6)$$

- The blocking probability of handoff calls is

$$B_h = \frac{\lambda_{hd} B_{hd} + \lambda_{hv} B_{hv}}{\Lambda_h} \quad (7)$$

- The dropping probability of handoff guaranteed calls is given by

$$D_G = 1 - (1 - B_n)^L \quad (8)$$

- The overall blocking probability and carried load (CL), respectively, are given by

$$B = \frac{\Lambda_n B_n + \lambda_{hd} B_{hd} + \lambda_{hd} B_{hv} + \Lambda_h D_G}{\Lambda}$$
$$\text{CL} = \frac{\Lambda_n(1-B_n) + \lambda_{hd}(1-B_{hd}) + \lambda_{hv}(1-B_{hv}) + \Lambda_h(1-D_G)}{\Lambda} \quad (9)$$

- The expected number of busy channels is

$$E[l] = \sum_{0 \le i \le c+r+N} iP_i \quad (10)$$

- The number of sub-rated channels is

$$E[l_s] = \sum_{c \le i \le c+r} 2(i-c)P_i \quad (11)$$

- The degradation of voice quality is given as

$$E[D] = E[l_s/l] = \sum_{c \le i \le c+r} \frac{2(i-c)P_i}{i} \tag{12}$$

- The average queue length of handoff voice calls is given by

$$Q_{hv} = \sum_{c+r \le i \le c+r+N} (i-(c+r))P_i \tag{13}$$

- Using Little's formula the average waiting time of handoff voice is obtained by

$$W_{hv} = Q_{hv}[(1-B_{hv})\lambda_{hv}]^{-1} \tag{14}$$

Adaptive algorithms

In this section, we propose two algorithms: (1) algorithm to assign optimal number of unreserved (s_j) channels and guard channels (r_j) to each cell in the cluster of K cells and (2) algorithm to compute the arrival rate of handoff calls in each cell.

Allocation of channels

We assume that there are total TC interference-free frequency channels available in the cellular cluster. The objective is to determine the optimal number of channels (s_j^*, r_j^*) in each of the cells of the cluster, which not only minimizes the overall blocking probability (B_h) of handoff calls in the cluster, but also ensures sufficient level of GoS for new calls. The problem is formulated as a nonlinear integer-programming problem (NIPP) as follows:

Problem

$$\text{Minimize } B_h = \sum_{j=1}^{K} \frac{\lambda_{j,hv}}{\Gamma} B_{j,hv}(s_j, r_j) + \sum_{j=1}^{K} \frac{\lambda_{j,hd}}{\Gamma} B_{j,hd}(s_j, r_j) \tag{15}$$

$$\text{subject to } B_{j,n}(s_j, r_j) \le B_{\text{target}}, \quad j = 1, 2, \ldots, K \tag{16}$$

$$\sum_{j=1}^{K} (s_j + r_j) \le \text{TC} \tag{17}$$

$s_j, r_j \ge 0,\ s_j,\ r_j (j = 1, 2, \ldots, K)$ being integers.

Here B_{target} is the minimum level of GoS to be satisfied by both types of calls. In addition $\Gamma = \sum_{j=1}^{K} \Lambda_{j,h}$. To solve this NIPP we suggest the following optimization algorithm:

Optimization Algorithm

For each cell j in the cluster, follow the following steps:

Step 0: Input Parameters:
Assign the values of $\lambda_{j,nd}, \lambda_{j,nv}, \lambda_{j,hd}, \lambda_{j,hv} (for\ j = 1,2,\ldots,K), \mu, \eta, TC, B_{Target}$.

Step 1: For j= 1, 2,…, K, initialize k=0 and $r_j^0 = 0$;

Step 2: For j= 1, 2,…, K, find the smallest integer s_j^0 such that
$B_{j,n}(s_j^0, 0) \le B_{Target}$;

Step 3: Compute $Z = TC - \sum_{j=1}^{K} s_j^0$;

Step 3.1: **If** (Z > 0) **then** fix $(s_j^0, r_j^0) = (s_j^0, 0)$ and proceed to step 4.

Step 3.2: **If** (Z = 0) **then** f ix the optimal allocation as $s_j^* = s_j^0\ and\ r_j^* = 0$ and end;

Step 3.3: **If** (Z < 0) **then** the problem is infeasible and the given channel set TC cannot satisfy the pre-specified GoS;

Step 4: For j= 1, 2,…, K, Increase r_j^k *i.e.* $r_j^k = r_j^k + 1$

Compute $B_{j,n}(s_j^k, r_j^k + 1)$;

If $(B_{j,n}(s_j^k, r_j^k + 1) \le B_{target})$

then compute $B_{j,hd}(s_j^k, r_j^k + 1)\ and\ B_{j,hv}(s_j^k, r_j^k + 1)$ and find the potential amount of decrease in B_h as

$$\alpha_j^k = \frac{\lambda_{j,hv}}{\Gamma}\{old\,B_{j,hv} - newB_{j,hv}\} + \frac{\lambda_{j,hd}}{\Gamma}\{old\,B_{j,hd} - newB_{j,hd}\};$$

and fix the indicator $I_k(j) = 0$.

else increase s_j^k *i.e.* $s_j^k = s_j^k + 1$;

compute $B_{j,hd}(s_j^k + 1, r_j^k)\ and\ B_{j,hv}(s_j^k + 1, r_j^k)$

$$\alpha_j^k = \frac{\lambda_{j,hv}}{\Gamma}\{old\,B_{j,hv} - newB_{j,hv}\} + \frac{\lambda_{j,hd}}{\Gamma}\{old\,B_{j,hd} - newB_{j,hd}\};$$

and fix $I_k(j) = 1$;

Step 5: Find l the index at which α_l^k is maximum;

If $(I_k(l) = 0)$ **then** $(s_j^{k+1}, r_j^{k+1}) = (s_j^{k+1}, r_j^k + 1)$;

If $(I_k(l) = 1)$ **then** $(s_j^{k+1}, r_j^{k+1}) = (s_j^k + 1, r_j^k)$;

Step 6: Increase k = k+1;

Step 6.1: **If** (k = Z) **then**, $s_j^* = s_j^{k+1}\ and\ r_j^* = r_j^{k+1}$ is an optimal solution;
end

Step 6.2: **If** (k < Z) **then** $\alpha_j^k = \alpha_j^{k+1}$ for all j except l selected in step 5, then proceed to next step;

Step 7: Compute $B_{l,n}(s_l^k, r_l^k + 1)$;

If $(B_{l,n}(s_l^k, r_l^k + 1) \le B_{target})$

then $\alpha_l^k = \frac{\lambda_{j,hv}}{\Gamma}\{old\,B_{j,hv} - newB_{j,hv}\} + \frac{\lambda_{j,hd}}{\Gamma}\{old\,B_{j,hd} - newB_{j,hd}\}$;

and fix $I_k(l) = 0$

else fix $\alpha_j^k = \frac{\lambda_{j,hv}}{\Gamma}\{old\,B_{j,hv} - newB_{j,hv}\} + \frac{\lambda_{j,hd}}{\Gamma}\{old\,B_{j,hd} - newB_{j,hd}\}$;

and $I_k(l) = 1$;

go to step 5;

Output: (s_j^*, r_j^*).

Computation of handoff traffic

We use the following iterative algorithm to compute the arrival rates of handoff voice and handoff data attempts in each cell. The blocking/dropping probabilities and arrival rates of new attempts and the handoff arrival rates are interdependent and are given by the following relation (see Lin et al. 1996):

Table 1 Optimal channel allocation taking $B_{\mathrm{target}} = 1\ \%$

TC	(s^*, r^*)	B_n	B_{hd}	B_{hv}	B_h	B	D_G	CL
150	(13, 1)	0.009861	0.000823	4.97E-06	0.00017	0.008323	0.057728	0.991677
	(16, 1)	0.008749	0.00078	5.38E-06	0.000156	0.007358	0.051359	0.992642
	(18, 0)	0.009209	0.009209	8.67E-04	0.002451	0.00811	0.053997	0.99189
	(21, 1)	0.009536	0.000946	8.10E-06	0.000183	0.008002	0.055872	0.991998
	(23, 0)	0.00939	0.00939	9.36E-04	0.002492	0.008257	0.055034	0.991743
	(26, 1)	0.009769	0.001032	1.00E-05	0.000196	0.008189	0.057203	0.991811
	(28, 0)	0.009283	0.009283	9.57E-04	0.002461	0.008156	0.054419	0.991844
160	(13, 1)	0.009861	8.23E-04	4.97E-06	1.70E-04	0.008323	0.057728	0.991677
	(16, 1)	0.008749	7.80E-04	5.38E-06	1.56E-04	0.007358	0.051359	0.992642
	(19, 1)	0.007604	7.01E-04	5.19E-06	1.37E–04	0.00639	0.044763	0.993611
	(21, 2)	0.009749	1.10E-04	7.37E-08	2.06E-05	0.008153	0.057084	0.991847
	(24, 2)	0.008438	9.81E-05	7.04E-08	1.81E-05	0.007055	0.049573	0.992945
	(26, 2)	0.009994	1.28E-04	1.08E-07	2.34E-05	0.008349	0.058488	0.991651
	(29, 2)	0.008702	1.13E-04	9.83E-08	2.05E-05	0.007268	0.051088	0.992732
170	(14, 2)	0.006265	4.42E-05	1.32E-08	8.90E-06	0.005272	0.037005	0.994728
	(16, 3)	0.008952	8.08E-06	2.14E-10	1.57E-06	0.007503	0.052524	0.992497
	(19, 3)	0.007782	7.86E-06	2.63E-10	1.49E-06	0.006517	0.045793	0.993483
	(21, 3)	0.009774	1.23E-05	5.81E-10	2.30E-06	0.008171	0.057227	0.991829
	(24, 3)	0.00846	1.12E-05	5.94E-10	2.06E-06	0.00707	0.049699	0.99293
	(27, 2)	0.007423	8.90E-05	6.81E-08	1.63E-05	0.006201	0.04372	0.993799
	(29, 3)	0.008726	1.37E-05	9.75E-10	2.48E-06	0.007286	0.05123	0.992714
180	(14, 1)	0.006277	3.17E-07	1.98E-13	6.38E-08	0.005281	0.037075	0.994719
	(16, 4)	0.008954	7.65E-07	1.05E-12	1.49E-07	0.007504	0.052535	0.992496
	(19, 4)	0.007784	7.82E-07	1.49E-12	1.48E–07	0.006518	0.045804	0.993482
	(21, 4)	0.009776	1.33E-06	4.00E-12	2.48E-07	0.008173	0.057243	0.991827
	(24, 4)	0.008463	1.24E-06	4.44E-12	2.28E-07	0.007072	0.049713	0.992928
	(27, 4)	0.007445	1.17E-06	4.92E-12	2.13E-07	0.006216	0.043847	0.993784
	(29, 5)	0.008730	1.85E-07	7.02E-14	3.35E-08	0.007288	0.051249	0.992712

Table 2 Optimal channel allocation for TC = 200

	(s^*, r^*)	B_n	B_{hd}	B_{hv}	B_h	B	D_G	CL
$B_{\mathrm{target}} = 0.01$	(14, 6)	0.006277	1.85E-09	1.55E-18	3.73E-10	0.005281	0.037075	0.994719
	(16, 7)	0.008954	4.97E-10	5.23E-20	9.65E-11	0.007504	0.052537	0.992496
	(19, 7)	0.007784	6.06E-10	1.27E-19	1.15E-10	0.006518	0.045805	0.993482
	(21, 7)	0.009777	1.34E-09	6.29E-19	2.49E-10	0.008173	0.057245	0.991827
	(24, 7)	0.008463	1.36E-09	9.47E-19	2.51E-10	0.007072	0.049715	0.992928
	(27, 7)	0.007445	1.41E-09	1.38E-18	2.56E-10	0.006217	0.043849	0.993783
	(29, 8)	0.00873	2.37E-10	2.10E-20	4.27E-11	0.007288	0.051249	0.992712
$B_{\mathrm{target}} = 0.015$	(13, 7)	0.009723	2.57E-10	7.82E-21	5.19E-11	0.008207	0.056938	0.991793
	(15, 8)	0.013011	7.63E-11	3.49E-22	1.48E-11	0.010926	0.075572	0.989074
	(18, 8)	0.011059	9.50E-11	9.79E-22	1.81E-11	0.009275	0.064549	0.990725
	(20, 8)	0.013152	2.05E-10	5.10E-21	3.82E-11	0.01102	0.076364	0.98898
	(23, 9)	0.011411	2.18E-11	4.39E-23	4.01E-12	0.009544	0.066541	0.990456
	(25, 9)	0.013026	3.94E-11	1.58E-22	7.18E-12	0.01089	0.075654	0.98911
	(27, 9)	0.014612	6.72E-11	5.01E-22	1.21E-11	0.012205	0.08453	0.987795

Table 2 continued

	(s^*, r^*)	B_n	B_{hd}	B_{hv}	B_h	B	D_G	CL
$B_{\text{target}} = 0.02$	(12, 8)	0.014412	3.37E-11	3.29E-23	6.82E-12	0.012215	0.083413	0.987785
	(14, 9)	0.018281	1.15E-11	2.10E-24	2.23E-12	0.015388	0.104794	0.984612
	(17, 9)	0.015253	1.47E-11	7.00E-24	2.80E–12	0.012816	0.088101	0.987184
	(19, 10)	0.017478	3.01E-12	1.62E-25	5.63E-13	0.014665	0.100393	0.985335
	(21, 10)	0.019654	6.70E-12	9.95E-25	1.23E-12	0.016454	0.112278	0.983546
	(24, 10)	0.016722	6.62E-12	1.60E-24	1.21E-12	0.013991	0.096228	0.986009
	(26, 10)	0.018223	1.10E-11	5.16E-24	2.00E-12	0.01524	0.104474	0.98476

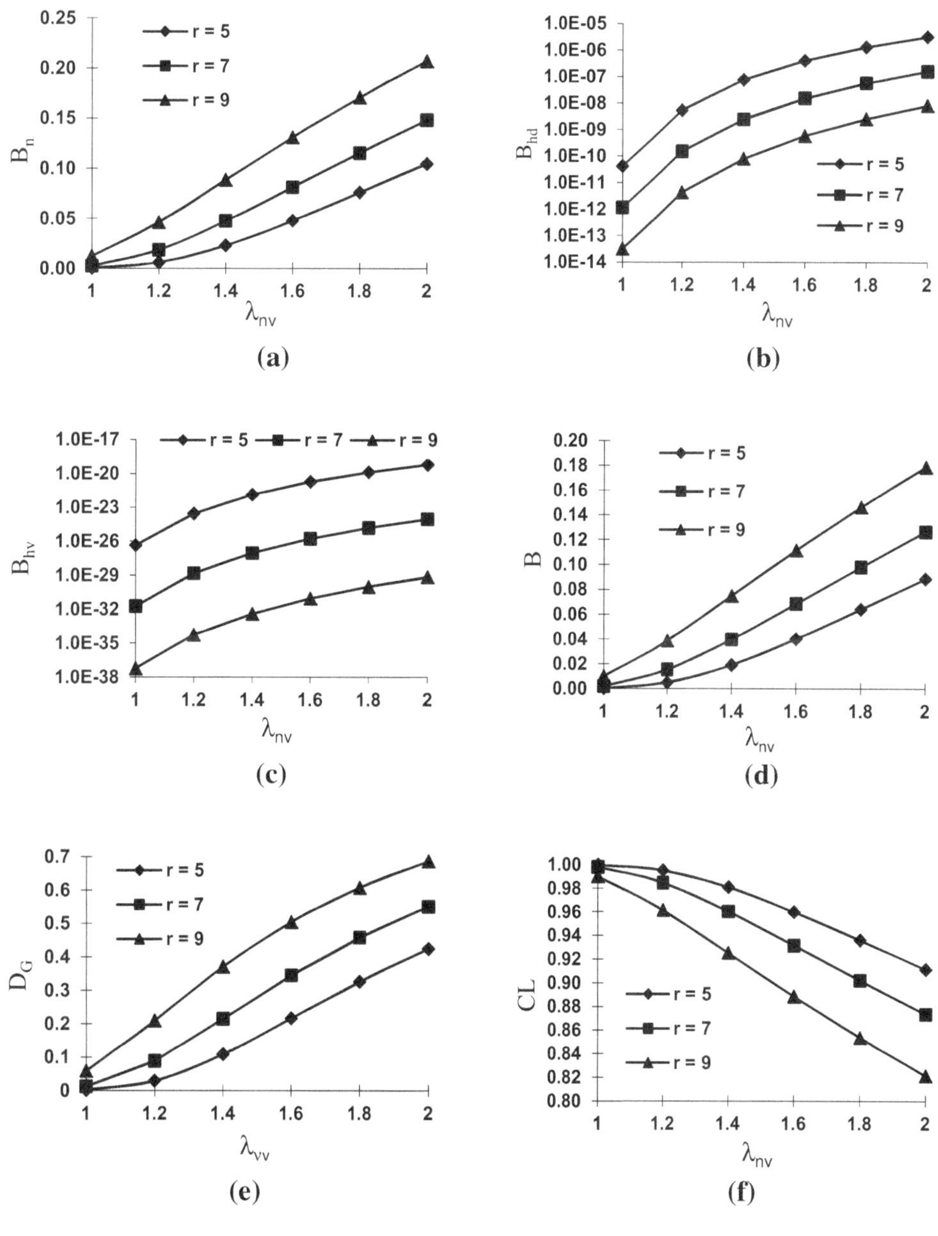

Fig. 2 Effect of λ_{nv} on **a** B_n, **b** B_{hd}, **c** B_{hv}, **d** B, **e** D_G, **f** CL for different values of r

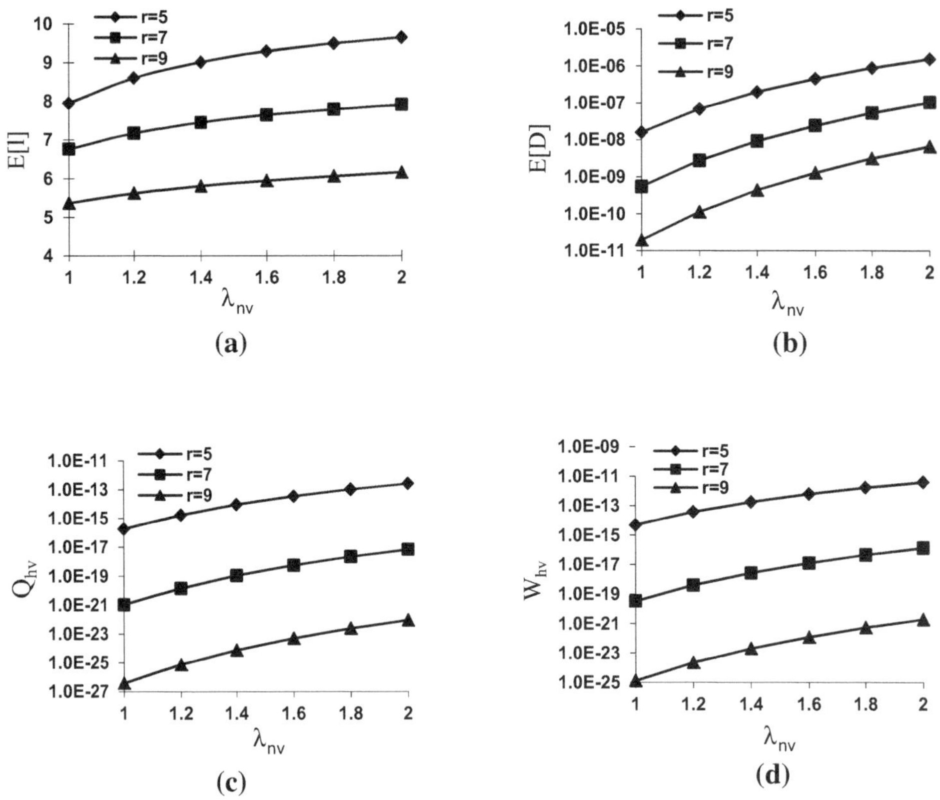

Fig. 3 Effect of λ_{nv} on **a** $E[l]$, **b** $E[D]$, **c** Q_{hv}, **d** W_{hv} for different values of r

$$\lambda_{j,hd} = \frac{\eta(1 - B_{j,n})}{\mu + \eta D_{j,hd}} \lambda_{j,nd} \quad \text{and} \quad \lambda_{j,hv} = \frac{\eta(1 - B_{j,n})}{\mu + \eta D_{j,hv}} \lambda_{j,nv}$$

Handoff Algorithm

Step 0: Initilize $\lambda_{j,hd}=0$ and $\lambda_{j,hv}=0$

Step 1: **While** $|\delta_j| < \varepsilon$ $(say\ \varepsilon = .00001)$ **do**

Step 1.1: **set** $B_{j,n}$, $B_{j,hd}$ and $B_{j,hv}$

set $\lambda_{j,hd} = \frac{\eta(1-B_{j,n})}{\mu+\eta D_{j,hd}}\lambda_{j,nd}$ and $\lambda_{j,hv} = \frac{\eta(1-B_{j,n})}{\mu+\eta D_{j,hv}}\lambda_{j,nv}$

Step 2: **Set** $\lambda_{j,h} = \lambda_{j,hd} + \lambda_{j,hv}$

Step 2.1: **set** $\delta_j = \left|\frac{old(\lambda_{j,h}) - new(\lambda_{j,h})}{new(\lambda_{j,h})}\right|$

end while

return ($B_{j,n}$, $B_{j,hv}$, $B_{j,hd}$, B_j, CL)

end Algorithm

Sensitivity analysis

In this section, the sensitivity analysis is carried out to examine the analytical results. Various performance indices for the proposed scheme are summarized in Tables 1, 2 and Figs. 2, 3, 4. For different values of TC, Table 1 provides the optimal number of channels (unreserved channels and reserved channels) to be allocated in each cell using optimal allocation algorithm by taking $B_{\text{target}} = 1\ \%$. The corresponding performance indices like blocking probability of new calls (B_n), blocking probability of handoff data calls (B_{hd}), blocking probability of handoff voice calls (B_{hv}), overall blocking probability of handoff calls (B_h), overall blocking probability of calls (B), dropping probability of handoff guaranteed calls (D_G) and CL in every cell, are also tabulated. For different values of B_{target}, Table 2 displays the results for various performance indices for TC $= 200$. For these results, a cluster of 7 cells is taken, i.e., $K = 7$ and the buffer size is taken as $N = 5$.

Figure 2a–f shows the effect of arrival rate of new voice calls (λ_{nv}) on B_n, B_{hd}, B_{hv}, B, and CL for different values of reserve channels (r). For this we choose the default parameters as $c = 15$, $N = 5$, $\lambda_{nd} = 1$, $\beta = 0.3$, $v = 0.2$, $\mu = 0.5$, and $\eta = 0.1$. Figure 2a, d, e depict that B_n, B and D_G first increases slowly then sharply with respect of λ_{nv}. B_n, B, and D_G also increase as r increases. In Fig. 2a, b B_{hd} and B_{hv} increase sharply with respect to λ_{nv} whereas B_{hd} and B_{hv} decrease on increasing r. Figure 2f shows that CL decreases sharply with λ_{nv} and r both.

Figures 3a–d and 4a–d shows the effect of arrival rate of new voice calls and new data calls (λ_{nv} and λ_{nd}) on expected number of busy channels ($E[l]$), average queue length of handoff voice calls (Q_{hv}) and waiting time of handoff voice calls (W_{hv}), respectively, for different values of r. Figure 3a depicts that $E[l]$ increases slowly as we

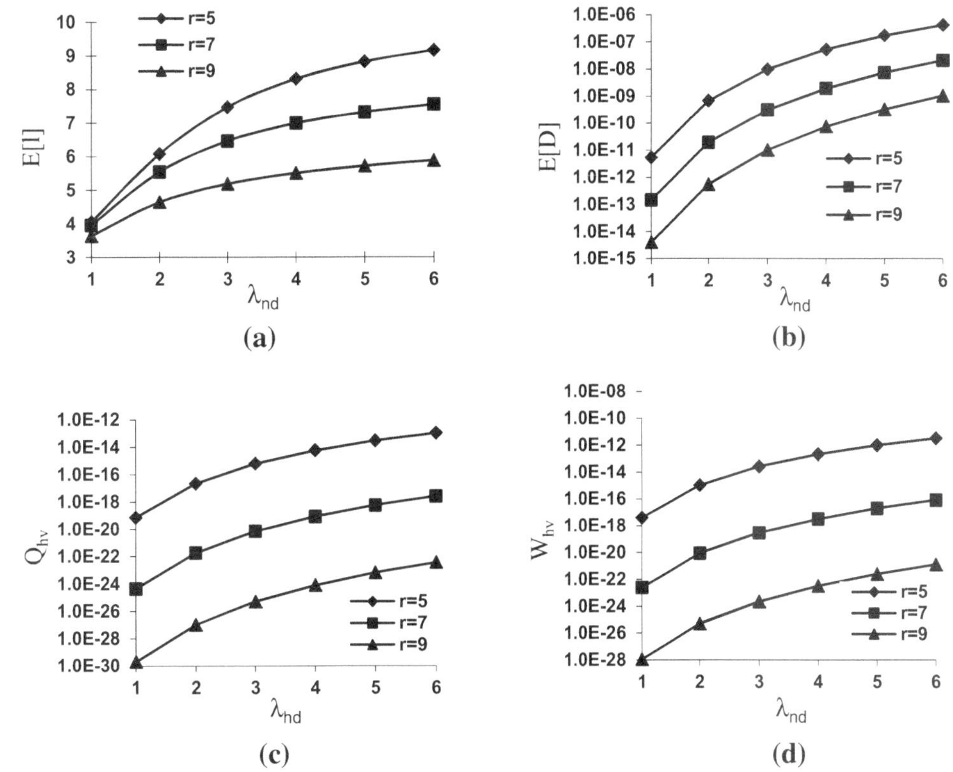

Fig. 4 Effect of λ_{nd} on **a** $E[l]$, **b** $E[D]$, **c** Q_{hv}, **d** W_{hv} for different values of r

increase λ_{nv} whereas in Fig. 4a $E[l]$ increases first sharply then gradually as λ_{nd} increases.

In Figs. 3b and 4b, it is noticed that $E[D]$ increases sharply on increasing λ_{nv} and λ_{nd}, respectively, whereas $E[D]$ decreases gradually on increasing r which is quite obvious because as the number of reserved channels increases, less degradation is achieved in handoff voice calls. Figures 3c, d and 4c, d shows that Q_{hv} and W_{hv} increase sharply as λ_{nv} and λ_{nd} increase whereas there is a sharp decrement in Q_{hv} and W_{hv} as r increases. The trend of Q_{hv} is quite at par as we expect.

Overall, we conclude that all performance measure except CL increase as arrival rate of new data calls and new voice calls increase which is the same as we expect and realize in real time system. The indices B_n, B, D_G and CL decrease as r increases whereas B_{hd}, B_{hv}, $E[l]$, $E[D]$, Q_{hv} and W_{hv} reveals the decreasing trend on increasing r.

Conclusion

The rapid growth in demand for mobile communication has led to intense research and development efforts towards a new generation of cellular systems. The new system must be able to provide GoS, support a wide range of services and improving the system capacity. Resource allocation and admission control are the two major issues which are faced by today's wireless/cellular industry. To control the admission of the incoming calls, the proposed priority channel assignment scheme may provide insights to the system designers and decision makers to utilize the bandwidth optimally. The optimal allocation algorithm suggested may be utilized to develop a controller which can optimally allocate the ordinary channels and guard channels in each cell of the cluster in cellular network. We have developed an iterative algorithm to computes the arrival rate of handoff calls in an efficient manner. The sensitivity analysis presented, exhibits the validity of the proposed algorithms.

References

Artalejo JR, Pla V (2009) On the impact of customer balking, impatience and retrials in telecommunication systems. Comput Math Appl 57(2):217–229

Choi YJ, Kim J, Bahk S (2006) QoS-aware selective feedback and optimal channel allocation in multiple shared channel environments. IEEE Trans Wirel Commun 5(11):3278–3286

Chowdhury KR, Nandiraju N, Chanda P, Agrawal DP, Zeng Q (2009) Channel allocation and medium access control for wireless sensor networks. Ad Hoc Netw 7(2):307–321

Galtier J (2001) Geographical reservation for guaranteed handover and routing in low earth orbit constellations. Telecommun Syst 18(1–2):101–121

Gross D, Harris MC (2003) Basic concepts of queueing theory. Wiley, Singapore

Haghighi AM, Mishev DP (2006) A parallel priority queueing system with finite buffers. J Parallel Distrib Comput 66:379–392

Hashemin SS, Ghomi SMTF (2012) Constrained consumable resource allocation in alternative stochastic networks via multi-objective decision making. J Ind Eng Int 8(18):1–9

Huang L, Kumar S, Kuo CCJ (2004) Adaptive resource allocation for multimedia QoS management in wireless networks. IEEE Trans Veh Technol 53(2):547–558

Islam MM, Murshed M (2007) Parametric mobility support dynamic resource reservation and call admission control scheme for cellular multimedia communications. Comput Commun 30(2):233–248

Jain M (2003) Prioritized handover request in cellular radio system with subrating for intrgrated voice/data traffic. IETE J Res 49(6):378–398

Jain M (2005) Channel allocation policy in cellular radio network. Appl Math Model 29(1):65–83

Jain M, Mittal R (2011) Performance analysis in a double layer cellular network with new call bounding scheme and directed retry. CSI J Comput Sci Eng 9(2):1–10

Jain M, Mittal R (2015) Call admission control for soft handoff coverage in CDMA cellular system. Int J Wirel Inf Netw 22:53–66

Jain M, Rakhee (2001) Queueing analysis for PCS with integrated traffic and subrating channel assignment scheme. J Comput Sci Inf 31(2):1–8

Jain M, Rakhee (2003) Priority based channel assignment scheme for PCS with integrated traffic. J Comput Sci Inf 33(1):8–18

Jain M, Priya K, Rakhee (2003a) Optimal channel allocation in cellular systems. J Inf Commun Technol 2(2):41–64

Jain M, Sharma GC, Saraswat VK (2003b) Channel assignment schemes for PCS with cutoff priority and subrating. Int J Inf Comput Sci 6(2):46–56

Jain M, Sharma GC, Mittal R (2008) Transient analysis of channel allocation in cellular radio network with balking and reneging. Comput Soc India 38(4):11–19

Jain M, Sharma GC, Mittal R (2013) Performance analysis of prioritized call admission control schemes for integrated traffic in wireless network. Int J Electron Commun Comput Eng 4(1):283–291

Jain M, Sharma GC, Mittal R (2014a) Call admission control and resource allocation of wireless networks. Int J Wirel Mob Comput 7(6):577–590

Jain M, Sharma GC, Mittal R (2014b) Queueing analysis and channel assignment scheme for cellular radio system with GPRS services. Int J Math Oper Res 6(6):704–731

Jain M, Sharma GC, Mittal R (2015) ANN model for multi channel infinite buffer queue under N-policy. Int J Oper Res 24(1):59–82

Jayarin PJ, Ravi T (2012) Optimal channel allocation mechanism to improve the QoS in wireless networks. Int J Comput Appl 45(20):30–34

Kia R, Shirazi H, Javadian N, Tavakkoli-Moghaddam R (2013) A multi-objective model for designing a group layout of a dynamic cellular manufacturing system. J Ind Eng Int 9(8):1–14

Krishna PV, Iyengar NCSN (2008) Optimal channel allocation algorithm with efficient channel reservation for cellular networks. Int J Commun Netw Distrib Syst 1(1):33–51

Kumar A, Bhushan A, Kumar M (2012) A study on minimization of 2G/3G handover failure. Int J Adv Res Comput Sci Softw Eng 2(4):4–7

Lin P (2003) Channel allocation for GPRS with buffering mechanisms. Wirel Netw 9:431–441

Lin YB, Anthony R, Harasty DJ (1996) The sub-rate channel assignment strategy for PCS handoffs. IEEE Trans Veh Technol 45(1):122–130

Mittal R, Jain M (2015) Maximum entropy analysis of $M^x/G/1$ retrial queue with k-phases of heterogeneous service and impatient calls under different vacation policies. Am J Math Manag Sci 34(2):117–146

Ning Z, Song Q, Guo L, Chen Z, Jamalipour A (2015) Integration of scheduling and network coding multi rate wireless mesh networks optimization models and algorithms. Adhoc Netw. doi:10.1016/j.adhoc.2015.08.013

Rashad S, Kantardzic M (2006) User mobility oriented predictive call admission control and resource reservation for next generation mobile networks. J Parallel Distrib Comput 66(7):971–988

Samanta RK, Bhattacharjee P, Sanyal G (2010) Performance analysis of cellular wireless network by queueing priority handoff calls. Int J Electr Electron Eng 3(8):472–477

Sharma G, Purohit GN (2011) Queueing modeling of handoff calls with subrating in wireless cellular netwrok. Appl Math Sci 5(21):1017–1024

Shet NSV, Chandrasekaran K, Shet KC (2010) Implementation of handoff through wireless access point techniques. J Telecommun 2(2):143–146

Tang S, Li W (2007) Performance analysis of a channel allocation scheme for multi-service mobile cellular networks. Int J Commun Syst 20:177–205

Wu X, He M, Wang F, Zheng J, Regentova E, Hao G (2009) Performance analysis of sub-rating for handoff calls in HCN. Int J Commun Netw Syst Sci 2(1):21–29

Wu X, Jaekel A, Bari A (2011) Optimal channel allocation for dynamic power control in cellular networks. Int J Comput Netw Commun 3(2):83–93

Yaghoubi S, Noori S, Mazdeh MM (2013) A heuristic method for consumable resource allocation in multi-class dynamic PERT networks. J Ind Eng Int 9(17):1–12

8

Robust optimization of a mathematical model to design a dynamic cell formation problem considering labor utilization

Moghadaseh Vafaeinezhad[1] · Reza Kia[1] · Parisa Shahnazari-Shahrezaei[1]

Abstract Cell formation (CF) problem is one of the most important decision problems in designing a cellular manufacturing system includes grouping machines into machine cells and parts into part families. Several factors should be considered in a cell formation problem. In this work, robust optimization of a mathematical model of a dynamic cell formation problem integrating CF, production planning and worker assignment is implemented with uncertain scenario-based data. The robust approach is used to reduce the effects of fluctuations of the uncertain parameters with regards to all possible future scenarios. In this research, miscellaneous cost parameters of the cell formation and demand fluctuations are subject to uncertainty and a mixed-integer nonlinear programming model is developed to formulate the related robust dynamic cell formation problem. The objective function seeks to minimize total costs including machine constant, machine procurement, machine relocation, machine operation, inter-cell and intra-cell movement, overtime, shifting labors between cells and inventory holding. Finally, a case study is carried out to display the robustness and effectiveness of the proposed model. The tradeoff between solution robustness and model robustness is also analyzed in the obtained results.

Keywords Dynamic cell formation problem · Scenario-based robust optimization · Mixed-integer nonlinear model · Worker assignment

✉ Reza Kia
rezakia@iaufb.ac.ir

1 Department of Industrial Engineering, Firoozkooh Branch, Islamic Azad University, Firoozkooh, Iran

Introduction

Today, global competitive environment has persuaded manufacturing practitioners to deliver low-cost and high-quality products. Some recently applied approaches have been put into practice to cope with the ever growing manufacturing costs, such as location, material handling system, and energy. One of these recent manufacturing approaches is Group Technology (GT). GT is one of the main building blocks to implementing Just-In-Time (JIT) philosophy. This approach is based upon grouping parts and machines together with respect to their similarities in production processes, functionalities, etc. The aspect of GT which associates with the configuration of manufacturing firms is cellular manufacturing system (CMS). The most outstanding benefit of CMS can be noted as reduction in some production factors, such as lot sizes, lead times, work-in-process inventories and setups, while higher level of investment is inevitable to implement this system. Designing of a CMS involves four main steps. The first step associates with cell formation problem which comprises assigning parts to their families and machines to their corresponding machine cells based on some features, such as similar geometric design or processing requirements. Second, intra-cell and inter-cell layouts are defined through Group Layout (GL). This step determines the location of machines and cells in the shop floor. Third, Group Scheduling (GS) is accomplished to schedule parts within part families. Finally, required resources such as labors and material handling devices are assigned to the manufacturing cells.

It has been clarified by Wu et al. (2007) that these four steps are interrelated and in other words, the solution for each step influences the other one. Thus, simultaneously solving these problems has to be applied by the researchers; that is, the matter not been paid attention enough. Nevertheless, due to the complexity and NP-complete nature of CF, GL, and GS decisions, most researchers have addressed two or three decisions sequentially or independently. However, the benefits gained from CMS implementation are highly affected by how theses stages of the CMS design have been performed in collaboration with each other.

Shorter product life cycles are an increasingly significant issue in CM. As a result, neglecting new products emerging at future imposes subsequent unplanned changes to the CMS design and causes production disruptions and unexpected costs. Hence, those changes should be incorporated in the design process. To come up with a solution to handle those changes, the dynamic cellular manufacturing system (DCMS) was introduced in which it is assumed that the product mix or volume changes of demands can be predicted in a multi-period planning horizon (Rheault et al. 1995).

Most DCMS models assume that the input parameters are deterministic and certain. However, in practical situations many parameters are uncertain and imprecise. DCMS design has to be implemented in many environments based on some parameters with uncertain values. However, there are few studies on designing cellular manufacturing systems under dynamic and uncertain conditions. These studies can be divided into four classes as fuzzy programming approach, stochastic programming approach, scenario-based programming approach, and robust optimization approach in terms of uncertainty expression type in the problem. Different robust optimization approaches have been introduced in the recent years to deal with the uncertainty of the data. In this study, a scenario-based robust optimization approach is used to cope with uncertainty and to find a solution that is robust with regard to data uncertainties in part demand, inter-cell and intra-cell movement cost, machine purchase cost, selling machine revenue, machine fixed/variable cost, machine relocation cost, inter-cell movement labor cost, process variable cost and inventory holding cost. It is the first time that this vast coverage of input parameters in a DCMS are considered uncertain to be handled by a robust optimization approach.

The aims of this study are twofold. The first one is to formulate a new mathematical model with an extensive coverage of important manufacturing features including batch intra-cell/inter-cell movement, production planning strategies (i.e., internal production, inventory holding, and lost sale as under-fulfilled demand), selling/purchase machine, labor movement, labor assignment, labor capacity, machine relocation, regular/overtime machine capacity, cell size limit, flexible operation sequence, machine/labor processing time, and uncertain scenario-based parameters (i.e., part demand and miscellaneous costs). The second aim is to develop a robust model based on the deterministic proposed model using scenario-based robust optimization approach. The important concern of the employed robust methodology is to obtain an optimal CM design that is robust with regard to data uncertainties in part demand and miscellaneous costs. The objective function of the integrated model is to minimize the total costs of machine constant, machine procurement, machine relocation, machine operation, inter-cell and intra-cell movement, overtime, shifting labors between cells and inventory holding. The main constraints are operator-machine-cell assignment, machine capacity, machine number equilibrium, labor capacity, cell size limit, and balancing inventory.

Recently, Kia et al. (2012) have formulated a mathematical model integrating the CF and GL decisions in a dynamic environment by considering some advantages including: (1) considering flexible configuration of cells, (2) calculating relocation cost based on the locations assigned to machines, (3) distance-based calculation of intra- and inter-cell material handling costs and (4) considering multi-rows layout of equal sized facilities. One disadvantage in their work was ignoring the assignment of operators to machines located in different cells. In another study, Bagheri and Bashiri (2014) investigated the simultaneous consideration of the cell formation problem with inter-cell layout and operator assignment problems in a dynamic environment by formulating a mathematical model with the objectives of minimization of inter–intra cell part trips, machine relocation cost and operator-related issues. A main drawback in both mentioned studies was that all parameters were considered deterministic despite the fact some of them should be predicted for the future periods in a dynamic environment with high level of uncertainty.

Generally, the presented study is an extension of the previous studies Kia et al. (2012), Bagheri and Bashiri (2014) by integrating the CF, production planning (PP) and worker assignment in a mathematical model with data uncertainties in most parameters of model including part demand and miscellaneous costs which is solved by a scenario-based robust optimization approach. The robust approach is used to reduce the effects of fluctuations of the uncertain parameters with regards to all possible future scenarios.

To investigate the effect of turbulence in the values of uncertain data on the model performance and obtained solutions, a robust model is developed. Then, a case study is carried out to demonstrate the validity of the employed robust approach and verify the integrated DCMS model. The obtained results of implementing the case study also illustrate the applicability of the proposed model in real industrial cases.

The remainder of this paper is organized as follows. In "Literature review" section, the literature review is carried out. The background of the robust optimization approach employed in this study is described in "Robust optimization" section. A mathematical model is formulated integrating CF, PP and worker assignment decisions in "Mathematical model and model description" section followed using some linearization procedures. In addition, a robust model is developed in this section. "A case study" section illustrates the case study that is implemented to investigate the features of the proposed model and assess the performance of the developed robust model. Finally, conclusion is given in "Conclusion" section.

Literature review

One of the most important issues which have received less attention in the literature body of DCMS is consideration of human-related issues. The first mathematical model developed for human-related aspects of DCMS was presented by Aryanezhad et al. (2009). They developed a new mathematical model to deal with DCMS and worker assignment problems, simultaneously. The objective function of this model contains system costs including machine purchase, operating, inter-cell material handling, machine relocation, worker hiring, training, salary and firing costs. Balakrishnan and Cheng (2005) presented a flexible framework for modeling cellular manufacturing when product demand changes during the planning horizon.

Most CMS models assume that the input parameters are deterministic and certain. However, in practical situations, many parameters such as parts demands, processing times and machines capacities are uncertain. Robust optimization as a strong technique was used to deal with uncertainty in the systems. Robust optimization can be very efficient and useful because of generation of the good and robust solutions for any possible occurrences of uncertain parameters (Mulvey et al. 1995). The concept of robust optimization in operation research was presented by Mulvey et al. (1995). They extended a robust counterpart approach with a nonlinear function that penalizes the constraint violations and addresses uncertainties via a set of discrete scenarios. Bai et al. (1997) demonstrated that the traditional stochastic linear program fails to determine a robust solution despite the presence of a cheap robust point. They evaluated properties of risk-averse utility functions in robust optimization. They discussed that a concave utility function should be incorporated in a model whenever the decision maker is risk averse. Ben-Tal and Nemirovski (1998) proposed a robust optimization approach to formulate continuous uncertain parameters. Ben-Tal and Nemirovski (1998), Ben-Tal and Nemirovski (2002) and Ben-Tal et al. (2002) developed robust theory of linear, quadratic and conic quadratic problems. Bertsimas and Sim (2002) and Bertsimas and Thiele (2003) proposed robust optimization methods for discrete optimization in continuous spaces.

Mirzapour Al-E-Hashem et al. (2011) studied multi-site aggregate production planning problems under uncertainty by defining multi-objective robust optimization models.

Mahdavi et al. (2010) proposed a mathematical model for solving dynamic cellular manufacturing problem considering two areas of cell configuration and assigning the operators to the machines. In the proposed model, some factors have been considered including machine capacity, multi-period planning horizon and the worker idleness time. Rafiei and Ghodsi (2013) designed a two-objective mathematical model for solving the operator assignment and cell configuration simultaneously. Minimizing total costs of machines purchase, machine relocation and overhead, parts intra-cell and inter-cell movements and the operator inter-cell movements were considered in the first objective function. The second objective function increased the utilization level of the operators.

In similar studies, Kia et al. (2013), Shirazi et al. (2014) presented multi-objective mixed-integer nonlinear programming models to combine the problems of dynamic cell formation and group layout. They utilized the multi-row layout for locating machines inside the cells with flexible size regarding the lot splitting feature and several other features (i.e., operation sequence, processing time, machine duplicates, and machine capacity).

Bashiri and Bagheri (2013) proposed a two-phase heuristic method for cell formation and operator assigning, where in the first phase, clustering technique and in the second phase, a mathematical model is used. Kia et al. (2011) presented a mathematical model for a multi-period CM system layout with fuzzy parameters. By taking the linear intra-cell machines layout, operation sequence, processing times and the machines capacity into account, the model intended to minimize the intra/inter-cell movements costs, the machines overhead costs and machines relocation costs.

Ghezavati et al. (2011) proposed a robust model for cell formation and group scheduling with supply chain approach. In this model, the uncertainty resulted from demand and parts processing time were expressed by stochastic scenarios with given probabilities. They formulated the problem with the objective to minimize delaying costs for parts delivery due time, the parts outsourcing costs to suppliers and the underutilization cost of machines and solved it by a hybrid meta-heuristic algorithm. Paydar et al. (2013) presented a mathematical model for integration of cell formation, machine layout and production planning. They considered customer demand and machine capacity uncertain and proposed a robust model. Forghani et al. (2012) suggested a robust model to determine cell formation and group layout where the parts demand is uncertain.

Sakhaii et al. (2015) developed a robust optimization approach for a new integrated MILP model to solve a DCMS with unreliable machines and a production planning problem simultaneously. They adopted a robust optimization approach immunized against even worst-case to cope with the parts processing time uncertainty. Hassannezhad et al. (2014) performed sensitivity analysis of modified self-adaptive differential evolution (MSDE) algorithm for basic parameters of cell formation problem. First, they presented a DCMS model. Then, two basic test CF problems were introduced to assess the performance of MSDE algorithm by diverse problems sizes.

Regarding this section, it could be concluded that no study has been done on simultaneous integrating of three problems as cell configuration, production planning and operator assigning so far with uncertainty considered in the most model parameters including part demands and cost parameters.

Robust optimization

Mulvey et al. (1995) presented a framework for robust optimization that involves two types of robustness: ''solution robustness'' (the solution is nearly optimal in all scenarios) and ''model robustness'' (the solution is nearly feasible in all scenarios). The robust optimization method extended by Mulvey et al. (1995), in fact, develops stochastic programming through replacing traditional expected cost minimization objective by one that explicitly addresses cost variability. The framework of robust optimization is briefly demonstrated by Feng and Rakesh (2010). The form of the robust optimization model is as follows:

$$\text{Min } c^T x + d^T y \tag{1}$$

$$Ax = b \tag{2}$$

$$Bx + Cy = e \tag{3}$$

$$x, y \geq 0 \tag{4}$$

where x defines the vector of decision variables that should be determined under the uncertainty of model parameters. B, C and e demonstrate random technological coefficient matrix and right- hand side vector, respectively. Assume a finite set of scenarios $\Omega = \{1, 2, \ldots, s\}$ to model the uncertain parameters; with each scenario $s \in \Omega$, we associate the subset $\{d_s; B_s; C_s; e_s\}$ and the probability of the scenario $p_s(\sum_{s=1}^{S} p_s = 1)$.

Note that a scenario is a series of data realizations over the planning horizon. In addition, control variable y, can be denoted as y_s for scenario s. δ_s represents the infeasibility of the model under scenario s, because of parameter uncertainty the model may be infeasible for some scenarios. If the model is feasible, δ_s will be equal to 0, otherwise; δ_s will receive a positive value according to Eq. (7). A robust optimization model is formulated as follows:

$$\text{Min } \sigma(x, y_1, \ldots, y_s) + \omega\rho(\delta_1, \delta_2, \ldots, \delta_s) \tag{5}$$

$$Ax = b \tag{6}$$

$$B_s x + C_s y_s + \delta_s = e_s \quad \text{for all} \quad s \in \Omega \tag{7}$$

$$x \geq 0, \quad y_s \geq 0 \quad \text{for all} \quad s \in \Omega \tag{8}$$

The first term presents solution robustness, a single choice for an aggregate objective in (1). The second term demonstrates model robustness, feasibility penalty function, which is used to penalize violation of the control constraint under some of the scenarios. Mulvey et al. (1995) used Eq. (9) to indicate solution robustness as follows:

$$\sigma(0) = \sum_{s \in \Omega} \psi_s p_s + \lambda \sum_{s \in \Omega} p_s \left(\psi_s - \sum_{s' \in \Omega} p_{s'} \psi_{s'} \right)^2 \tag{9}$$

As can be seen, there is a quadratic term in Eq. (9). Yu and Li (2000) proposed an absolute deviation instead of the quadratic term, because the computational effort required due to the quadratic term is less, shown as follows:

$$\sigma(0) = \sum_{s \in \Omega} \psi_s p_s + \lambda \sum_{s \in \Omega} p_s \left| \psi_s - \sum_{s' \in \Omega} p_{s'} \psi_{s'} \right| \tag{10}$$

Mathematical model and model description

In this section, a new mixed-integer nonlinear programming model of a DCMS integrating CF, PP and worker assignment is presented to minimize total costs including machine constant, machine procurement, machine relocation, machine operation, inter-cell and intra-cell movement, overtime, shifting labors between cells and inventory holding respecting to the following assumptions.

Assumptions

1. Each part type has several operations which must be processed according to their sequence data.
2. Process time and manual workload time required for performing operations of a part type on various machine types are known and deterministic.
3. Part demands in each period are uncertain and defined in scenarios.

4. Time-capacity in regular time and overtime for each machine type are known and deterministic over the planning horizon.
5. Purchasing price and revenue from selling of each machine type are uncertain.
6. Constant cost of each machine type is uncertain. It covers overall service and maintenance cost. It is burdened for each machine even when a machine is idle.
7. Variable cost of each machine type in regular time and overtime is uncertain. It covers operating cost depending on the workload allocated to the machine.
8. Holding inventory is allowed and its related cost is uncertain.
9. In each period, the number of cells and the maximum cell size is known.
10. All machine types are multipurpose. Therefore, each operation of each part can be processed by more than one machine which brings flexibility for processing routes. However, each operation is allowed to be assigned to only one machine. In addition, there is no changeover cost for performing different operations by a machine.
11. Total number of labors is constant for all periods. Firing and hiring are not allowed.
12. Relocation cost of each machine between cells and shifting cost of operators between cells during successive periods are uncertain.
13. Batch sizes are fixed for moving parts between and within cells during planning horizon. However, inter-cell and intra-cell batches have different sizes. It is supposed that inter-cell and intra-cell transferring of batches has uncertain costs.

Indices

c Index for cells ($c = 1,\ldots,C$).
m Index for machine types ($m = 1,\ldots,M$).
p Index for part types ($p = 1,\ldots,P$).
h Index for time periods ($h = 1,\ldots,H$).
j Index for operations of part p ($j = 1,\ldots,Op$).
s Index for scenarios ($s = 1,\ldots,S$).

Input parameters

L Total number of labors.
D_{phs} Demand for part p in period h under scenario s.
ϑ_{phs} 1 if part p is planned to be produced in period h under scenario s; 0 otherwise.
B_p^{inter} Batch size for inter-cell movements of part p.
B_p^{intra} Batch size for intra-cell movements of part p.
γ_s^{inter} Inter-cell movement cost per batch under scenario s.
γ_s^{intra} Intra-cell movement cost per batch under scenario s. For justification of CMS, it is assumed that $(\gamma_s^{\text{intra}}/B_p^{\text{intra}}) < (\gamma_s^{\text{inter}}/B_p^{\text{inter}})$.
φ_{ms} Marginal cost to purchase machine type m under scenario s.
h_{phs} Inventory cost for holding part p at the end of period h under scenario s.
W_{ms} Marginal revenue from selling machine type m under scenario s.
α_{ms} Constant cost of machine type m in each period under scenario s.
ρ_{hs} Constant cost of inter-cell labor movement in period h under scenario s.
β_{ms} Variable cost of machine type m for each unit time in regular time under scenario s.
δ_{ms} Relocation cost of machine type m under scenario s.
T_{mh} Time-capacity of machine type m in period h in regular time.
T'_{mh} Time-capacity of machine type m in period h in overtime.
θ_{mhs} Variable cost of processing on machine type m per hour in overtime in period h under scenario s.
UB Maximal cell size.
t_{jpm} Processing time required to perform operation j of part type p by machine type m.
t'_{jpm} Manual workload time required to perform operation j of part type p by machine type m.
a_{jpm} 1 if operation j of part p can be processed by machine type m; 0 otherwise.
p_s Occurrence probability of scenario s.
WT Available time capacity per worker.

Decision variables

N_{mch} Number of machine type m allocated to cell c in period h.
k^+_{mch} Number of machine type m added in cell c in period h.
k^-_{mch} Number of machine type m removed from cell c in period h.
I^+_{mh} Number of machine type m purchased in period h.
I^-_{mh} Number of machine type m sold in period h.
X_{jpmchs} 1 if operation j of part type p is processed by machine type m in cell c in period h under scenario s; 0 otherwise.
L_{ch} Number of labors assigned to cell c in period h.
T'_{mch} Extra time needed for machine type m allocated to cell c in period h.
δ_{phs} the under-fulfillment of demand of part type p in period h under scenario s.

I_{phs} The inventory level of part p at the end of time period h under scenario s.

Q_{phs} Number of demand of part type p produced in period h under scenarios s.

Problem formulation

The objective function consists of nine components, given in Eqs. (1.1)–(1.9), seeks to minimize the sum of miscel-

$$\text{Min } Z = \sum_{h=1}^{H}\sum_{m=1}^{M}\sum_{c=1}^{C} N_{mch}\cdot\alpha_{ms} \quad (1.1)$$

$$+\sum_{h=1}^{H}\sum_{m=1}^{M} I_{mh}^{+}\cdot\varphi_{ms} - \sum_{h=1}^{H}\sum_{m=1}^{M} I_{mh}^{-}\cdot w_{ms} \quad (1.2)$$

$$+\sum_{h=1}^{H}\sum_{c=1}^{C}\sum_{p=1}^{P}\sum_{j=1}^{O_p}\sum_{m=1}^{M} \beta_{ms}\cdot Q_{phs}\cdot t_{jpm}\cdot X_{jpmchs} \quad (1.3)$$

$$+1/2\sum_{h=1}^{H}\sum_{p=1}^{P}\gamma_s^{\text{inter}}\cdot\left[\frac{Q_{phs}}{B_p^{\text{inter}}}\right]\sum_{j=1}^{O_p-1}\sum_{c=1}^{C}\left|\sum_{m=1}^{M}X_{(j+1)pmchs}-\sum_{m=1}^{M}X_{jpmchs}\right| \quad (1.4)$$

$$+1/2\sum_{h=1}^{H}\sum_{p=1}^{P}\gamma_s^{\text{intra}}\cdot\left[\frac{Q_{phs}}{B_p^{\text{intra}}}\right]\sum_{j=1}^{O_p-1}\sum_{c=1}^{C}\left(\sum_{m=1}^{M}\left|X_{(j+1)pmchs}-X_{jpmchs}\right|-\left|\sum_{m=1}^{M}X_{(j+1)pmchs}-\sum_{m=1}^{M}X_{jpmchs}\right|\right) \quad (1.5)$$

$$+\sum_{h=1}^{H}\sum_{m=1}^{M}\sum_{c=1}^{C}T'_{mch}\cdot\theta_{mhs} \quad (1.6)$$

$$+1/2\sum_{h=1}^{H}\sum_{c=1}^{C}\rho_{hs}\cdot\left(\left|L_{c(h+1)}-L_{ch}\right|\right) \quad (1.7)$$

$$+1/2\sum_{h=1}^{H}\sum_{m=1}^{M}\sum_{c=1}^{C}\delta_{ms}\cdot\left(k_{mch}^{+}+k_{mch}^{-}\right) \quad (1.8)$$

$$+\sum_{h=1}^{H}\sum_{p=1}^{P}h_{phs}\cdot I_{phs} \quad (1.9)$$

s.t:

$$\sum_{c=1}^{C}\sum_{m=1}^{M}X_{jpmchs}\cdot a_{jpm}=\vartheta_{phs} \qquad \forall j,p,h,s \quad (2)$$

$$X_{jpmchs}\le a_{jpm} \qquad \forall j,p,m,c,h,s \quad (3)$$

$$\sum_{p=1}^{P}\sum_{j=1}^{O_p}X_{jpmchs}\cdot Q_{phs}\cdot t_{jpm}\le T_{mh}\cdot N_{mch}+T'_{mch} \qquad \forall m,c,h,s \quad (4)$$

$$\sum_{c=1}^{C}N_{mch}-\sum_{c=1}^{C}N_{mc(h-1)}=I_{mh}^{+}-I_{mh}^{-} \qquad \forall m,h \quad (5)$$

$$N_{mc(h-1)}+k_{mch}^{+}-k_{mch}^{-}=N_{mch} \qquad \forall m,c,h \quad (6)$$

$$\sum_{c=1}^{C}T'_{mch}\le T'_{mh} \qquad \forall m,h \quad (7)$$

$$\sum_{c=1}^{C}L_{ch}\le L \qquad \forall h \quad (8)$$

$$\sum_{m=1}^{M}N_{mch}\le UB \qquad \forall c,h \quad (9)$$

$$\sum_{j=1}^{O_p}\sum_{p=1}^{P}\sum_{m=1}^{M}X_{jpmchs}\cdot Q_{phs}\cdot t'_{jpm}\le WT\cdot L_{ch} \qquad \forall c,h,s \quad (10)$$

$$D_{phs}=Q_{phs}-I_{phs}+I_{p(h-1)s} \qquad \forall p,h,s \quad (11)$$

$$Q_{phs}\le M\vartheta_{phs} \qquad \forall p,h,s \quad (12)$$

X_{jpmchs} in binary $\quad \forall j,p,m,c,h,s \quad$ (13)

$L_{ch}, N_{mch}, k_{mch}^{+}, k_{mch}^{-}, I_{mh}^{+}, I_{mh}^{-}$ are positive and integer $\quad \forall m,c,h \quad$ (14)

$Q_{phs}, I_{phs}, T'_{mch}\ge 0$ are positive and continuous $\quad \forall p,m,c,h,s \quad$ (15)

laneous costs. Term (1.1) demonstrates sum of constant cost of all machines which have been used over the planning horizon for entire cells. Term (1.2) shows the total purchase cost minus selling income for entire machines during all periods. Term (1.3) indicates the variable cost of processing operations by different machines in whole cells and periods. Terms (1.4) and (1.5) calculate inter-cell and intra-cell movement costs, respectively. Term (1.6) represents the total costs for overtime working of machines which is required to produce the partial fraction of demand. Term (1.7) demonstrates the total costs of shifting labors between cells over the planning horizon. Various parameters such as labors training, wage rate of skilled labors and labors transference among the cells affect this expenditure. Term (1.8) indicates the cost of machines relocations. Finally, the last term of the objective function considers inventory holding costs. It is worth mentioning that all components (1.1)–(9) in the objective function are calculated under scenario s.

The first constraint introduced in Eq. (2) ensures that each operation of part p is allocated to only one machine capable of processing that part operation and one cell in period h on condition that part p is planned to be produced in that period. Equation (3) guarantees that an operation of a part is assigned to a machine provided that the machine is capable of processing that part operation. Equation (4) guarantees that machine capacity is not exceeded. Equation (5) calculates the number of each machine type bought or sold during each period. Equation (6) shows that the number of machines type m in cell c at the current period h equals to the number of that machines moved into cell c, plus the number of the same machine type present in the previous period and minus the number of machines removed from that cell. Equation (7) shows that summation of the extra time dedicated to all cells per machine type m cannot exceed the total capacity of machine type m in period h in overtime. Equation (8) ensures the number of labors allocated to all cells in each period is equal to the total number of available labors. Equation (9) determines the number of machines assigned to a cell in each period is less than the upper cell size limit. Equation (10) guarantees that available time capacity per worker is not exceeded. Equation (11) shows the balancing inventory constraint between periods for each part type at each period. It means that the inventory level of each part at the end of each period is equal to the quantity of production plus the inventory level of the part at the end of the previous period minus the part demand volume in the current period. Equation (12), complementary to Eq. (2), ensures that a portion of the part demand can be produced at the given period if its operations are assigned in the constraint given in Eq. (2). Logical binary, non-negativity integer or continuous necessities for the decision variables are determined in Eqs. (13), (14) and (15).

Linearization of the proposed model

The proposed model is a mixed-integer nonlinear programming model because of absolute terms in Eqs. (1.4), (1.5) and (1.7) and the product of decision variables in Eqs. (1.3), (4) and (10).

The linearization process for absolute terms (1.4), (1.5) and (1.7) is accomplished by transforming the absolute terms into the linear form as follows:

To linearize term (1.4), non- negative variables Z^1_{jpchs} and Z^2_{jpchs} are introduced and term (1.4) is rewritten as follows:

$$1/2\gamma_s^{\text{inter}} \cdot \sum_{h=1}^{H}\sum_{p=1}^{P}\left[\frac{Q_{phs}}{B_p^{\text{inter}}}\right]\sum_{j=1}^{O_p-1}\sum_{c=1}^{C}\left(Z^1_{jpchs}+Z^2_{jpchs}\right) \tag{11}$$

where the following constraint must be added to the original model.

$$Z^1_{jpchs} - Z^2_{jpchs} = \sum_{m=1}^{M} X_{(j+1)pmchs} - \sum_{m=1}^{M} X_{jpmchs} \quad \forall j,p,c,h,s \tag{12}$$

Likewise, to transform the term (1.5) to the linear form, non- negative variables Y^1_{jpmchs} and Y^2_{jpmchs} are introduced and this term is rewritten as follows:

$$1/2\gamma_s^{\text{intra}}\sum_{h=1}^{H}\sum_{p=1}^{P}\left[\frac{Q_{phs}}{B_p^{\text{intra}}}\right]\sum_{j=1}^{O_p-1}\sum_{c=1}^{C} \times \left(\sum_{m=1}^{M}\left(Y^1_{jpmchs}+Y^2_{jpmchs}\right) - \left(Z^1_{jpchs}+Z^2_{jpchs}\right)\right) \tag{13}$$

where the following constraint must be added to the original model.

$$Y^1_{jpmchs} - Y^2_{jpmchs} = X_{(j+1)pmchs} - X_{jpmchs} \quad \forall j,p,m,c,h,s \tag{14}$$

Equation (11) is still nonlinear term. In the next step, to transform Eq. (11) to the linear form, non-negative variable φ^1_{jpchs} is introduced, and this equation is rewritten as follows:

$$1/2\gamma_s^{\text{inter}}\sum_{h=1}^{H}\sum_{p=1}^{P}\sum_{j=1}^{O_p-1}\sum_{c=1}^{C}\left[\frac{\varphi^1_{jpchs}}{B_p^{\text{inter}}}\right] \tag{15}$$

where the following constraints set must be added to the original model.

$$\varphi^1_{jpchs} \geq Q_{phs} - M\left(1 - Z^1_{jpchs} - Z^2_{jpchs}\right) \quad \forall j,p,c,h,s \tag{16}$$

$$\varphi^1_{jpchs} \leq Q_{phs} + M\left(1 - Z^1_{jpchs} - Z^2_{jpchs}\right) \quad \forall j,p,c,h,s \quad (17)$$

Likewise, to transform Eq. (13) to the linear form, non-negative variable φ^2_{jpchs} is introduced, and this equation is rewritten as follows:

$$1/2\gamma_s^{\text{intra}} \sum_{h=1}^{H}\sum_{p=1}^{P}\sum_{j=1}^{O_p-1}\sum_{c=1}^{C}\left[\frac{\varphi^2_{jpchs}}{B_p^{\text{intra}}}\right] \quad (18)$$

where the following constraints must be added to the original model.

$$\varphi^2_{jpchs} \geq Q_{phs} - M\left\{1 - \sum_{m=1}^{M}\left(Y^1_{jpmchs} + Y^2_{jpmchs}\right) + \left(Z^1_{jpchs} + Z^2_{jpchs}\right)\right\} \quad \forall j,p,c,h,s \quad (19)$$

$$\varphi^2_{jpchs} \leq Q_{phs} + M\left\{1 - \sum_{m=1}^{M}\left(Y^1_{jpmchs} + Y^2_{jpmchs}\right) + \left(Z^1_{jpchs} + Z^2_{jpchs}\right)\right\} \quad \forall j,p,c,h,s \quad (20)$$

To transform product terms in Eqs. (1.3), (4) and (10) to the linear forms, non-negative variable φ_{jpmchs} is introduced and replaced by $X_{jpmchs} \times Q_{phs}$ in the aforementioned terms. Then, the following constraints must be added to the original model.

$$\varphi_{jpmchs} \geq Q_{phs} - M\left(1 - X_{jpmchs}\right) \quad \forall j,p,m,c,h,s \quad (21)$$

$$\varphi_{jpmchs} \leq Q_{phs} + M\left(1 - X_{jpmchs}\right) \quad \forall j,p,m,c,h,s \quad (22)$$

The absolute term Eq. (1.7) is transformed into the linear form as follows:

$$1/2\sum_{h=1}^{H}\sum_{c=1}^{C}\rho_{hs}\left(W^1_{ch} + W^2_{ch}\right) \quad (23)$$

where the following constraint must be added to the original model:

$$W^1_{ch} - W^2_{ch} = L_{c(h+1)} - L_{ch} \quad \forall c,h \quad (24)$$

The final linear model is written as follows:

$$\text{Min } Z = \sum_{h=1}^{H}\sum_{c=1}^{C}\sum_{p=1}^{P}\sum_{j=1}^{O_p}\sum_{m=1}^{M}\beta_{ms} \cdot t_{jpm} \cdot \varphi_{jpmchs} + \text{Eq. (1.1)} + \text{Eq. (1.2)} + \text{Eq. (1.6)} + \text{Eq. (1.8)} + \text{Eq. (1.9)} + \text{Eq. (15)} + \text{Eq. (18)} + \text{Eq. (23)}$$

s.t:

Equations (2) and (3)

$$\sum_{p=1}^{P}\sum_{j=1}^{O_p}\varphi_{jpmchs}t_{jpm} \leq T_{mh}N_{mch} + T'_{mch} \quad \forall m,c,h,s \quad (25)$$

Equations (5)–(9)

$$\sum_{j=1}^{O_p}\sum_{p=1}^{P}\sum_{m=1}^{M}\varphi_{jpmchs}t'_{jpm} \leq WTL_{ch} \quad \forall c,h,s \quad (26)$$

Equations (11)–(15), (12), (14), (16), (17), (19–22) and (24)

$$\varphi_{jpmchs}, Z^1_{jpchs}, Z^2_{jpchs}, \varphi^1_{jpchs}, Y^1_{jpmchs}, Y^2_{jpmchs}, \varphi^2_{jpchs}, W^1_{ch}, W^2_{ch} \geq 0$$

Robust optimization formulation

In this paper, a robust optimization approach based on Mulvey's model is employed in which uncertainty is represented by a set of discrete scenarios. The extended robust optimization model for the mentioned problem can be stated as follows:

$$\begin{aligned} TC_s = & \sum_{h=1}^{H}\sum_{m=1}^{M}\sum_{c=1}^{C}N_{mch}\alpha_{ms} + \sum_{h=1}^{H}\sum_{m=1}^{M}I^+_{mh}\varphi_{ms} - \sum_{h=1}^{H}\sum_{m=1}^{M}I^-_{mh}w_{ms} \\ & + \sum_{h=1}^{H}\sum_{c=1}^{C}\sum_{p=1}^{P}\sum_{j=1}^{O_p}\sum_{m=1}^{M}\beta_{ms}\varphi_{jpmchs}t_{jpm} \\ & + 1/2\gamma_s^{\text{inter}}\sum_{h=1}^{H}\sum_{p=1}^{P}\sum_{j=1}^{O_p-1}\sum_{c=1}^{C}\left[\frac{\varphi^1_{jpchs}}{B_p^{\text{inter}}}\right] \\ & + 1/2\gamma_s^{\text{intra}}\sum_{h=1}^{H}\sum_{p=1}^{P}\sum_{j=1}^{O_p-1}\sum_{c=1}^{C}\left[\frac{\varphi^2_{jpchs}}{B_p^{\text{intra}}}\right] \\ & + \sum_{h=1}^{H}\sum_{m=1}^{M}\sum_{c=1}^{C}T'_{mch}\theta_{mhs} + 1/2\sum_{h=1}^{H}\sum_{c=1}^{C}\rho_{hs}\left(W^1_{ch} + W^2_{ch}\right) \\ & + 1/2\sum_{h=1}^{H}\sum_{m=1}^{M}\sum_{c=1}^{C}\delta_{ms}\left(k^+_{mch} + k^-_{mch}\right) + \sum_{h=1}^{H}\sum_{p=1}^{P}h_{phs}I_{phs} \end{aligned} \quad (27)$$

$$\text{Min } Z = \sum_{s=1}^{S}p_sTC_s + \lambda_1\sum_{s=1}^{S}p_s\left|TC_s - \sum_{s'}^{S}p_{s'}TC_{s'}\right| + \omega\sum_{s=1}^{S}\sum_{m=1}^{M}\sum_{h=1}^{H}p_s\delta_{phs} \quad (28)$$

s.t

$$D_{phs} = \delta_{phs} + Q_{phs} - I_{phs} + I_{p(h-1)s} \quad \forall p,h,s \quad (29)$$

Equations (2), (3), (5)–(9), (12)–(15), (12), (14), (16), (17), (19)–(22), (24), (25), (26).

The first and second terms in the objective function (28) are the expected value and variance of the objective function (27), respectively, and they measure solution robustness. The third term in (28) measures the model robustness with regards to infeasibility associated with control constraints (29) under scenario s. Equation (29) is a control constraint that is used to specify the level of inventory and the under-fulfillment of part demand via violation level δ_{phs} under scenario s. It is noted that if the total quantity of products produced in period h plus previous inventory at period h-1 is greater than market demand D_{phs}, then the inventory at period h will be equal to $I_{phs} = I_{p(h-1)s} + Q_{phs} - D_{phs}$ and under minimization, the violation level $\delta_{phs} = 0$; whereas if $I_{p(h-1)s} + Q_{phs}$ is less than market demand D_{phs}, then $I_{phs} = 0$, and $\delta_{phs} = D_{phs} - Q_{phs} - I_{p(h-1)s}$, demonstrating under-fulfillment of part demand, thus an infeasible solution is obtained.

Although Eq. (28) is a nonlinear function, the absolute term is transformed into the linear form as follows:

$$\min = \sum_{s=1}^{S} p_s TC_s + \lambda_1 \sum_{s=1}^{S} p_s(p_s + q_s) + \omega \sum_{s=1}^{S} \sum_{m=1}^{M} \sum_{h=1}^{H} p_s \delta_{phs}. \tag{30}$$

s.t:

$$p_s - q_s = TC_s - \sum_{s'}^{S} p_{s'} TC_{s'} \qquad \forall s \tag{31}$$

$\delta_{phs} \geq 0$, Eqs. (2), (3), (5)–(9), (12)–(15), (12), (14), (16), (17), (19)–(22), (24), (25), (26), (29).

A case study

Case data description

A case study is conducted for a typical equipment manufacturer located in the Mazandaran province in the north of Iran. Badeleh Machinery Company was pioneered in 1988 with a factory for producing different kinds of tanked and trailed sprayers. Parallel with an increment in production rate, there came a variety of other types of machines, thus an increase in the factory's area, as far as 15,000 meters for production section with another 15,000 meters of area left for future developments, in which 70 people consisting of workers and specialists work seven days a week. Regarding the customized demand in such case study, different scenarios in different season could be defined. Eight part types (farm equipment) consisting of (1) sprinkler, (2) Rot cultivator, (3) Stalk-Shredder, (4) chipper, (5) Roller Chisel, (6) Borers with hydraulic inverter, (7) Borers with hydraulic inverter, and (8) Rear Hydraulic Crane Arm are produced in the company. To validate the proposed model and investigate the credibility of the employed robust optimization approach, the case study is solved using GAMS 22.0 software (solver CPLEX). First, the input data are described. Next, the obtained results are analyzed. This case study suggested in an uncertain environment includes 8 parts ($p1,\ldots,p8$), six types of machines ($m1,\ldots,m6$), three time periods ($h1$, $h2$, $h3$) and three types of cells ($c1$, $c2$, $c3$). For each part, three operations ($j1$, $j2$, $j3$) have to be processed sequentially considering processing times. The maximum available time for each worker in a time period is 40 h and the number of workers is 70. Besides, it has been assumed that the future economic scenarios will fit four probable scenarios that, respectively, are boom, good, fair and poor with the related probabilities 0.45, 0.25, 0.2, and 0.15.

Demand for part type p in period h under scenario s is shown Table 1. Batch size for inter and intra-cell movement of part p are shown Table 2. Inter-cell and intra-cell movement costs per batch under scenario s are shown Table 3. Purchase cost of machine type m under scenario s is shown Table 4. Marginal revenue from selling machine type m under scenario s is shown Table 5. Constant cost of machine type m in each period under scenario s is shown Table 6. Variable cost of machine type m for each unit time in regular time is shown Table 7. Relocation cost of machine type m under scenario s is shown Table 8. Fixed cost of inter-cell labor moving in period h under scenario s is shown Table 9. Time-capacity of machine type m in regular and overtime are shown Table 10. Variable cost of processing on machine type m in overtime in period h under scenario s is shown Table 11. Processing time required

Table 1 Demand for eight part types in two periods under four scenarios

Dphs	Scenario	*P1*	*P2*	*P3*	*P4*	*P5*	*P6*	*P7*	*P8*
*h*1	Boom	550	800	0	500	0	450	0	800
	Good	0	0	250	300	0	200	300	0
	Fair	350	500	0	0	200	0	250	250
	Poor	0	0	100	100	100	100	100	100
*h*2	Boom	700	800	0	500	0	800	0	950
	Good	0	0	500	300	0	500	300	0
	Fair	500	400	0	0	300	0	200	350
	Poor	0	0	200	100	100	100	100	100
*h*3	Boom	400	650	0	500	0	700	0	750
	Good	0	0	300	300	0	300	400	0
	Fair	200	400	0	0	250	0	300	200
	Poor	0	0	100	100	100	200	200	100

Table 2 Batch size for inter-cell and intra-cell movement of four part types

	*P*1	*P*2	*P*3	*P*4	*P*5	*P*6	*P*7	*P*8
B^{inter}	35	25	20	40	45	30	35	40
B^{intra}	7	5	4	8	9	5	7	8

Table 3 Inter-cell and intra-cell movement cost per batch under four scenarios

	Boom	Good	Fair	Poor
γ_s^{inter}	50	40	30	20
γ_s^{intra}	8	7	6	5

Table 4 Purchase cost of six machine types under four scenarios

φ_{ms}	Boom	Good	Fair	Poor
$M1$	14,000	12,000	11,000	10,000
$M2$	14,000	12,000	11,000	10,000
$M3$	15,000	13,000	12,000	11,000
$M4$	14,000	13,000	12,000	11,000
$M5$	15,000	13,000	12,000	11,000
$M6$	16,000	13,000	12,000	11,000

Table 5 Marginal revenue from selling six machine types under four scenarios

ω_{ms}	Boom	Good	Fair	Poor
$M1$	9800	8100	7700	7000
$M2$	9800	8100	7700	7000
$M3$	10,500	8700	8400	7700
$M4$	9800	8100	7700	7000
$M5$	9100	8000	7700	7000
$M6$	11,200	9000	8400	7700

Table 6 Constant cost of six machine types in each period under 4 scenarios

α_{ms}	Boom	Good	Fair	or
$M1$	1400	1200	1100	1000
$M2$	1400	1200	1100	1000
$M3$	1500	1400	1200	1100
$M4$	1400	1300	1200	1100
$M5$	1300	1200	1100	1000
$M6$	1600	1300	1200	1100

Table 7 Variable cost of six machine types for each unit time in regular time

β_{ms}	Boom	Good	Fair	Poor
$M1$	9	8	7	5
$M2$	9	8	7	6
$M3$	8	7	6	5
$M4$	8	7	6	5
$M5$	9	8	7	6
$M6$	8	6	5	4

Table 8 Relocation cost of six machine types under four scenarios

δ_{ms}	Boom	Good	Fair	Poor
$M1$	650	600	550	500
$M2$	700	650	600	550
$M3$	750	700	650	600
$M4$	700	650	600	550
$M5$	650	600	550	500
$M6$	800	750	700	650

Table 9 Fixed cost of inter-cell moving of a labor in three periods under four scenarios

ρ_{hs}	Boom	Good	Fair	Poor
$H1$	200	150	100	70
$H2$	200	150	100	70
$H3$	200	150	100	70

Table 10 Time-capacity of six machine types in regular and over time

	T_{mh}	T'_{mh}
$M1$	500	200
$M2$	500	200
$M3$	500	200
$M4$	500	200
$M5$	500	200
$M6$	500	200

Table 11 Variable cost of processing on six machine types in overtime in three periods under four scenarios

θ_{mhs}	Scenario	$M1$	$M2$	$M3$	$M4$	$M5$	$M6$
h1	Boom	15	11	17	12	10	20
	Good	14	10	16	11	9	19
	Fair	13	9	15	10	8	18
	Poor	10	8	10	9	7	10
h2	Boom	15	11	17	12	10	20
	Good	14	10	15	11	9	19
	Fair	13	9	13	10	8	18
	Poor	10	8	10	9	7	10
h3	Boom	15	11	17	12	10	20
	Good	13	10	12	11	9	19
	Fair	12	9	11	10	8	18
	Poor	10	8	10	9	7	10

to perform operation j of part type p on machine type m is shown Table 12. Manual workload time required to perform operation j of part type p on machine type m is shown Table 13. Inventory holding cost for part type p in period h under scenario s is shown Table 14.

Table 12 Processing time required to perform the operations of eight part types on six machine types

t^{jpm}	P1			P2			P3			P4			P5			P6			P7			P8		
	J1	J2	J3	J1	J2	J3	J1	J2	J3	J1	J2	J3	J1	J2	J3	J1	J2	J3	J1	J2	J3	J1	J2	J3
M1	0	0	0	0.76	0	0.39	0	0	0	0	0.83	0	0	0	0	0	0.57	0	0	0	0	0	0	0.54
M2	0	0	0	0	0	0	0.99	0	0	0	0	0.74	0	0	0	0.72	0	0.47	0.44	0	0.28	0	0.17	0
M3	0.73	0.93	0	0	0	0	0	0	0	0.45	0	0	0	0	0	0	0	0	0	0.97	0	0	0	0.15
M4	0	0	0.46	0	0.81	0	0	0	0	0	0	0	0	0.26	0	0.55	0	0	0	0.47	0	0.84	0.86	0.78
M5	0	0	0	0	0	0	0	0.48	0	0	0	0	0.12	0	0.75	0	0	0.12	0.76	0	0.86	0.2	0	0
M6	0	0	0	0	0	0	0.36	0	0.00	0	0.78	0	0	0.45	0.59	0.81	0.48	0	0	0	0	0	0	0

Table 13 Manual workload time required to perform the operations of eight part types on six machine types

t^{jpm}	P1			P2			P3			P4			P5			P6			P7			P8		
	J1	J2	J3	J1	J2	J3	J1	J2	J3	J1	J2	J3	J1	J2	J3	J1	J2	J3	J1	J2	J3	J1	J2	J3
M1	0	0	0	0.076	0	0.039	0	0	0	0	0.083	0	0	0	0	0	0.057	0	0	0	0	0	0	0.054
M2	0	0	0	0	0	0	0.099	0	0	0	0	0.074	0	0	0	0.072	0	0.047	0.044	0	0.028	0	0.017	0
M3	0.073	0.093	0	0	0	0	0	0	0	0.045	0	0	0	0	0	0	0	0	0	0.097	0	0	0	0.015
M4	0	0	0.046	0	0.081	0	0	0	0	0	0	0	0	0.026	0	0.055	0	0	0	0.047	0	0.084	0.086	0.078
M5	0	0	0	0	0	0	0	0.048	0	0	0	0	0.012	0	0.075	0	0	0.012	0.076	0	0.086	0.2	0	0
M6	0	0	0	0	0	0	0.036	0	0	0	0.078	0	0	0.045	0.059	0.081	0.048	0	0	0	0	0	0	0

Table 14 Cost of inventory holding for eight part types in three periods under four scenarios

H_{phs}	Scenario	$P1$	$P2$	$P3$	$P4$	$P5$	$P7$	$P8$
$h1$	Boom	17	17	25	20	19	33	25
	Good	15	15	21	18	14	25	22
	Fair	13	13	17	15	13	22	19
	Poor	10	10	10	10	10	10	10
$h2$	Boom	18	22	22	22	17	32	22
	Good	15	15	18	19	12	27	19
	Fair	14	13	15	14	11	22	13
	Poor	11	10	12	11	10	12	11
$h3$	Boom	17	18	20	17	20	17	29
	Good	13	13	15	15	15	15	22
	Fair	12	12	11	12	13	13	17
	Poor	10	10	10	10	11	10	10

Table 15 Sensitivity analysis for model Z

ω	Z
0	2850
100	264,416.1
200	281,446.5
300	292,060.5
400	298,182.5
500	301,576.6
600	301,941
700	308,054.6
800	308,636.1

Results analysis

As stated in "Robust optimization" section, robustness means that the model output should not be highly sensitive to the exact values of the model input parameters, and if the model remains feasible for each certain scenario, the model is robust. According to the objective function Z [Eq. (28)], the model robustness is calculated through third term in objective function. Because of the uncertainty of the demand parameter and the cost parameters related to the cell formation, the model might be infeasible for some various scenarios. Thus, third term of objective function (28) that is the penalty function for infeasibility penalizes the violation of the control constraint (29). The violation of the control constraint means an infeasible solution is obtained under some scenarios. In fact, δ_{phs} is the violation vector showing the infeasibility level in control constraint (29) under a given scenario. If the under-fulfilled demand (δ_{phs}) equals zero, the model is feasible, otherwise, δ_{phs} will be positive. Table 15 presents sensitivity analysis for the robustness of Model Z with different values for parameter ω.

It is seen from Table 15 that the objective function Z is sensitive in return for various values of ω, (δ_{phs}) obtains a positive value and the objective function Z is positive under some scenarios. At the point $\omega = 0$, the part under-fulfilled demand (δ_{phs}) obtains the maximum value since no production occurs and this way, it acquires a positive value in a descending manner until at the point $\omega = 800$, the part under-fulfilled demand (δ_{phs}) equals zero and the model becomes feasible. Figure 1 depicts sensitivity analysis for the model robustness and objective function value Z. As Fig. 1 illustrates, the value of Z increases as ω increases and the objective function value Z goes up. In fact, Fig. 1 shows that model Z has penalized the violation of control constraint (29) under some scenarios and as ω increases, the objective function value gets higher because the infeasibility penalty function acquires a positive value.

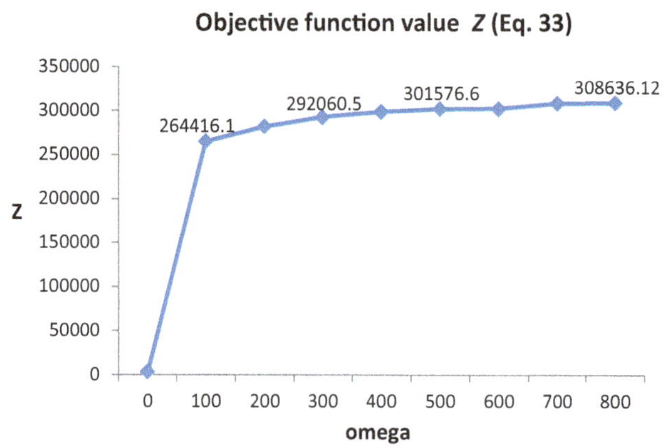

Fig. 1 Sensitivity analysis for the model robustness and the objective function value Z

Here, the model solution is analyzed considering $\omega = 300$. The computational results are given in Tables 16 and 17. Table 16 depicts the under-fulfilled demand of part type p in the period h under scenario s. As can be seen, the under-fulfilled demand of parts 1 and 2 obtain positive values in periods 1 and 2 for a boom scenario. Since the infeasibility penalty function (28) obtains positive value, it penalizes control constraint violation under some scenarios. While the demand for part 2 in period 1 under the boom scenario is 800, the optimal production value is 783 and the under-fulfilled demand is 17. Similarly, the demand for part 1 in period 2 is 700, the optimal production value is 625 and the under-fulfillment demand is 75 under the boom scenario. The demand for part 2 in period 2 is 800, the optimal production is 683 and the under-fulfilled demand is seven under the boom scenario. That is, violation of the control constraint (29) for the boom scenario in parts 1 and 2 in periods 1 and 2 happened at $\omega = 300$.

Table 17 illustrates the total costs based on Eq. (27) including costs of machine constant, machine variable,

Table 16 The under-fulfilled demand of eight part types in three periods under four scenarios

δ_{phs}	Scenario	$P1$	$P2$	$P3$	$P4$	$P5$	$P6$	$P7$	$P8$
$h1$	Boom	0	17	0	0	0	0	0	0
	Good	0	0	0	0	0	0	0	0
	Fair	0	0	0	0	0	0	0	0
	Poor	0	0	0	0	0	0	0	0
$h2$	Boom	75	117	0	0	0	0	0	0
	Good	0	0	0	0	0	0	0	0
	Fair	0	0	0	0	0	0	0	0
	Poor	0	0	0	0	0	0	0	0
$h3$	Boom	0	0	0	0	0	0	0	0
	Good	0	0	0	0	0	0	0	0
	Fair	0	0	0	0	0	0	0	0
	Poor	0	0	0	0	0	0	0	0

machine purchase, intra and inter-cell movement, part inventory holding and overtime under various scenarios.

According to Table 1, it is clear that the part demands and the scenario-based cell formation cost parameters are incremental from poor scenario to the boom one. As can be seen in Table 17, all cost components have increased from the poor scenario to the boom one, except the inventory holding cost. Since the part under-fulfilled demand has obtained positive value for parts 1 and 2 according to Table 16 in the boom scenario, violation occurred and according to the control constraint (29), the part inventory amount and its related cost is zero in the boom scenario. However, from the boom scenario to the poor one the part inventory has increased and similarly, the part inventory cost has increased as well. Since in the good, fair and poor scenarios, δ_{phs} equals zero, the part inventory level gets a positive level and the inventory holding cost gets a positive level as well.

Figure 2 depicts the cells configuration in three periods for the main model of the DCMS under boom scenario. The part operation assignments to machines and the machines assignments to cells are also shown in Fig. 2. For example, in the first period, 2 units of machines types 5 and 3 have been assigned to the cells 1 and 3, respectively.

In period 1, operations 1 and 2 of part 1 are processed inside cell 1 by machines 5 and 4, respectively, and operation 3 inside cell 2 by machine 6. Then, there is need for an intra-cell movement for operations 1 and 2 and an inter-cell movement for operations 2 and 3. In period 1, eight inter-cell movements and two intra-cell movement are performed for the parts processing. In period 2, seven inter-cell movements and three intra-cell movement are performed for the parts processing.

Figure 3 shows the cells configuration in periods 1, 2 and 3 for DCMS model solved by the robust optimization approach respect to 4 scenarios. Here, compared with the cell configurations obtained for the main model under boom scenario, there are some similarities and some differences. For example, in period 1, operations 1 and 2 of part 1 are processed inside cell 1 by machines 5 and 2, and operation 3 inside cell 2 by machine 6 as shown in Fig. 3. In period 2, according to Fig. 3, three inter-cell movements and seven intra-cell movement are performed for the parts processing. In period 3, five inter-cell movements and five intra-cell movement are performed for the parts processing. Totally, the number of inter-cell movements and the number of machines decrease; as a result, the relocation cost, machine constant cost and inter-cell movement cost become lower.

Tradeoff between solution robustness and model robustness

Tradeoff between solution robustness (expected total costs) and model robustness (expected under-fulfillment) can be found using different values of ω in the objective function (28). Robust optimization approach allows for infeasibility in the control constraints by means of penalties. When ω is considered equal to zero, δ_{phs} in constraint (29) is equal to D_{phs} due to the minimization of objective function (28). In fact, the total under-fulfillment obtains its highest value, and obviously this decision cannot be upheld. Therefore, it is necessary to evaluate the proposed robust optimization model with various values of ω. Tradeoff between feasibility and costs is illustrated in Fig. 4. As the value of ω increases, the expected total costs representing solution robustness increases exponentially, and the expected under-fulfilled demand representing model robustness drops. This means that for larger value of ω, the obtained solution is approaching 'almost' feasible for any realization

Table 17 Cost components of total costs [Eq. (27)] in four scenarios

	Total costs	Machine constant	Machine variable	Purchasing machine	Inter-cell movement	Intra-cell movement	Inventory holding	Overtime
Boom	275,040.3	30,000	87,218	100,000	12,571.7	1862.1	0	28,813.4
Good	189,211.5	27,000	28,888.1	88,000	4422.7	1689.9	388.5	25,826. 1
Fair	179,160.1	24,300	25,414.8	81,000	3313	1226.5	634.9	22,807.2
Poor	134,714.6	22,000	9528	74,000	1266.6	1250	1569.6	19,819.9

Period h1			C1		C2	C3	
			P1	P2	P4	P6	P7
C1	2	M5	1				
	1	M4	2				
	1	M2		2	3	1	
C2	1	M6	3		2		3
	3	M1		1		3	
C3	2	M3		3			
	1	M4					
	1	M6			1		2
	1	M5				2	1

Period h2	units		C1		C2	C3	
			P1	P2	P4	P6	P7
C1	2	M1	1			2	
	1	M2			3		
	1	M3		2		1	
C2	1	M4			1		
	3	M2	2			3	
	2	M1		1			
	1	M6		3			1
C3	2	M2	3				3
	1	M4					2
	1	M6			2		

Period h3	units		C1		C2	C3	
			P1	P2	P4	P6	P7
C1	2	M5	1			2	
	1	M4	2		3		
	1	M2		2			
C2	1	M6	3		1		
	3	M1		1		3	
C3	2	M3		3			3
	1	M4				1	
	1	M6			2		1
		M5					2

Fig. 2 Cell configurations for the main DCMS model under boom scenario

Period h1			C1		C2	C3	
			P1	P2	P4	P6	P7
C1	2	M5	1	1			
	1	M2	2	2,3	3		
C2	1	M6	3		2		
	3	M1			1		1
C3	2	M3				3	
	1	M4				1	3
	1	M6				2	2

Period h2			C1		C2	C3	
			P1	P2	P4	P6	P7
C1	2	M1	1				
	1	M2	2	3	3		
	1	M3	3	2			
C2	1	M4			1		
	3	M2			2	3	
	2	M1		1			
C3	2	M2				1	3
	1	M4				2	1,2

Period h3			C1			C3	
			P1	P2	P4	P6	P7
C1	2	M5	1			2	
	1	M4	2	3			
	1	M2		2	3		
C3	2	M3	3				
	1	M4		1	1	1	3
	1	M6				3	1
	2	M5			2		2

Fig. 3 Cell configurations for the DCMS model by the proposed robust optimization approach

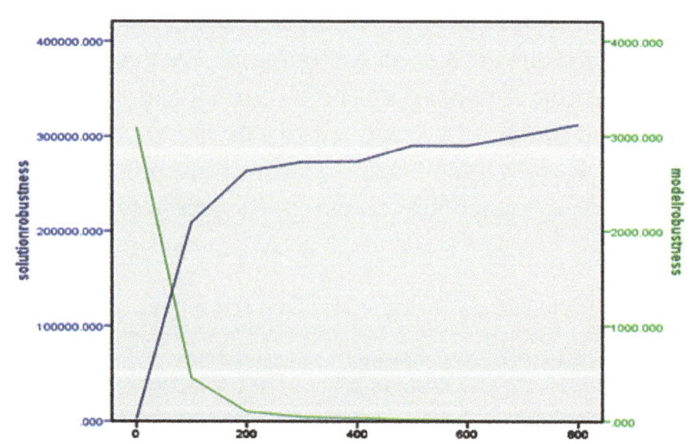

Fig. 4 Trade-off between expected total costs and expected under-fulfillment

of scenario s through the payment of more total costs. In addition, e expected under-fulfillment will eventually drop to zero with an increase in value of ω to 800.

Comparing the effectiveness of robust model and mean-value based model

To illustrate the robust dynamic cell formation that could be obtained by the proposed MIP model, expected values of uncertain parameters are used in the primary mixed-integer linear programming model presented in "Linearization of

Table 18 Total objective function values obtained by the robust and mean-value based models

Problem number	1	2	3	4	5	6	7	8	9	10
Robust model	257,361.8	286,914.8	294,127.3	285,993.8	271,391.6	301,591.6	298,255.4	308,646.9	279,743.5	309,443
Mean-value based model	316,642.7	394,204.2	389,787.9	421,225.4	410,668.2	432,788.6	371,073.8	439,419.8	406,668.8	434,360.9

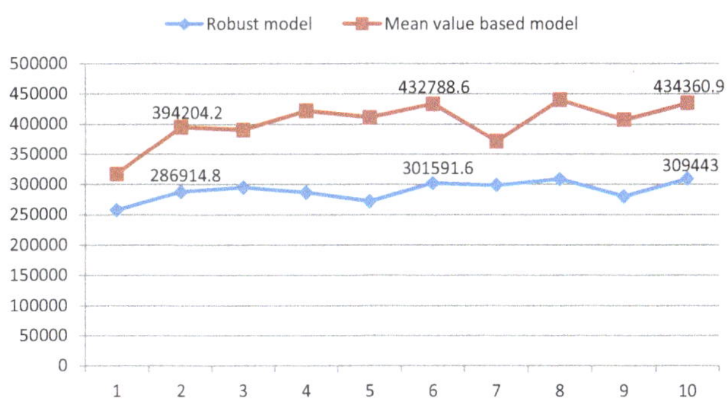

Fig. 5 Comparison of total objective function between robust and mean-value based models

the proposed model" section as certain value parameters, hereafter called mean-value based model. The results of these two models (i.e., robust model and mean-value based model) are compared with each other at the following. Robust optimization is used to attain a robust solution against the fluctuation of uncertain parameters in the future. Note that at the inception of planning horizon, some parameters are uncertain, and only in the execution time of the plan, the real values of uncertain parameters will be realized. For this purpose, we simulate some real and conceivable scenarios that may occur after executing the cell formation in the future. We consider 10 random occurrences for the uncertain parameters and compute the objective function Z of each instance for the dynamic cell formation problem obtained by the robust and mean-value based models.

The objective function values for the scenarios with probabilities 0.15, 0.2, 0.25 and 0.45 are shown in the Table 18 and Fig. 5. As shown in Fig. 5, the objective function values of dynamic cell formation problem obtained by the proposed MIP model are robust against the amount of uncertain parameters in the future and yield a series of solutions that are less sensitive to realizations of the uncertain data. In other words, the violation of results attained by the robust optimization model is less than that by mean-value based model.

In fact, the values of the objective function Z for different scenarios are closer to each other than these values for the mean-value based model. The curve of values in the proposed method follows a more robust incline, but the fluctuation in the curve of values for the classical approach is very high. This achievement indicates that the proposed approach is efficient for any systems that the robustness of solution is important in addition to objective function value Z of production for their managers. Indeed, for such systems having a solution with minimum total objective is not adequate, but the fluctuation in real scenarios in future should be handled. Therefore, numerical results show the robustness and effectiveness of the proposed model.

Conclusion

In this study, a mathematical model based on a robust optimization approach has been presented in dynamic cell formation problem with uncertain data to integrate CF, PP and worker assignment. The robust optimization approach reduces the effect of the fluctuations of uncertain parameters under certain scenarios. In this study, the majority of cell formation parameters including cost parameters and part demand fluctuation were considered uncertain.

Next, sensitivity analysis has been presented for solution robustness and model robustness. Since the objective function has been influenced by ω, the relationship between the model robustness and solution robustness has been analyzed only for the objective function value.

The computational experiments obtained from a set of real-world data for an Iranian farm tanked and trailed sprayers manufacturer illustrated that the proposed robust model is more practical for handling uncertain parameters in the production environments. The tradeoff between optimality and infeasibility was used for obtaining robust solution based on the opinion of decision-makers. The results showed the robustness and effectiveness of the model in real-world cell formation problem.

In addition, the results obtained by the robust MIP model indicated the advantages of robust optimization in generating more robust cell configurations with less cost over the considering expected value of uncertain parameters in a deterministic mean-value based model. In fact, in such systems designed here as the mean-value based model, having only solution with the minimum value of the objective function and lower costs is not sufficient rather the fluctuations in the related scenarios have to be lowered in future.

The future studies in the following of the present study can be pursued in multi-objective DCMS modeling, employing the other robust optimization methods, taking into account the setup time, defining the processing times and time-capacity of machines as uncertain, consideration of machine layout, allowing partial or total subcontracting, workload balancing among the cells, and using meta-heuristics to tackle large-sized problems.

References

Aryanezhad MB, Deljoo V, Mirzapour Al-e-hashem SMJ (2009) Dynamic cell formation and the worker assignment problem: a new model. Int J Adv Manuf Technol 41:329–342

Bagheri M, Bashiri M (2014) A new mathematical model towards the integration of cell formation with operator assignment and inter-cell layout problems in a dynamic environment. Appl Math Model 38(4):1237–1254

Bai D, Carpenter T, Mulvey J (1997) Making a case for robust optimization models. Manag Sci 43:895–907

Balakrishnan J, Cheng CH (2005) Dynamic cellular manufacturing under multi period planning horizons. J Manuf Technol Manag 16(5):516–530

Bashiri M, Bagheri M (2013) A two stage heuristic solution approach for resource assignment during a cell formation problem. Int J Eng Trans C Asp 26(9):943

Ben-Tal A, Nemirovski A (1998) A Robust convex optimization. Math Oper Res 23:769–805

Ben-Tal A, Nemirovski A (2002) A robust optimization—methodology and applications. Math Program 92:453–480

Ben-Tal A, Nemirovski A, Roos C (2002) Robust solutions of uncertain quadratic and conic-quadratic problems. SIAM J Optim 13:535–560

Bertsimas D, Sim M (2002) Robust discrete optimization Technical Report, Sloan School of Management and Operations Research Center, MIT Cambridge

Bertsimas D, Thiele A (2003) A robust optimization approach to supply chain management. Technical Report Working Paper MIT

Feng P, Rakesh N (2010) Robust supply chain design under uncertain demand in agile manufacturing. Comput Oper Res 37(4):668–683

Forghani K, Sobhanallahi M, Mirzazadeh A, Mohammadi M (2012) A mathematical model in cellular manufacturing system considering subcontracting approach under constraints. Manag Sci Lett 2(7):2393–2408

Ghezavati V, Sadjadi S, Dehghan Nayeri M (2011) Integrating strategic and tactical decisions to robust designing of cellular manufacturing under uncertainty: fixed suppliers in supply chain. Int J Comput Intell Syst 4(5):837–854

Hassannezhad M, Cantamessa M, Montagna F, Mehmood F (2014) Sensitivity analysis of dynamic cell formation problem through meta-heuristic. Procedia Technol 12:186–195

Kia R, Paydar MM, Jondabeh MA, Javadian N, Nejatbakhsh Y (2011) A fuzzy linear programming approach to layout design of dynamic cellular manufacturing systems with route selection and cell reconfiguration. Int J Manag Sci Eng Manag 6(3):219–230

Kia R, Baboli A, Javadian N, Tavakkoli-Moghaddam R, Kazemi M, Khorrami J (2012) Solving a group layout design model of a dynamic cellular manufacturing system with alternative process routings, lot splitting and flexible reconfiguration by simulated annealing. Comput Oper Res 39:2642–2658

Kia R, Shirazi H, Javadian N, Tavakkoli-Moghaddam R (2013) A multi-objective model for designing a group layout of a dynamic cellular manufacturing system. J Ind Eng Int 9:8

Mahdavi I, Aalaei A, Paydar MM, Solimanpur M (2010) Designing a mathematical model for dynamic cellular manufacturing systems considering production planning and worker assignment. Comput Math Appl 60(4):1014–1025

Mirzapour Al-E-Hashem SMJ, Malekly H, Aryanezhad MB (2011) A multi-objective robust optimization model for multi-product multi-site aggregate production planning in a supply chain under uncertainty. Int J Prod Econ 134(1):28–42

Mulvey JM, Vanderbei RJ, Zenios SA (1995) Robust optimization of large-scale systems. Math Oper Res 43:264–281

Paydar MM, Saidi-Mehrabad M, Teimoury E (2013) A robust optimisation model for generalised cell formation problem considering machine layout and supplier selection. Int J Comput Integr Manuf 27(8):772–786

Rafiei H, Ghodsi R (2013) A bi-objective mathematical model toward dynamic cell formation considering labor utilization. Appl Math Model 37(4):2308–2316

Rheault M, Drolet J, Abdulnour G (1995) Physically reconfigurable virtual cells: a dynamic model for a highly dynamic environment. Comput Ind Eng 29(1–4):221–225

Sakhaii M, Tavakkoli-Moghaddam R, Bagheri M, Vatani B (2015) A robust optimization approach for an integrated dynamic cellular manufacturing system and production planning with unreliable machines. Appl Math Model. doi:10.1016/j.apm.2015.05.005

Shirazi H, Kia R, Javadian N, Tavakkoli-Moghaddam R (2014) An archived multi-objective simulated annealing for a dynamic cellular manufacturing system. J Ind Eng Int 10:58

Wu X, Chu C-H, Wang Y, Yue D (2007) Genetic algorithms for integrating cell formation with machine layout and scheduling. Comput Ind Eng 53(2):277–289

Yu CS, Li HL (2000) A robust optimization model for stochastic logistic problems. Int J Prod Econ 64(1–3):385–397

9

A reliability-based maintenance technicians' workloads optimisation model with stochastic consideration

D. E. Ighravwe[1,2] · S. A. Oke[1] · K. A. Adebiyi[2]

Abstract The growing interest in technicians' workloads research is probably associated with the recent surge in competition. This was prompted by unprecedented technological development that triggers changes in customer tastes and preferences for industrial goods. In a quest for business improvement, this worldwide intense competition in industries has stimulated theories and practical frameworks that seek to optimise performance in workplaces. In line with this drive, the present paper proposes an optimisation model which considers technicians' reliability that complements factory information obtained. The information used emerged from technicians' productivity and earned-values using the concept of multi-objective modelling approach. Since technicians are expected to carry out routine and stochastic maintenance work, we consider these workloads as constraints. The influence of training, fatigue and experiential knowledge of technicians on workload management was considered. These workloads were combined with maintenance policy in optimising reliability, productivity and earned-values using the goal programming approach. Practical datasets were utilised in studying the applicability of the proposed model in practice. It was observed that our model was able to generate information that practicing maintenance engineers can apply in making more informed decisions on technicians' management.

Keywords Experiential knowledge · Stochastic workloads · Goal programming · Technician's reliability · Technician's fatigue

Introduction

Maintenance workload optimisation scheduling problem has witnessed the application of different modelling approaches in improving equipment availability (Safaei et al. 2008). Performance metrics, methodologies and tools used to ascertain the control of technicians' workload during the plant's operating hours have evolved for decades. The pressure and growing interest in technicians' workload practices and research seems to be associated strongly with the recent surge in competition globally. It was also triggered by the increasingly difficulty in economic growth of industries. Current research trends in this area therefore match technicians' characteristics with workloads and other considerations. There are immense benefits in understanding the reliability of technicians (Gregoriades and Sutcliffe 2008). This will help in the integration of technicians' reliability into technicians' work performance evaluation. All these efforts are to minimise maintenance job rework and reduce overtime. It also aids making the overhaul activity more effective and reduce equipment breakdown to an acceptable level. This study is motivated by lack of information on how to optimise the technicians' costs, workloads, sizes and reliability of technicians. Most theoretical accounts of reliability have been skewed towards machines.

An organisation with an effective and efficient group of technicians experiences several benefits. These benefits are

✉ S. A. Oke
sa_oke@yahoo.com

1 Department of Mechanical Engineering, Faculty of Engineering, University of Lagos, Room 10, Mezzanine Complex, Akoka-Yaba, Lagos, Nigeria

2 Department of Mechanical Engineering, Faculty of Engineering, Ladoke Akintola University of Technology, Ogbomosho, Nigeria

improved productivity, elevated sales, enhanced customer service, reduced labour costs, improved employee engagement and satisfaction (Bartels and Richey 2008). Focusing on the maintenance department as a major area for driving the above-mentioned benefits is important because of its interrelationships with other departments. Maintenance problems in manufacturing systems are diverse. Technicians' workload and reliability problems, which are key problems in solving other problems in maintenance system, are least considered in literature. Several contributions on spare parts management, machine availability (Safaei et al. 2008) and maintenance budget (Mansour 2011; Ighravwe and Oke 2014; Ighravwe et al. 2015) have been documented. Within the last few decades, researchers and industrial practitioners have started considering human factors in maintenance systems with a view of ensuring robust maintenance management systems.

Dealing with the maximisation of a multi-objective problem that is nonlinear in formulation, a single objective model with weights for each objective was considered. Such a model can be solved easily using metaheuristics (evolutionary algorithms and swarm algorithms) in order to generate compromise solutions. Currently, there exists sparse documentation on big-bang–big-crunch (BB–BC) and EP algorithms as solution methods for technicians' parameters optimisation.

Studying fatigue during the design of technicians' workload and reliability for planned operations will help in improving work-plans. Another aspect of technician's problem, which has been sparingly considered in the literature, is experiential knowledge of the technicians who engage in rework activities.

The objective of this study was to develop a mixed-integer optimisation model that optimises technicians' sizes, workloads (service times), reliability, availability, performance (efficiency) and quality of workdone. A performance comparative analysis of EP and BB–BC algorithms when used in solving the proposed model was also carried out. The selection of these algorithms was motivated by their low computation time.

Literature review

Maintenance system operations are often executed using maintenance teams (Hedjazi 2015). These teams are formed based on the skills of technicians (Hedjazi 2015) and the types of maintenance activities (Ighravwe and Oke 2014; Ighravwe et al. 2015) in an organisation. Tohidi and Tarokh (2005) reported that when forming teams, there is the need to ensure proper communication systems. These help in improving the productivity of teams. Thus, the interest of researchers and practitioners in maintenance systems is on the analysis of types of maintenance activities, policies, technicians' attributes (productivity, service time and skills) and cost of operating maintenance systems (Lai et al. 2015). The core reason for studying maintenance departmental performance is to improve organisation's profitability (Alsyouf 2007). This has stimulated different scientific methods in analysing maintenance systems.

Maintenance workload management helps in determining equipment downtime. When workload is optimally shared among the team members, low amounts of equipment downtime will be experienced. For example, the concern of Oladokun et al. (2006) was how to predict equipment downtime. In their study, a predictive model was designed to predict equipment downtime using machine breakdown periods and maintenance repair time as dependent variable. The issue of how to determine maintenance repair times was not exhaustively addressed in their study. Rana and Purohit (2012) investigated the application of critical path method in addressing the problems of technician's productivity, maintenance time and tasks balancing. One drawback of their study is the assumption that technicians are capable of doing any maintenance work in an organisation. In practice, this assumption is feasible for small-scale organisations. However, for large-scale systems, the assumption may be violated. This study relaxed the assumption of Rana and Purohit (2012) by considering technicians for different sections (mechanical, electrical and instrumentation). The scope of Ighravwe and Oke's (2014) study limited maintenance activities to routine maintenance. Ighravwe and Oke (2014) modelled maintenance time utilisation problem by considering the allocated and actual maintenance time used by technicians. No mention of technician's training and fatigue experienced during the execution of maintenance activities was mentioned in their work. In the work of He et al. (2014), the issue of technician's reliability and cost were addressed using nonlinear programming techniques. Their proposed model can be applied in a maintenance system. However, consideration was not given to the impact of technician's fatigue, experiential knowledge and training on technician's reliability.

Huge amount of funds are usually invested in maintenance systems of organisations. This is necessary avert machine breakdowns, which could be more costly. Fajardo and Drekic (2015) pointed out that cost-effective maintenance checks could be achieved during close-down periods. The problem of technician cost and equipment availability was studied by Safaei et al. (2008). The application of simulated annealing as a solution method for handling multi-objective technicians' problem was demonstrated. They considered the management of regular, outsourcing and overtime maintenance activities. Mansour's (2011) studied maintenance cost minimisation problem by proposing a mixed-integer programming model that incorporates equipment complexity. Their work investigated the suitability of genetic algorithm as a solution method for technicians' parametric optimisation.

Kaufman and Lewis (2007) considered the application of repair and replacement models for maintenance workload management. In their study, information on service rates and failures, fixed maintenance times and random replacement times was generated. A model that minimised the total cost of technicians used for maintenance activities was developed by De Bruecker et al. (2015) using mixed-integer programming approach. In their study, a model enhancement heuristic was proposed as solution method for their model. The heuristic considered stochastic service levels.

Knapp and Mahajan (1998) conducted a comparative analysis between centralised and decentralised organisational structure as it affects maintenance systems. Consideration was given to technician-type (in-house and subcontracted technicians) and training levels of technicians. The results of their study demonstrated how optimal allocation of technicians can be achieved. A study which considered organisational policies on technicians' capacity was presented by Mjema (2002). The challenge of managing the switches of technicians for mechanical and electrical workloads was considered. Their analysis was based on the technicians' utilisation and through-put time, work-order requirement and prioritisation rules.

Manzini et al. (2015) presented a nonlinear model which minimises cost of preventive and corrective maintenance as well as technicians' workload and spare parts cost. These maintenance costs were optimised under total expected and probabilistic costs. Hervet and Chardy (2012) reported the use of mixed-integer programming approach in addressing the problem of preventive maintenance workforce needs during the design of passive optical network. Jarugumill (2011) studied workforce problem by considering regular and overtime activities as well as workers' skills.

Research methodology

This section presents the proposed model ("Model formulation" section) and discussed the solution methods used in solving the proposed model ("Solution methods" section). The development of the proposed model was based on the following assumptions:

1. Technicians in the same group can be categorised differently, and these categories are known in advance;
2. Available maintenance work is carried out by in-house maintenance crew only; and
3. Inventories for maintenance work are available and released when needed.

Some of the notations used in formulating the proposed model are given as follows:

Indices

i Maintenance activity
j Maintenance section
k Technicians category
t Planning period
M Total number of types of maintenance activities
N Total number of maintenance sections
K Total number of technicians categories
T Total number of sub-planning periods

Decision variables

x_{ijkt} Number of technicians required for maintenance activity i from maintenance section j belonging to technician category k at period t
R_{ijkt} Reliability of a technician required for maintenance activity i from maintenance section j belonging to technician category k at period t
δ_{ijkt} Amount of maintenance time required by technician to carry out maintenance activity i from maintenance section j belonging to technician category k at period t (h)

Parameters

v_{ijkt} Earned-value of a technician that carries out maintenance activity i from maintenance section j belonging to technician category k at period t (₦)
c_{ijkt} Unit cost of technician that carries out maintenance activity i from maintenance section j belonging to technician category k at period t (₦)
r_{it} Expected value of technicians' reliability for maintenance activity i at period t
χ_{it} Total cost of technicians required to carry out maintenance activity i at period t (₦)
χ_t Total cost of technicians required to carry out available maintenance activities at period t (₦)
a_{ijt} Actual number of days a technician in section j from technicians category k at period t is available in a maintenance system (days)
p_{ijt} Actual performance of a technician in section j from technician's category k at period t
q_{ijt} Actual quality of workdone by a technician in section j from technician's category k at period t (kg)

$\hat{a}_{ijt}$ Expected days a technician in section j from technician's category k at period t is available at a maintenance system (days)

$\hat{p}_{ijt}$ Expected performance of a worker in section j from technician's category k at period t

$\hat{q}_{ijt}$ Expected quality of workdone by a technicians in section j from technician's category k at period t (kg)

$b_{ijk\,1}$ The minimum value of fatigue experienced during maintenance activity i by a technician from maintenance section j belonging to technicians' category k (h)

$b_{ijk\,2}$ The maximum value of fatigue experienced during maintenance activity i by a technician from maintenance section j belonging to technicians' category k (h)

$c_{jk\,1}$ The minimum value of extra fatigue on technicians for overtime activities by technicians from maintenance section j belonging to technician's category k (h)

$c_{jk\,2}$ The maximum value of extra fatigue on technicians for overtime activities by technicians from maintenance section j belonging to technician's category k (h)

$e_{jk\,1}$ The minimum value of experiential on technicians for reworked activities by technicians from maintenance section j belonging to technician's category k (h)

$e_{jk\,2}$ The maximum value of experiential on technicians for reworked activities by technicians from maintenance section j belonging to technician's category k (h)

$V_{ijk\,2}$ The minimum value of training impact on technicians for maintenance activity i by technicians from maintenance section j belonging to technician's category k (h)

$V_{ijk\,2}$ The maximum value of training impact on technicians for maintenance activity i by technicians from maintenance section j belonging to technician's category k (h)

$f(b)$ Probability density function of fatigue on technicians for maintenance activities

$f(v)$ Probability density function of training impact on technicians service times

$f(e)$ Probability density function of extra fatigue experience by technicians during overtime maintenance activity

$f(c)$ Probability density function of experience gain by technicians during reworked maintenance activity

Model formulation

This study presented two technicians' objectives (earned-value and reliability) which were subjected to technician's cost, service time, reliability, availability, performance and quality of workdone.

The cost of keeping a particular level of technician in a system depends largely on the expected contributions from the technicians. Using the concept of earned-value from engaging technicians in planned and unplanned maintenance activities in an organisation, the total earned-value from scheduling the different technicians for stochastic maintenance activities in a maintenance system was expressed as Eq. (1). The technician's earned-value is a function of the size of technicians assigned to carry out maintenance activities (x_{ijkt}), time spent on maintenance activities (δ_{ijkt}) and the unit earned-value expected from each technician (v_{ijkt}). Technician's earned-value could be defined as a measure of the value of utilising a technician for maintenance activities with respect to maintenance time.

$$\text{Max}\, f_1 = \sum_{t=1}^{T}\sum_{i=1}^{M}\sum_{j=1}^{N}\sum_{K=1}^{K} v_{ijkt}\delta_{ijkt}x_{ijkt} \tag{1}$$

By looking beyond the concept of earned-value of technicians, technicians scheduling analysis may be improved on using the average expected technician reliability (R_{ijk}) over a planning period. The study of technician's reliability provides a means of estimating the degree to which the expected earned-value of technicians will be achieved. Since technicians work in groups, we considered their reliability as being parallel. The expected average technicians' reliability is expressed as Eq. (2).

$$\text{Max}\, f_2 = \frac{1}{T}\sum_{t=1}^{T}\left[\prod_{i=1}^{M} 1 - \left(1 - \prod_{j=1}^{N}\left(\prod_{k=1}^{K}\left[1-\left(1-R_{ijkt}\right)^{x_{ijkt}}\right]\right)\right)\right] \tag{2}$$

The volume of maintenance work from overtime, overtime and rework maintenance activities in a system varies from one period to another. This variation may be attributed to equipment usage, equipment age, organisation's maintenance policy and quantity of spare parts used. Furthermore, factors such as fatigue, training and experiential knowledge affect technicians performance when restoring equipment to acceptable functional state.

The issue of routine maintenance activities has been studied in the literature (Mansour 2011). However, the concerns of researchers have been on equipment. Information on factors which affect technicians' performance is sparse in maintenance literature. This study considered the issue of fatigue and training on technician's service time. Since it is often difficult to measure fatigue in qualitative terms, a continuous function was used to capture the amounts of fatigue a technician experience during the maintenance activities. This is possible by considering two extreme values for technician's fatigue for a particular kind of maintenance activity. The expected reduction in maintenance time of technicians in a section is expressed as Eq. (3).

$$\begin{cases} L_{1jt} \leq \sum\limits_{k=1}^{K}\left(\delta_{1jkt} - \int\limits_{b_{ijk1}}^{b_{ijk2}} b\cdot f(b)\mathrm{d}b\right)x_{1jkt} \leq \bar{L}_{1jt} & t=1 \\ L_{1jt} \leq \sum\limits_{k=1}^{K}\left(\left(\delta_{1jk(t-1)} + \int\limits_{v_{ijk1}}^{v_{ijk2}} v\cdot f(v)\mathrm{d}v - \int\limits_{b_{ijk1}}^{b_{ijk2}} b\cdot f(b)\mathrm{d}b\right)\right)x_{1jkt} \leq \bar{L}_{1jt} & \text{Otherwise} \end{cases} \quad \forall(j,t) \tag{3}$$

$$\begin{cases} L_{2jt} \leq \sum\limits_{k=1}^{K}\left(\delta_{2jkt} - \int\limits_{b_{ijk1}}^{b_{ijk2}} b\cdot f(b)\mathrm{d}b - \int\limits_{c_{jk1}}^{c_{jk2}} c\cdot f(c)\mathrm{d}c\right)x_{2jkt} \leq \bar{L}_{2jt} & t=1 \\ L_{2jt} \leq \sum\limits_{k=1}^{K}\left(\delta_{2jk(t-1)} + \int\limits_{v_{ijk1}}^{v_{ijk2}} v\cdot f(v)\mathrm{d}v - \int\limits_{b_{ijk1}}^{b_{ijk2}} b\cdot f(b)\mathrm{d}b - \int\limits_{c_{jk1}}^{c_{jk2}} c\cdot f(c)\mathrm{d}c\right)x_{2jt} \leq \bar{L}_{2jkt} & \text{Otherwise} \end{cases} \quad \forall(j,t) \tag{4}$$

where L_{1jt} and $\bar{L}_{1jt}$ are the minimum and maximum amounts of routine maintenance tasks available for maintenance section j at period t, respectively.

The extension of production activities beyond the scheduled periods makes it necessary for technicians to be on ground. The technicians are expected to carry out maintenance activities that are required during overtime production activities. The need for overtime activities may be attributed to a change in production volume and maintenance-related problems. Shiftan and Wilson (1994) considered this problem under a deterministic condition. This study deviated from Shiftan and Wilson (1994) approach by introducing stochastic element into overtime maintenance activities. These stochastic elements involve the fatigue experienced during normal and overtime periods as well as reduction in maintenance time as a result of training technicians (Eq. 4). The determination of the number of overtime technicians will improve technicians' cost control in systems (Aghdaghi and Jolai 2008).

where L_{2jt} and $\bar{L}_{2jt}$ are the minimum and maximum amounts of overtime maintenance tasks available for maintenance section j at period t, respectively.

The failures of machines to produce the expected number of defective products after maintenance often result in re-maintenance (rework) of such machines. This problem results from poor diagnosis of the causes of breakdown of installed machines or the use of inferior spare parts. Whenever this problem occurs, technicians who are responsible for the initial maintenance of such machines could have gained experience on the possible causes of the poor machine performance. By the combining experiential knowledge, training and fatigue, the constraint for the expected time for sectional rework activities is given as Eq. (5).

$$\begin{cases} L_{3jt} \leq \sum\limits_{k=1}^{K}\left(\delta_{3jkt} + \int\limits_{e_{jk1}}^{e_{jk2}} e\cdot f(e)\mathrm{d}e - \int\limits_{b_{ijk1}}^{b_{ijk2}} b\cdot f(b)\mathrm{d}b\right)x_{3jkt} \leq \bar{L}_{3jt} & t=1 \\ L_{3jt} \leq \sum\limits_{k=1}^{K}\left(\delta_{3jk(t-1)} + \int\limits_{v_{ijk1}}^{v_{ijk2}} v\cdot f(v)\mathrm{d}v + \int\limits_{e_{jk1}}^{e_{jk2}} e\cdot f(e)\mathrm{d}e - \int\limits_{b_{ijk1}}^{b_{ijk2}} b\cdot f(b)\mathrm{d}b\right)x_{3jkt} \leq \bar{L}_{3jt} & \text{Otherwise} \end{cases} \quad \forall(j,t) \tag{5}$$

where L_{3jt} and $\bar{L}_{3jt}$ are the minimum and maximum amounts of rework maintenance tasks available for maintenance section j at period t, respectively.

Using the expected value and standard deviation of technician in a unit for the classes of technicians in each section, the minimum number of technicians required for each class under the various units in a maintenance department can be expressed as follows:

$$\delta_{ijkt}x_{ijkt} \geq \mu_{x_{ijkt}} - \alpha_r \left(\frac{1}{N} \left(\sum_{i=1}^{N} x_{ijkt}^2 P(x_{ijkt}) - 2\mu_{x_{ijkt}} \times \sum_{i=1}^{N} x_{ijkt}^2 P(x_{ijk}) + \mu_{x_{ijkt}}^2 \sum_{i=1}^{N} P(x_{ijkt}) \right) \right)^{1/2} \quad \forall (i,j,k,t) \tag{6}$$

where $\mu_{x_{ijkt}}$ is mean value of maintenance task i from maintenance section j for technician k at period t. $P(x_{ijkt})$ is the probability of occurrence of maintenance task i from maintenance section j for technician k at period t and α_r contribution of variance in maintenance tasks to $\mu_{x_{ijkt}}$.

The average number of technicians required in a planning horizon should not be less than the specified number of technicians (Eq. 7). This allows variations in the number of technicians in a section from one period to another.

$$\frac{\sum_{t=1}^{T} x_{ijkt}}{T} \leq x_{ijk,\text{ave}} \quad \forall (i,j,k) \tag{7}$$

where $x_{ijk,\text{ave}}$ is the average number of technicians expected to be scheduled to carried out maintenance task i from maintenance section j belonging to technician's category i at period t.

The sum of technicians in each period is expected to be within a specified range. This constraint helped in controlling the amount of technicians within a particular category (Eq. 8).

$$xx_{\min} \leq \sum_{i=1}^{M} \sum_{j=1}^{N} \sum_{k=1}^{K} x_{ijkt} \leq xx_{\max} \ \forall t \tag{8}$$

In Ighravwe et al. (2015), technicians' overall effectiveness (TOE) was considered. To control the expected value of TOE, there is the need to consider constraining the minimum values for sectional technicians' availability, quality of workdone and performance. The expected performance of each section in a maintenance department could be considered as the expected technician's efficiency. We applied uniform distribution concept (Wu 2008) to estimate the expected technicians' availability, quality of workdone and performance. Technicians' availability is expressed as Eq. (9), while technicians' performance was expressed as Eq. (10). The quality of workdone constraints is expressed as Eq. (11).

$$\frac{\sum_{k=1}^{K} a_{ijkt}x_{ijkt}}{\sum_{k=1}^{K} \hat{a}_{ijkt}x_{ijkt}} \geq (\bar{b}_j - \bar{a}_j)(1 - \bar{\alpha}_j) + \bar{a}_j \quad \forall (i,j,t) \tag{9}$$

$$\frac{\sum_{k=1}^{K} p_{ijkt}x_{ijkt}}{\sum_{k=1}^{K} \hat{p}_{ijkt}x_{ijkt}} \geq (\widehat{b}_j - \widehat{a}_j)(1 - \widehat{\alpha}_j) + \widehat{a}_j \quad \forall (i,j,t) \tag{10}$$

$$\frac{\sum_{k=1}^{K} q_{ijkt}x_{ijkt}}{\sum_{k=1}^{K} \hat{q}_{ijkt}x_{ijkt}} \geq (\tilde{b}_j - \tilde{a}_j)(1 - \tilde{\alpha}_j) + \tilde{a}_j \quad \forall (i,j,t) \tag{11}$$

where $\widehat{a}_j$ and $\widehat{b}_j$ are the minimum and maximum values of technicians' performance for maintenance task i from technicians in section j at period t, respectively. $\bar{a}_j$ and $\bar{b}_j$ are the minimum and maximum values of technicians' availability for maintenance task i from technicians in section j at period t, respectively. $\tilde{a}_j$ and $\tilde{b}_j$ are the minimum and maximum values of technicians' quality of work for maintenance task i from technicians in section j at period t, respectively. $\bar{\alpha}_j$, $\widehat{\alpha}_j$ and $\tilde{\alpha}_j$ are the confident levels for technicians' availability, performance and quality of workdone for maintenance task i expected from technicians in section j.

We relaxed the expression for TOE as defined by Ighravwe et al. (2015), to a stochastic constraint using the concept of normal distribution (Wu 2008) as Eq. (12).

$$\frac{\sum_{j=1}^{N} \sum_{k=1}^{K} a_{ijkt}x_{ijkt}}{\sum_{j=1}^{N} \sum_{k=1}^{K} \hat{a}_{ijkt}x_{ijkt}} \cdot \frac{\sum_{j=1}^{N} \sum_{k=1}^{K} p_{ijkt}x_{ijkt}}{\sum_{j=1}^{N} \sum_{k=1}^{K} \hat{p}_{ijkt}x_{ijkt}} \cdot \frac{\sum_{j=1}^{N} \sum_{k=1}^{K} q_{ijkt}x_{ijkt}}{\sum_{j=1}^{N} \sum_{k=1}^{K} \hat{q}_{ijkt}x_{ijkt}} \geq \mu_i + \Phi^{-1}(1 - \alpha_i)\,\sigma_i \quad \forall (i,t) \tag{12}$$

where μ_i, σ_i and α_i are the mean, standard deviation and confident level for TOE expected for maintenance task i from the technicians in a maintenance system, respectively.

Beyond specifying the expected TOE of technicians, there is the need for the cost implication consideration that will be associated with scheduling a particular level of technicians for each maintenance task at the different planning periods. The expression of the cost for each maintenance activity is considered (Eq. 13). The total cost of technicians required for the maintenance activities at each period is expressed as Eq. (14).

$$\sum_{j=1}^{N} \sum_{K=1}^{K} C_{ijkt}\delta_{ijkt}x_{ijkt} \leq \chi_{it} \quad \forall (i,t) \tag{13}$$

$$\sum_{i=1}^{M} \sum_{j=1}^{N} \sum_{K=1}^{K} C_{ijkt}\delta_{ijkt}x_{ijkt} \leq \chi_t \quad \forall (t) \tag{14}$$

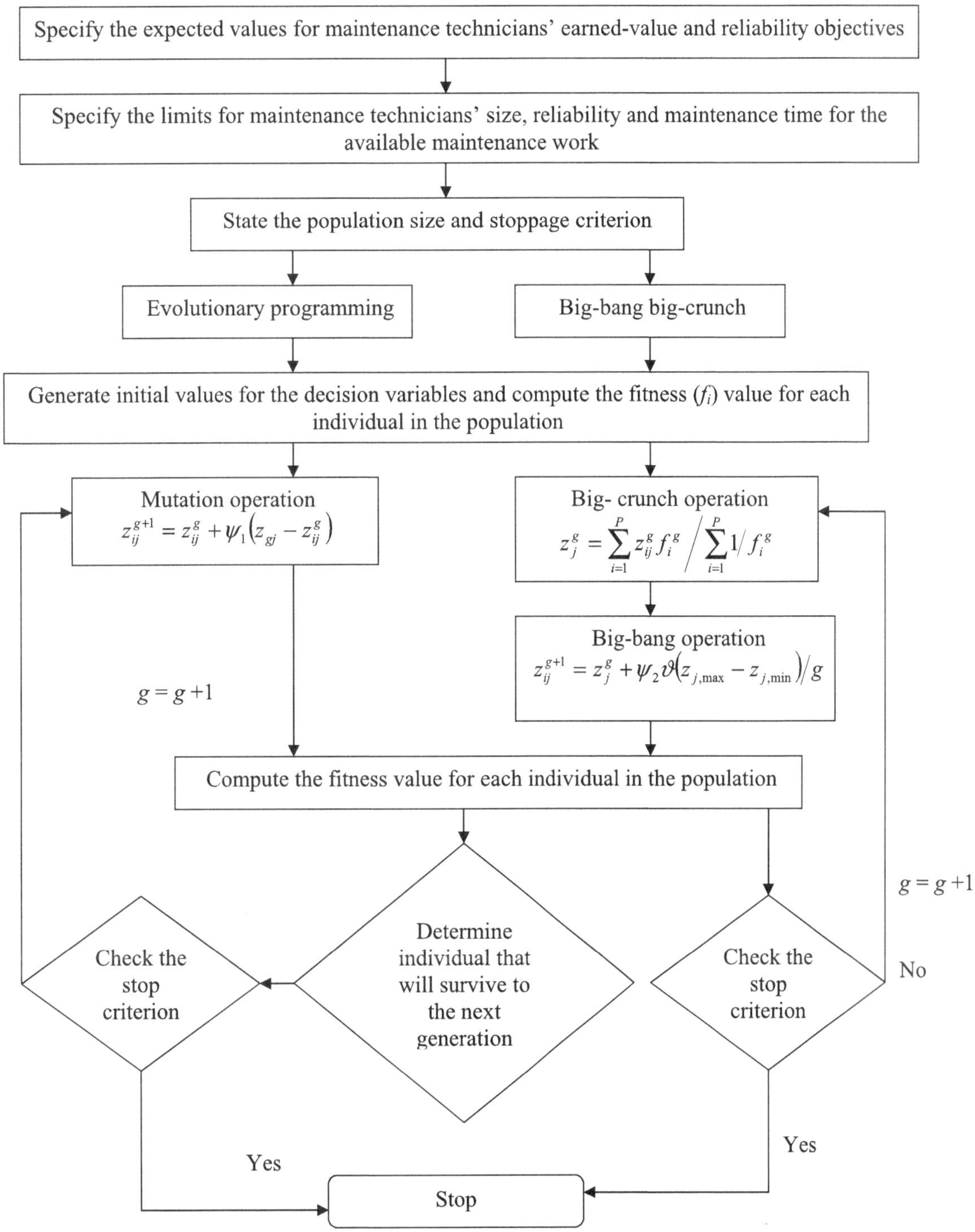

Fig. 1 Flow chart for the solution methods (Wong and Yuryevich 1997; Sakthivel and Mary 2013)

To further constraint the proposed model, the issue of the expected reliability from each maintenance section is considered. The reliability of each section in a maintenance department is expressed as Eq. (15).

$$1 - \prod_{j=1}^{N} \left(\prod_{k=1}^{K} \left[1 - \left(1 - R_{ijk} \right)^{x_{ijkt}} \right] \right) \geq r_{it} \quad \forall (i,t) \qquad (15)$$

Solution methods

The handling of the multi-objective is carried out using the weights of each of the objectives and deviational variables for the objective functions (Wu 2008). The new objective function is given as Eq. (16).

$$\text{Minimise } \frac{w_1(d_1^+ + d_1^-)}{f_{1,\max} - f_1} + \frac{w_2(d_2^+ + d_2^-)}{f_{2,\max} - f_2} \quad (16)$$

where $f_{1,\max}$ and $f_{2,\max}$ are the maximum values for the technicians earned-value and reliability, respectively.

The summary of the proposed mixed-integer programming model is presented as follows:

$$\text{Minimise } \frac{w_1(d_1^+ + d_1^-)}{f_{1,\max} - f_1} + \frac{w_2(d_2^+ + d_2^-)}{f_{2,\max} - f_2}$$

Subject to the following constraints:

$$f_1 + d_1^- = f_{1,\max} \quad (17)$$

$$f_2 + d_2^- = f_{2,\max} \quad (18)$$

Equations (3)–(15).

Non-negativity constraints

The flow chart for the evolutionary programming and big-bang–big-crunch algorithms is depicted in Fig. 1. The termination of each of these algorithms was taken as the maximum epoch. EP algorithms are different from other evolutionary algorithms (genetic algorithm, differential evolution and genetic programming) because it does not require crossover operation. The first EP algorithm was proposed by Fogel (1962). The quest to improve the mutation operation in EP algorithms has led to different versions of the EP algorithms.

The potentials of EP and BB–BC algorithms in generating optimal solution for computational problems are due to their stochastic-populate-based capacity. For the EP algorithm, the mutation introduced randomness into a current solution. In the BB–BC algorithm (Osman and Eksin 2006), randomness to current solution is introduced through the generation of centre of mass (big-crunch) and the new variable (big-bang). In Fig. 1, z_{ij}^g is the value of solution i decision variable j at epoch g and z_j^g the centre of mass for decision variable j at epoch g. The variable z_{gj} is the value of global solution variable j in a current epoch. Some authors have considered z_{gj} as the centre of mass during the implementation of BB–BC algorithm. The variable f_i^g is the quality of solution i at epoch g. In the EP section, ψ_1 is a uniform random number which lies between (0, 1). This variable helps in controlling the influence of the difference between current global and local optimal solutions in a population at a particular epoch g. The variable ψ_2 in the BB–BC algorithm is a random variable which lies at ± 1, and ϑ is a constant parameter that helps in controlling the search capacity of the BB–BC algorithm. The variables $z_{j,\min}$ and $z_{j,}$ are the minimum and maximum values of decision variable j.

Table 1 Simulated maintenance time for the different maintenance activities

Technician	$t = 1$	$t = 2$	$t = 3$	$t = 4$
Planned maintenance (h)				
x_{111t}	3132.95	2950.99	3182.18	2889.25
x_{112t}	2545.74	2490.33	2338.70	2371.24
x_{121t}	1472.62	1395.36	1389.03	1358.50
x_{122t}	1082.14	1060.97	1040.93	1152.72
x_{131t}	581.39	629.68	579.78	627.42
x_{132t}	592.05	574.70	533.63	545.71
Overtime maintenance (h)				
x_{211t}	1253.18	1180.40	1272.87	1155.70
x_{212t}	1018.30	996.13	935.48	948.50
x_{221t}	589.05	558.15	555.61	543.40
x_{222t}	432.85	424.39	416.37	461.09
x_{231t}	232.56	251.87	231.91	250.97
x_{232t}	236.82	229.88	213.45	218.29
Rework maintenance (h)				
x_{311t}	939.89	885.30	954.65	866.78
x_{312t}	763.72	747.10	701.61	711.37
x_{321t}	441.79	418.61	416.71	407.55
x_{322t}	324.64	318.29	312.28	345.82
x_{331t}	174.42	188.91	173.94	188.23
x_{332t}	177.62	172.41	160.09	163.71

Model application

The proposed model and algorithms (BB–BC and EP) were coded using C# programming language on a Windows 8 computer with installed memory (RAM) of 4.00 GB, 1.80 GHz processor and 64 bit operating system. To demonstrate the applicability of the proposed model, datasets from Ighravwe and Oke (2014) were used and complemented with simulated data. The datasets that were simulated are technician's reliability, earned-value, quality of workdone, availability and performance as well as the amount of rework and overtime maintenance activities. The amount planned maintenance work in Ighravwe and Oke (2014) was increased by 20 %. This enables us to generate upper bounds for the various planned maintenance activities.

The minimum value for the amounts of overtime was about 40 % of minimum planned maintenance workload. We considered the amount of rework maintenance activities as 30 % of the amounts of minimum planned maintenance workload. Table 1 shows the amount of planned, rework and overtime maintenance activities for the system. By observing the total amounts of workloads for the different sections, available maintenance time and

Table 2 Results of each objective minimum and maximum bounds solution method-wise

Priority	EP algorithm		BB–BC algorithm	
	Goal 1 (₦)	Goal 2	Goal 1 (₦)	Goal 2
Goal 1	1,539,809.6	0.98	1,368,011.02	0.97
Goal 2	1,485,545.60	0.97	1,296,020.80	0.94

Table 3 Optimal values for the objective functions

Objectives	EP algorithm	BB–BC algorithm
Technicians' earned-value (₦)	1,435,598.40	1,415,878.40
Technicians' reliability	0.96	0.98
Computation time (s)	1.4500	0.9246

technician's categories, the bounds for the numbers of technicians in the different maintenance sections were estimated. The unit cost for overtime maintenance activities was about 110 % of unit cost of engaging them in normal maintenance activities. The unit cost of rework maintenance activities was about 30 % of the unit cost of engaging them in normal maintenance activities.

The proposed model was solved using pre-emptive goal programming approach. This enabled us to determine the bounds for the objective functions. We considered the technicians' earned-valued goal as being a higher priority than the technicians' reliability goal. However, some organisations may consider the technicians' reliability goal as being a higher priority than the technicians' earned-valued goal. To address priority determination, priority scored performance measurement can be used (Tarokh and Nazemi 2006). The minimum value of the expected reliability of the total scheduled technicians was 85 %. During the testing of the proposed model, the minimum acceptable reliability of the technicians was 60 %. The maximum acceptable reliability of the technicians was 95 %.

For the EP algorithm, the selection between parents and off-springs was based on Boltzmann–Gibbs approach in simulated annealing (Engelbrecht 2007). The total number of epochs for the EP and BB–BC algorithms was 200. The limiting parameter (ψ) for the BB–BC algorithm was 0.2. The Pareto solutions obtained for at the initial implementation of EP and BB–BC algorithms are in Table 2.

Based on the objective functions results in Table 2, the differences between the upper and lower bounds for the technicians' earned-value objective are ₦54,264.00 and ₦71,990.22 for the EP and BB–BC algorithms, respectively. For the reliability objective, the differences between the upper and lower bounds for the EP and BB–BC algorithms are 0.1 and 0.3, respectively. Equal importance was given to the objective functions (i.e. $w_1 = 0.5$ and $w_2 = 0.5$). The values of the Pareto solutions obtained for the objective functions are in Table 3.

From the values for the two objective functions obtained from the two solution methods presented in Table 2, the differences between the upper and lower bounds for the maintenance technicians' earned-value objective are ₦54,264.00 and ₦71,990.22, for the EP and BB–BC algorithms, respectively. For the reliability objective, the differences between the upper and lower bounds for the EP and BB–BC algorithms are 0.1 and 0.3, respectively. To obtain optimal solution from the proposed model using the EP and BB–BC algorithms, equal importance was assigned to maintenance technicians earned-value and reliability objectives (i.e. $w_1 = 0.5$ and $w_2 = 0.5$). By using Eq. (16) as objective function for the proposed model and Eqs. (17) and (18) as soft constraints, the values of the Pareto solutions obtained for the two objective functions are presented in Table 3.

Based on the information in Table 3, the BB–BC algorithm performed better than the EP algorithm. By using the BB–BC algorithm as solution method for the proposed model, the number of maintenance technicians required to be scheduled to carry out the available maintenance work are shown in Table 4. In addition, the breakdown of maintenance time and reliability associated with each scheduled technicians for the various maintenance activities are in Table 4.

Discussion of results

The total number of technicians required for routine maintenance was 121 technicians, while 101 technicians were required for overtime maintenance activities. Rework maintenance activities required 84 technicians (Table 3). For routine maintenance activities, the system required 66 mechanical technicians and 32 electrical technicians. A total of 23 instrumentation technicians were required for routine maintenance activities. During the scheduling of technicians for overtime maintenance activities, the required mechanical technicians were 50. A total of 19 instrumentation technicians were required for overtime maintenance activities. The number of electrical technicians required for overtime electrical maintenance activities was 32 technicians.

The number of instrumentation technicians required for rework maintenance activities was the same as the number of workers for overhaul maintenance (19 technicians). The number of technicians for rework mechanical maintenance

Table 4 Optimal distribution of selected technicians' variables for different maintenance activities

Technicians	Technicians' size				Maintenance time				Technicians' reliability			
	$t = 1$	$t = 2$	$t = 3$	$t = 4$	$t = 1$	$t = 2$	$t = 3$	$t = 4$	$t = 1$	$t = 2$	$t = 3$	$t = 4$
Routine maintenance (h)												
x_{111t}	11	10	11	10	269	269	269	269	0.82	0.85	0.85	0.81
x_{112t}	6	6	6	6	229	226	229	228	0.83	0.85	0.84	0.83
x_{121t}	5	5	5	5	269	269	266	266	0.83	0.84	0.83	0.84
x_{122t}	3	3	3	3	223	225	224	226	0.81	0.83	0.86	0.84
x_{131t}	2	3	3	3	269	268	269	267	0.84	0.86	0.80	0.82
x_{132t}	3	3	3	3	232	225	225	219	0.83	0.86	0.82	0.85
Overtime maintenance (h)												
x_{211t}	7	7	6	6	211	206	216	202	0.81	0.85	0.82	0.82
x_{212t}	6	6	6	6	176	176	176	178	0.82	0.82	0.83	0.84
x_{221t}	5	5	5	5	208	216	204	213	0.84	0.84	0.85	0.82
x_{222t}	3	3	3	3	171	172	181	172	0.79	0.84	0.83	0.85
x_{231t}	2	2	3	2	211	212	209	208	0.80	0.83	0.83	0.83
x_{232t}	2	3	3	2	177	177	177	174	0.84	0.84	0.87	0.78
Rework maintenance (h)												
x_{311t}	6	6	5	5	178	179	179	178	0.84	0.84	0.84	0.84
x_{312t}	5	5	4	5	177	173	174	179	0.83	0.83	0.82	0.81
x_{321t}	3	3	3	3	177	179	178	173	0.83	0.82	0.82	0.80
x_{322t}	3	3	3	3	176	176	177	174	0.84	0.83	0.84	0.84
x_{331t}	2	3	3	2	178	179	177	179	0.82	0.83	0.85	0.82
x_{332t}	3	2	2	2	173	175	174	175	0.83	0.80	0.79	0.83

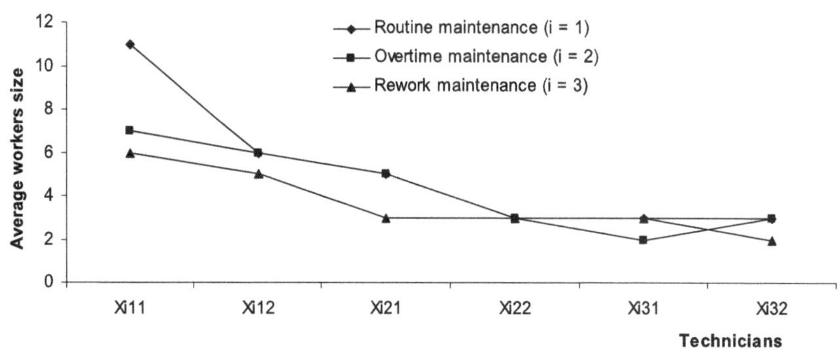

Fig. 2 Average technicians' size

was 41 technicians. The company's electrical section required about 24 technicians for rework maintenance activities.

The importance of this information is that it will assist maintenance managers in preparing logistics required to carry out successful maintenance tasks. For instance, in situations where production facilities are not on the same site, adequate arrangement for technician and spare parts transportation logistics will be made (Yuceer 2013). Another benefit of the above information is that business owners will have the opportunity to know how much of their working capital will be devoted to technicians' expenses. For instance, the highest number of technicians required for the routine and the overtime maintenance tasks was at period 3. For the routine maintenance tasks, 31 technicians are required, while 26 technicians are required to cover the available overtime maintenance tasks. The highest number of technicians for rework maintenance tasks was at periods 1 and 2.

To further simplify the above analysis, the pattern for the average number of the different categories of technicians is shown in Fig. 2. This provides an overview of the

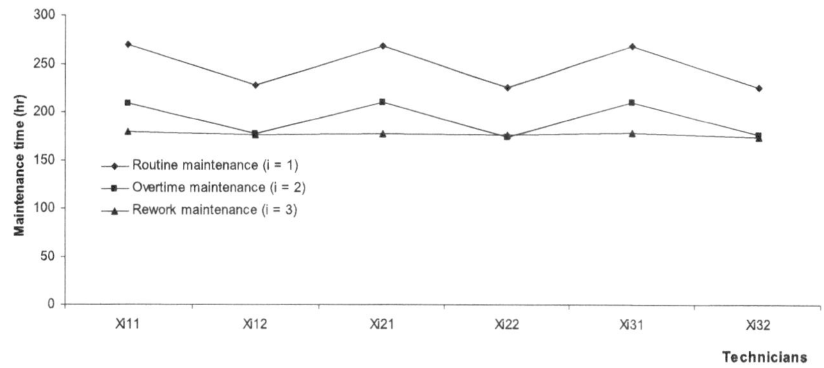

Fig. 3 Average technicians' time

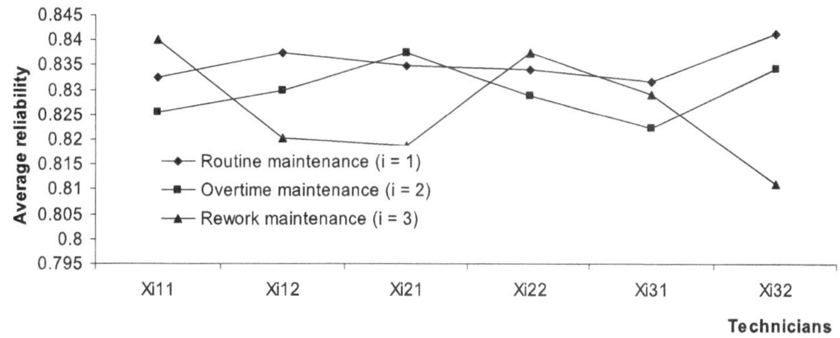

Fig. 4 Average technicians' reliability

trend of the required average number of the technicians for the system.

The organisation will need about ₦107, 694,516 for the technicians' expense for four periods. The total amount of funds required to cater for the technicians' expenses at period 1 was ₦8, 274,106, while a sum of ₦8, 212,776 was required at period 2. At period 3, ₦8, 322,675 was required for technicians' expenses. The technicians' expense at period 4 was less than that of period 3 with about 4 % (₦335, 716). The company's management can decide to budget funds for each period or for all the periods. Since resources are scare, the decision markers will prefer to plan for technicians' expenses at each period.

Beyond the issue of cost management, how the allocated time for maintenance activities (time management) will be utilised is equal of importance to the decision markers. The amounts of maintenance time for rework maintenance activities are relative the same for the maintenance sections (Fig. 3). There are differences in the amounts of maintenance times for routine and overtime maintenance activities required for the maintenance sections (Fig. 3). The combination of the different maintenance times will help in analysing production line availability whenever there is fluctuation in product demand. At period 1, the unavailability of the production line was 3704 h, while the production line was unavailable for 3702 h at period 2. The amount of time of the production line unavailability for period 3 was the same as that of period 2 (3704 h). Period 4 had the least production line unavailability (3680 h).

The question of how reliable the scheduled technicians are when carrying out assignment based on the allocated maintenance times is displayed in Table 3. By providing answer to this question, the problem of how to manage a maintenance system will be reduced to other areas such as spare management and design of technicians' motivational schemes. A summary of the average technician's reliability for the different technician's categories is in Table 4. The average reliability expected from the technicians during routine maintenance is relatively stable when compared with the average reliability during rework activities (Fig. 4).

Conclusions

This study has succeeded in proposing a multi-objective technician's scheduling model using mixed-integer programming approach. The integer aspect of the proposed model deals with the technicians' size and time, while the technicians' reliability is considered as a real-value decision variable. We demonstrated how the proposed model can be used in obtaining Pareto solutions for technicians' earned-value and reliability. Also, the proposed model results for different planning periods technicians size, maintenance time and reliability for the different kinds of maintenance task and technician's categories are presented. The computational results from EP and BB–BC algorithms as solution methods are compared, and the suitability of these algorithms has been shown from the results obtained. It was observed that the BB–BC algorithm performed better than the EP algorithm. The value of Pareto optimal solution obtained using the BB–BC algorithm from is ₦1, 415,878.40 and 98 % for the total technicians earned-value and reliability.

The main benefit of the proposed model is that it can be used to select technicians' matrix for the different kinds of maintenance tasks in a system. Another benefit of the model is that the information it generates can be used in estimating the productivity of technicians assigned for the different maintenance activities.

The contributions of this study are as follows: First, the introduction of training, fatigue and experimental knowledge in designing technicians' schedule plan is presented. Second, an integrated platform for determining the number of technicians for routine, rework and overtime maintenance tasks was contributed to technicians' allocation literature. Third, the use of technicians' earned-value in a technician's optimisation was added to literature. Lastly, the application of BB–BC algorithm as a promising solution method for technician's optimisation models has been explored.

However, we acknowledge some limitations of the proposed model. The proposed model lacks the capacity to deal with the problem of technician's transfers from one section to another. Also, our model cannot be used to address workers' preferences for particular types of maintenance activities and the effects of technician's absenteeism on work-plans.

The use of swarm algorithm in obtaining Pareto solution when applying the proposed model can be investigated as a further study. Also, the application of system dynamic modelling approach in establishing the interrelations among technicians' variables (time, size and reliability) and other maintenance parameters (spare parts) can be investigated as a further study.

References

Aghdaghi M, Jolai F (2008) A goal programming model for vehicle routing problem with backhauls and soft time windows. J Ind Eng Int 4(6):7–18

Alsyouf I (2007) The role of maintenance in improving companies' productivity and profitability. Int J Prod Econ 105(1):70–78

Bartels S, Richey J (2008) Workforce management—impacting the bottom-line, an oracle white paper. Oracle Corporation, Redwood Shores, CA, pp 1–10

De Bruecker P, Van den Bergh J, Belien J, Demeulemeester E (2015) A model enhancement heuristic for building robust air craft maintenance personnel rosters with stochastic constraints. Eur J Oper Res 246:661–673

Engelbrecht AP (2007) Computational intelligence: an introduction. Wiley, London

Fajardo VA, Drekic S (2015) Controlling the workload of M/G/1 queues via the q-policy. Eur J Oper Res 243:607–617

Fogel LJ (1962) Autonomous automata. Ind Res Mag 4(2):14–19

Gregoriades A, Sutcliffe A (2008) Workload prediction for improved design and reliability of complex systems. Reliab Eng Syst Saf 93:530–549

He M, Hu Q, Wu X, Jing P (2014) A combination algorithm for selecting functional logistics service vendors based on SQP and BNB. J Chem Pharm Res 6(5):2013–2018

Hedjazi D (2015) Scheduling a maintenance activity under skills constraints to minimise total weighted tardiness and late tasks. Int J Ind Eng Comput 6:135–144

Hervet C, Chardy M (2012) Passive optical network design under operations administration and maintenance considerations. J Appl Oper Res 4(3):152–172

Ighravwe DE, Oke SA (2014) A non-zero integer non-linear programming model for maintenance workforce sizing. Int J Prod Econ 150:204–214

Ighravwe DE, Oke SA, Adebiyi KA (2015) Maintenance workload optimisation with accident occurrence considerations and absenteeism from work using genetic algorithms. Int J Manag Sci Eng Manag. doi:10.1080/17509653.2015.1065208

Jarugumill S (2011) Integrated workforce planning considering regular and overtime decisions. In: Doolen T, Van Aken E (eds) Proceedings of the 2011 industrial engineering research conference

Kaufman DL, Lewis ME (2007) Machine maintenance with workload considerations. Naval Res Logist 54(7):750–766

Knapp GM, Mahajan M (1998) Optimisation of maintenance organisation and manpower in process industries. J Qual Maint Eng 4(3):168–183

Lai Y-C, Fan D-C, Huang K-W (2015) Optimising rolling stock assignment and maintenance plan for passenger railway operations. Comput Ind Eng 85:284–295

Mansour MAA-F (2011) Solving the periodic maintenance scheduling problem via genetic algorithm to balance workforce levels and maintenance cost. Ame J Eng Appl Sci 4(2):223–234

Manzini R, Accorsi R, Cennerazzo T, Ferrari E, Maranesi F (2015) The scheduling of maintenance. A resource-constraints mixed integer linear programming model. Comput Ind Eng 87:561–568

Mjema EAM (2002) An analysis of personnel capacity requirement in maintenance department by using simulation method. J Qual Maint Eng 8(3):253–273

Oladokun VO, Charles-Owaba OE, Nwaouzru CS (2006) An application of artificial neural network to maintenance management. J Ind Eng Int 2(3):19–26

Osman KE, Eksin I (2006) New optimisation method: big-bang big-crunch. Adv Eng Softw 37:106–111

Rana RS, Purohit R (2012) Balancing of maintenance task during maintenance of four wheeler. Int J Eng Res Appl 2(3):496–504

Safaei N, Banjevic D, Jardine AKS (2008) Multi-objective simulated annealing for a maintenance workforce scheduling problem: a case study. In: Tan CM (ed) Simulated annealing. I-Tech Education and Publishing, Vienna, pp 27–48

Sakthivel S, Mary D (2013) Big bang-big crunch algorithm for voltage stability limits improvement by coordinated control of svc settings. Res J Appl Sci Eng Technol 6(7):1209–1217

Shiftan Y, Wilson NHM (1994) Absence, overtime, and reliability relationships in transit workforce planning. Transp Res Part A: Policy and Pract 28(3):245–258

Tarokh MJ, Nazemi E (2006) Performance measurement in industrial organisations case study: Zarbal Complex. J Ind Eng Int 2(3):54–69

Tohidi H, Tarokh MJ (2005) Team size effect on teamwork productivity using information technology. J Ind Eng Int 1(1):37–42

Wong KP, Yuryevich PD (1997) Evolutionary programming-based algorithms for environmentally-constrained economic dispatch. IEEE Trans Power Syst 13(2):301–316

Wu Z (2008) Hybrid multi-objective optimisation models for managing pavement assets. Ph.D. Thesis, Virginia Polytechnic Institute and State University

Yuceer U (2013) An employee transporting problem. J Ind Eng Int 9:3

10

A mixed integer bi-level DEA model for bank branch performance evaluation by Stackelberg approach

Morteza Shafiee[1] · Farhad Hosseinzadeh Lotfi[2] · Hilda Saleh[3] · Mehdi Ghaderi[1]

Abstract One of the most complicated decision making problems for managers is the evaluation of bank performance, which involves various criteria. There are many studies about bank efficiency evaluation by network DEA in the literature review. These studies do not focus on multi-level network. Wu (Eur J Oper Res 207:856–864, 2010) proposed a bi-level structure for cost efficiency at the first time. In this model, multi-level programming and cost efficiency were used. He used a nonlinear programming to solve the model. In this paper, we have focused on multi-level structure and proposed a bi-level DEA model. We then used a liner programming to solve our model. In other hand, we significantly improved the way to achieve the optimum solution in comparison with the work by Wu (2010) by converting the NP-hard nonlinear programing into a mixed integer linear programming. This study uses a bi-level programming data envelopment analysis model that embodies internal structure with Stackelberg-game relationships to evaluate the performance of banking chain. The perspective of decentralized decisions is taken in this paper to cope with complex interactions in banking chain. The results derived from bi-level programming DEA can provide valuable insights and detailed information for managers to help them evaluate the performance of the banking chain as a whole using Stackelberg-game relationships. Finally, this model was applied in the Iranian bank to evaluate cost efficiency.

✉ Morteza Shafiee
m_shafiee277@yahoo.com

1 Department of Industrial Management, Shiraz Branch, Islamic Azad University, Shiraz, Iran

2 Department of Mathematics, Science and Research Branch, Islamic Azad University, Tehran, Iran

3 Department of Mathematics, Central Tehran Branch, Islamic Azad University, Tehran, Iran

Keywords Bi-level programming · DEA · Mixed integer programming · Stackelberg equilibrium · Game theory · Decentralized decision making structure · Bank performance evaluation

Introduction

Banks, as one of the most complex industries in this rapidly changing high-tech world of computers and telecommunications, need to be flexible enough to respond rapidly to change and also to keep up with stiff competition. In such competitive environment, continuous improvement is critical for any successful organization. Therefore, improving performance is widely recognized as essential to gaining an extra competitive edge. Due to working in a competitive business environment, banking chain has a multi-dimensional and complex structure. Banking chain plays an important role in the economic cycle of each country, so to gain/remain on sustainable competitive edge performance evaluation has become a critical role which management of every financial institution may play. In this regard, bank branch performance, as the focus of empirical application in this paper, has become more difficult due to size variety, offering different services to different costumers, and multi-dimensional structure. Generally two methods are used to measure bank branch operational efficiency; parametric and non-parametric. The drawback of parametric techniques is a number of inherent limitations which make them unsuitable for fully reflecting the increasingly complex nature of banking chains. For example, regression analysis as one of the best parametric techniques is a central tendency method and is only suitable for modeling single input-multiple outputs or multiple inputs-single output systems. Data envelopment analysis (DEA), as a non-parametric method, is an excellent efficiency analysis tool that creates efficient production frontier to compare the DMU under evaluation relative to the best decision making units (DMUs) if they are operating under the same conditions. Dealing with multiple inputs and multiple outputs is an appealing feature which gives DEA an edge over other analytical tools. Standard DEA evaluates the relative efficiency of DMUs in presence of multiple inputs and multiple outputs, but it does not provide sufficient details for managerial decisions. In real world scenarios, companies (DMUs) are mainly comprised of various divisions/levels that are linked together and have a great deal of interactions between them and follow multi-level Stackelberg relationships. Complex hierarchical structure of banking chains makes their managers deal with many challenges. The most important challenges that managers should cope with are: variable costs and demand under the multi-level Stackelberg relationships with interactions among levels, and coordinating these relationships to provide high quality services for customers. Bank managers can improve their cost efficiency through remaining or improving the quality of their services in order to create competitive advantage. Banking chain cost efficiency evaluates the ability of banks in producing current outputs at minimal level of cost which provides managers some insights behind the total cost in operations. The problems in the way of evaluation of banking branch performance naturally exhibit a multi-level decision making models which are connected in a hierarchical way. Since in the multi-level decentralized decision companies the individual set of decision variables is often controlled by each level which have their own, often mutually conflicting, objectives the evaluation of banking chain performance is based on multi-perspective. Therefore, performance evaluation in such particular multi-level decentralized decision structures can be modeled by bi-level programming DEA approach. Bi-level programming data envelopment analysis (BLPDEA) approach goes to the black box and embodies internal structure and interior interactions of system when it has a hierarchical structure. BLPDEA also can provide a valuable insight and detailed information to manager when evaluating the performance of a system with Stackelberg-game relationships (Wu 2010). The main purpose of this paper is to illustrate a seldom utilized non-parametric analysis technique called BLPDEA which addresses the problem of cost efficiency evaluation under the Stackelberg leader–follower relationships in the context of banking. Bi-level DEA model is a NP-hard nonlinear programming. Since nonlinear models are reducing the validity of the model, in this paper, we recommended the Mixed integer programming method which converts the bi-level mathematical programming to a linear mathematical programming and significantly improves the process to achieve the optimum solution.

The remainder of this paper is organized as follows: In the next section, "Literature review" section, literature is reviewed. Section "Fundamentals of DEA, DEA cost efficiency and bi-level programming" introduces the concepts of cost efficiency DEA and bi-level programming. In "Proposal model" section bi-level programming DEA model proposed by Wu (2010) is presented. In "The framework of the efficiency evaluation of the banks" section, we demonstrate empirical use of bi-level programming DEA model in the form of a case study (Iranian bank). Conclusions are presented in "Empirical study" section.

Literature review

In comparison with techniques of assessing organization performance, the method of DEA proposed by Charnes et al. (1978) is a better way to organize and analyze data because it allows efficiency to change over time and requires no prior assumption on the specification of the efficient frontier. Thus, DEA is an excellent approach for the performance analysis of banking industry. In many real world scenarios, DMUs have a two-stage network process and due to this reason, DEA has been extended to examine the efficiency of two-stage processes, where all of the outputs from the first stage are intermediate measures that make up the inputs of the second stage. Wang (1997) present a two-stage process in the banking industry where the banks use inputs (of the first stage) including fixed assets, labor, and information technology (IT) investments to generate deposits. The banks then use the deposits (intermediate measure) to generate loans and profits (as the outputs). Momen et al. (2012) measured the operational risk of Iranian banks based on Loss Distribution Approach. Bhattacharya et al. (1997) used a two-stage DEA approach to examine the impact of liberalization on the efficiency of the Indian banking industry. In the first stage, a technical efficiency score was calculated, whereas in the second stage a stochastic frontier analysis was used to attribute variation in efficiency scores of three sources: temporal, ownership and noise component. Seiford and Zhu (1999) examined the performance of the top 55 US banks using a two-stage DEA approach. Results indicated that relatively large banks exhibit better performance on profitability, whereas smaller banks tend to perform better with respect to marketability. Sexton and Lewis (2003) proposed a two-stage process for evaluating Major League Baseball performance. Khalili-Damghani and Taghavifard (2012) proposed a generic process in which just-in-time (JIT) practices are changed into agility indices, and agility indices are converted into performance measurement in supply chain in form of a conceptual model. Then, a three-stage data envelopment analysis (TSDEA) model proposed to measure the relative efficiency of aforementioned process and sub-processes. Babazadeh et al. (2012) designed a net work and applied a mixed integer linear programming to evaluate supply chain efficiency. Khalili-Damghani and Tavana (2013) proposed a new network DEA (NDEA) model for measuring the performance of agility in supply chains. Their proposed fuzzy NDEA model was linear and independent of the α-cut variables. Sanieemonfared and Safi (2013) applied a novel DEA network structure to measure relative efficiency of academic colleges in both teaching quality and research productivity. Kao and Hwang (2008) developed a different approach where the entire two-stage process can be decomposed into the product of the efficiencies of the two sub-processes. As a result, both the overall efficiency and each stage's efficiency are obtained. Tone and Tsutsui (2009) extended the SBM model into a network framework to deal with intermediate products formally. Avkiran (2009), as the first empirical study of NSBM, relies on actual aggregate data of domestic commercial banks in the UAE, and applied the non-oriented network slacks-based measure for evaluating the profit efficiency. Fukuyama and Weber (2010) extended the slacks-based inefficiency measure for evaluating a two-stage system with bad outputs in a Japanese bank. Paradi et al. (2011) developed a two-stage DEA approach for simultaneously benchmarking the performance of operating units. Li et al. (2012) extended the findings of Liang et al. (2008) and proposed a centralized and non-cooperative model to evaluate the efficiency of two-stage process to further decompose the overall efficiency for complex network structure. Khalili-Damghani and Hosseinzadeh Lotfi (2012) developed a fuzzy two-stage data envelopment analysis (FTSDEA). In this paper, each decision making unit supposed to make up of two serially connected sub-DMUs. Tavana and Khalili-Damghani (2014), proposed an efficient two-stage fuzzy DEA model to calculate the efficiency scores for a DMU and its sub-DMUs. They used the Stackelberg (leader–follower) game theory approach to prioritize and sequentially decompose the efficiency score of the DMU into a set of efficiency scores for its sub-DMUs.

Bi-level decision making or bi-level programming techniques, first introduced by Von Stackelberg (1952), have been developed for mainly solving decentralized decision process with decision makers in a hierarchical organization. Decision maker at the upper level is termed as the leader, and in the lower level, the followers which have their own and perhaps mutually conflicting objective. Bi-level programming has been applied in a great deal of fields. Ryu et al. (2004) presented a bi-level programming framework to capture conflicting interests of multiple elements in the context of supply chain planning problems. Huijun et al. (2008) proposed a bi-level programming model to describe and solve the location problem in which both the benefits of costumers and logistics planers are taken into account. Sakawa et al. (2002) dealt with a transportation problem in a housing material manufacturer and derived a satisfactory solution to the problem. Roghanian et al. (2007) considered a probabilistic bi-level linear multi-objective programming problem and its application in enterprise-wide supply chain planning problem. Arora and Gupta (2009) presented an interactive fuzzy goal programming approach for bi-level programming problem with the characteristics of dynamic programming. Hongjie et al. (2011) established two inventory control models of deteriorating items respectively according to time-based and the quantity-based integrated delivery strategies for suppliers under VMI model based on bi-level programming. Lachhwani and Poonia (2012) developed a procedure for solving multilevel fractional programming problems in a large hierarchical decentralized organization by fuzzy goal programming approach. In this paper, fuzzy goal programming approach is used for achieving the highest degree of each of the membership goal by minimizing negative deviational variables. Alimardani et al. (2013) developed a continuous review policy for inventory control in a three-echelon supply chain including retailers, a central warehouse with limited storage space, and two independent manufacturing plants which offer two kinds of product to the customer. Arianezhad et al. (2013) presented a new two-echelon model to control the inventory of perishable goods. The main purpose of the model is to minimize the maintenance cost of the entire chain. Due to the complexity of the model, they used genetic algorithm under MATLAB to solve and confirm the accuracy of the model's performance.

Cost efficiency model is used to show the ability of DMUs to produce current outputs at minimal level of cost and how DEA can be used to identify types of inefficiency, which can emerge for treatment when information on costs is known exactly (Cooper et al. 2007). Tohidi and Khodadadi (2013) introduced a new model to evaluate cost efficiency of DMUs with negative data. They also demonstrated that proposed cost efficiency is a product of a locative and range directional measure efficiencies. Bahri and Tarokh (2012) focused on "seller–buyer" supply chain model with exponential distribution lead time and showed that their method can minimize the costs compared with systems that ignore the relation between seller and buyer.

Wu (2010) developed an innovative quantitative approach to evaluate the performance of multi-level decision network structure by integrating cost DEA into the bi-level programming framework and create bi-level programming DEA model. To show applicability of bi-level programming DEA model, Wu (2010) have demonstrated applications of the model in two practical examples: a banking chain and a manufacturing supply chain.

Fundamentals of DEA, DEA cost efficiency and bi-level programming

Data envelopment analysis

DEA is a linear programming based methodology which can calculate multiple inputs and outputs and can also evaluate DMUs both qualitatively and quantitatively. DMU, which can be related to different firms or the condition of the same firm over time, stands for decision making unit.

DEA was first proposed by Charnes et al. (1978). The evolutionary form of CCR model was suggested in 1984 by Banker et al. In subsequent years, several models were developed by a large number of researchers. Orientation, disposability, diversification, and return to scale are different aspects that can be seen in these models.

There are numerous studies on efficiency evaluation with the DEA model in several filed. For example, in banking such as Elyasiani and Mehdian (1990), Kao and Liu (2004), Camanho and Dyson (1999, 2005), and Cook and Hababou (2001). In supply chain efficiency evaluation, such as Khalili-Damghani and Sadi-Nezhad (2014), Khalili-Damghani and Hosseinzadel Lofit (2012), Khalili-Damghani and Taghavifard (2012, 2013), Abtahi and Khalili-Damghani (2011), Khalili-Damghani and Tavana (2013), and Khalili-Damghani et al. (2011, 2012). In other companies such as Tavana et al. (2014), Khalili-Damghani et al. (2015), Khalili-Damghani and Taghavifard (2012) and Tavana et al. (2013).

DEA cost efficiency

Many different types of DEA models with different aims have been developed. The aim of the majority of DEA models is focused on the technical-physical aspects of production for use in situations with unknown unit price and cost information, or where their uses are limited because of variability in the prices and costs that might need to be considered. Cost efficiency model is used to show the ability of DMUs to produce current outputs at minimal level of cost and how DEA can be used to identify types of inefficiency which can emerge for treatment when information on costs is known exactly (Cooper et al. 2007).

Suppose there are n DMUs under evaluation, each indexed by $(j = 1, \ldots, n)$, and $X = (x_1, \ldots, x_m)^T$ is the input vector which produces the output vector $Y = (y_1, \ldots, y_s)^T$ under the production possibility set. Then, the DEA cost efficiency model of 0th DMU, $(0 \in \{1, \ldots, n\})$, can be formulated as following linear programming:

$$\begin{aligned} & cx^* = \min \sum_{i=1}^{m} c_i x_i \\ & s.\,to \\ & \sum_{j=1}^{n} x_{ij}\lambda_j \le x_i, \quad i = 1, \ldots, m; \\ & \sum_{j=1}^{n} y_{rj}\lambda_j \ge y_{r0}, \quad r = 1, \ldots, s; \\ & \lambda_j \ge 0, \; x_i \ge 0. \end{aligned} \tag{1}$$

where (x_i, λ_j) are decision variables and (c_i) is the unit cost of input i which may vary from one DMU to another. This model allows substitutions in inputs. The objective function of model is to minimize the total cost of 0th DMU.

Based on an optimal solution (x^*, λ^*) of the above LP, the cost efficiency of DMU_0 is defined as:

$$E_c = \frac{CX^*}{CX_0} \tag{2}$$

where X_0 is the existing input vector of DMU_0.

Bi-level programming

Bi-level programming which is motivated by Von Stackelberg's game theory (1952) refers to situations where there are two decision makers in an organization which are inter-connected in a hierarchical structure. In such situations, the decision maker who first makes decision is termed as the leader and the other who knows the decision of opponent then makes a decision is termed as the follower. These two decision makers have independent, perhaps mutually conflicting, objectives. In the context of bi-level programming, the leader first specifies a decision and then the follower with the full knowledge of the leader's decision determines a decision so as to optimize his/her objective function. Accordingly, the leader also makes a decision so as to optimize his/her objective function. The obtained solution of the above mentioned procedure is a Stackelberg equilibrium solution (Sakawa and Nishizaki 2009). A bi-level linear programming problem for obtaining the Stackelberg solution is formulated as follows:

$$\begin{aligned} & \underset{x}{\textbf{minimize}}\; z_1(x, y) = c_1 x + d_1 y \\ & \textbf{where } y \textbf{ solves} \\ & \underset{y}{\textbf{minimize}}\; z_2(x, y) = c_2 x + d_2 y \\ & \textbf{subject to} \quad Ax + By \le b; \\ & x \ge 0, y \ge 0. \end{aligned} \tag{3}$$

where c_i $i = 1, 2$ are n_1-dimensional row coefficient vector, d_i, $i = 1, 2$ are n_2-dimensional row coefficient vector, A is an $m \times n_1$ coefficient matrix, B is an $m \times n_2$ coefficient matrix, b is an m-dimensional column constant vector. $z_1(x,y)$ and $z_2(x,y)$ are the objective function of the leader and the follower, respectively. x and y are a set of decision variables which are controlled by the leader and follower, respectively (Sakawa and Nishizaki 2009).

Sakawa and Nishizaki (2009) gave the following definitions on basis of bi-level programming:

Definition 1 (Sakawa and Nishizaki 2009) S is the feasible region of the bi-level linear programming problem:

$$S = \{(x,y)|Ax + By \leq b, x \geq 0, y \geq 0\}.$$

Definition 2 (Sakawa and Nishizaki 2009) $S(x)$ is the decision space of the follower after x is specified by the leader:

$$S(x) = \{y \geq 0 | By \leq b - Ax, x \geq 0\}.$$

Definition 3 (Sakawa and Nishizaki 2009) S_X is the decision space of the leader:

$$S_X = \{x \geq 0 | \text{there is a } y \text{ such that } Ax + By \leq b, y \geq 0\}.$$

Definition 4 (Sakawa and Nishizaki 2009) $R(x)$ is the set of rational responses of the follower for x specified by the leader:

$$R(x) = \left\{ y \geq 0 | y \in \arg\min_{y \in S(x)} z_2(x,y) \right\}.$$

Definition 5 (Sakawa and Nishizaki 2009) Inducible region:

$$\text{IR} = \{(x,y)|(x,y) \in S, y \in R(x)\}.$$

Definition 6 (Sakawa and Nishizaki 2009) Stackelberg solution:

$$\left\{ (x,y)|(x,y) \in \arg\min_{(x,y) \in \text{IR}} z_1(x,y) \right\}.$$

In the bi-level programming, follower optimization problem is considered as one of the constraints of a bi-level optimization problem. So, by applying the Kuhn–Tucker approach as a popular way in solving BLP, the follower's problem can be replaced by the Kuhn–Tucker conditions of the follower's problem. Then the leader's problem with constraints involving the optimality conditions of the follower's problem is solved (Sakawa and Nishizaki 2009). Using Kuhn–Tucker conditions, the bi-level linear programming problem (3) can be rewritten as the following equivalent single-level nonlinear programming problem:

$$\begin{aligned}
&\textbf{\textit{minimize}} \quad z_1(\boldsymbol{x},\boldsymbol{y}) = \boldsymbol{c}_1\boldsymbol{x} + \boldsymbol{d}_1\boldsymbol{y} \\
&\textbf{\textit{subject to}} \quad \boldsymbol{uB} - \boldsymbol{v} = -\boldsymbol{d}_2; \\
&\boldsymbol{u}(\boldsymbol{Ax} + \boldsymbol{By} - \boldsymbol{b}) - \boldsymbol{vy} = 0; \\
&\boldsymbol{Ax} + \boldsymbol{By} \leq \boldsymbol{b}; \\
&\boldsymbol{x} \geq 0, \boldsymbol{y} \geq 0, \boldsymbol{u}^T \geq 0, \boldsymbol{v}^T \geq 0,
\end{aligned} \tag{4}$$

where u is an m-dimensional row vector and v is an n_2-dimensional row vector. u and v are the dual variable associated with constraints $Ax + By \leq b$ and $y \geq 0$, respectively.

By introducing zero-one vectors $w_1 = (w_{11}, \ldots, w_{1m})$ and $w_2 = (w_{21}, \ldots, w_{2n_2})$, the non-linear programming problem (4) is transformed into a following linear mixed zero-one programming problem, and it can be solved by a zero-one mixed integer solver:

$$\begin{aligned}
&\textbf{\textit{minimize}} \quad z_1(\boldsymbol{x},\boldsymbol{y}) = \boldsymbol{c}_1\boldsymbol{x} + \boldsymbol{d}_1\boldsymbol{y} \\
&\textbf{\textit{subject to}} \quad 0 \leq \boldsymbol{u}^T \leq \boldsymbol{M}\boldsymbol{w}_1^T; \\
&0 \leq \boldsymbol{b} - \boldsymbol{Ax} - \boldsymbol{By} \leq \boldsymbol{M}(\boldsymbol{e} - \boldsymbol{w}_1^T); \\
&0 \leq (\boldsymbol{uB} + \boldsymbol{d}_2)^T \leq \boldsymbol{M}\boldsymbol{w}_2^T; \\
&0 \leq \boldsymbol{y} \leq \boldsymbol{M}(\boldsymbol{e} - \boldsymbol{w}_2^T); \\
&\boldsymbol{x} \geq 0,
\end{aligned} \tag{5}$$

where e is an m-dimensional vector of ones, and M is a large positive constant. The readers who are interested in bi-level programming may refer to Sakawa and Nishizaki (2009).

Proposal model

Banking branch performance evaluation problems naturally exhibit a two-level decision modeling which are connected in a hierarchical way. In a bank chain structure, the funds from costumers in the form of deposits are collected within the first level and in the second level the deposits from previous level are taken to make profit. Since the funds collected from the first level determines on the investment decision in the second level, the banking branch performance evaluation problem can be modeled as a leader-follower Stackelberg problem (Wu 2010). The conceptual bi-level DEA model with shared resource is depicted in Fig. 1.

Consider the n banking branch, which was indexed by $(j = 1, \ldots, n)$, is involving two levels L_1, and L_2. This L_1–L_2 chain has been addressed using bi-level programming structure, where the first level is termed as a leader and the second level as a follower. These two bank branch chains performance evaluation problem for specific DMU_0 can be

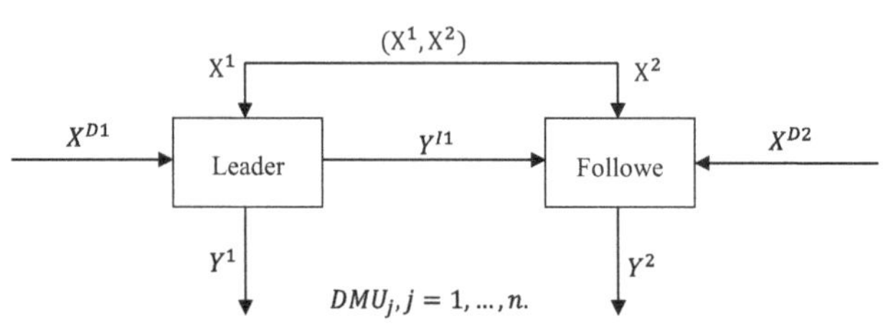

Fig. 1 Bi-level programming DEA model with shared resource (Wu 2010)

mathematically modeled in the bi-level programming DEA model, which considers the hierarchical structure of the bank branch including decision maker in each level who makes a decision to control a set of decision variables independently, as follows:

$$\begin{aligned}
&\min_{\overline{X}^1,\overline{X}^{D1},\lambda} \left(C^{1^T}\overline{X}^1 + C^{2^T}\overline{X}^{D1}\right) + \left(C^{1^T}\overline{X}^2 + D^{1^T}\overline{X}^{D2} + D^{2^T}\overline{Y}^{I1}\right)\\
&s.to\\
&\sum_{j=1}^{n} X_j^1\lambda_j \le \overline{X}^1;\\
&\sum_{j=1}^{n} X_j^{D1}\lambda_j \le \overline{X}^{D1};\\
&\sum_{j=1}^{n} Y_j^1\lambda_j \ge Y_0^1;\\
&\sum_{j=1}^{n} Y_j^{I1}\lambda_j \ge Y_0^{I1};\\
&\overline{X}^1 + \overline{X}^2 \le E(const.); \qquad (6)\\
&\min_{\overline{X}^2,\overline{X}^{D2},\overline{Y}^{I1},\pi} \left(C^{1^T}\overline{X}^2 + D^{1^T}\overline{X}^{D2} + D^{2^T}\overline{Y}^{I1}\right)\\
&s.to\\
&\sum_{j=1}^{n} X_j^2\pi_j \le \overline{X}^2;\\
&\sum_{j=1}^{n} X_j^{D2}\pi_j \le \overline{X}^{D2};\\
&\sum_{j=1}^{n} Y_j^{I1}\pi_j \le \overline{Y}^{I1};\\
&\sum_{j=1}^{n} Y_j^2\pi_j \ge Y_0^2;\\
&\overline{X}^1, \overline{X}^2, \overline{X}^{D1}, \overline{X}^{D2}, \overline{Y}^{I1},\ \lambda, \pi \ge 0,
\end{aligned}$$

where the X^1 and X^2 are m_1-dimensional row vectors of the shared input of the leader and the follower, respectively, X^{D1} is an m_2-dimensional row vector of the direct input of the leader, X^{D2} is the m_3-dimensional row vector of the direct input to the follower, Y^{I1} is an m_4-dimensional row vector which is the intermediate output to the leader and the intermediate input to the follower, Y^1 is an m_5-dimensional row vector of the direct output of the leader, and Y^2 is an m_6-dimensional row vector of the direct output to the follower. $C^{1T}, C^{2T}, D^{1T}, D^{2T}$ are the input unit cost vectors associated with the shared input, the direct input to the leader, the direct input to the follower, and the intermediate input to the follower, respectively. λ and π are the nonnegative multiplier used to aggregate existing leader and follower activities, respectively (Wu 2010).

To solve the bi-level model problem Wu (2010) converts bi-level mathematical programming to a non-linear mathematical programming and then uses trial and error process to achieve the optimum solution, which makes the process lengthy and time-consuming. Since nonlinear models are reducing the validity of the model, in this paper, we recommended the Mixed integer programming method, which converts the bi-level mathematical programming to a linear mathematical programming and significantly improves the process to achieve the optimum solution.

The bi-level programming DEA model (6) is transformed into the mixed integer single-level linear programming DEA model as follows:

$$\begin{aligned}
&\min\left(C^{1^T}\overline{X}^{1^T} + C^{2^T}\overline{X}^{D1^T}\right) + \left(C^{1^T}\overline{X}^2 + D^{1^T}\overline{X}^{D2} + D^{2^T}\overline{Y}^{I1^T}\right)\\
&s.to\\
&\sum_{j=1}^{n} X_j^1\lambda_j \le \overline{X}^1;\\
&\sum_{j=1}^{n} X_j^{D1}\lambda_j \le \overline{X}^{D1};\\
&\sum_{j=1}^{n} Y_j^1\lambda_j \ge Y_0^1;\\
&0 \le \overline{X}^2 - \left(\sum_{j=1}^{n} X_j^2\pi_j\right) \le Mw_1^T;\\
&0 \le U^1 \le M\left(e - w_1^T\right);\\
&0 \le \overline{X}^{D2} - \left(\sum_{j=1}^{n} X_j^{D2}\pi_j\right) \le Mw_2^T;\\
&0 \le U^2 \le M\left(e - w_2^T\right);\\
&0 \le \overline{Y}^{I1} - \left(\sum_{j=1}^{n} Y_j^{I1}\pi_j\right) \le Mw_3^T;\\
&0 \le U^3 \le M\left(e - w_3^T\right);\\
&0 \le \sum_{j=1}^{n} Y_j^2\pi_j - Y_0^2 \le Mw_4^T; \qquad (7)\\
&0 \le U^4 \le M\left(e - w_4^T\right);\\
&0 \le E - \overline{X}^1 - \overline{X}^2 \le Mw_5^T;\\
&0 \le U^5 \le M\left(e - w_5^T\right);\\
&U^1 - U^5 + V^1 = C^1;\\
&U^2 + V^2 = D^1;\\
&U^3 + V^3 = D^2;\\
&-X_j^2U^1 - X_j^{D2}U^2 - Y_j^{I1}U^3 + Y_j^2U^4 + V^4 = 0;
\end{aligned}$$

$0 \le \overline{X}^2 \le Mw_6^T;$
$0 \le V^1 \le M(e - w_6^T);$
$0 \le \overline{X}^{D2} \le Mw_7^T;$
$0 \le V^2 \le M(e - w_7^T);$
$0 \le \overline{Y}^{I1} \le Mw_8^T;$
$0 \le V^3 \le M(e - w_8^T);$
$0 \le \pi_j \le Mw_9^T;$
$0 \le V^4 \le M(e - w_9^T);$
$\overline{X}^1, \overline{X}^{D1}, \lambda \ge 0,$

where the U^1 and V^1 are the m_1-dimensional dual vectors correspond to shared input constraints and variables to the follower, respectively, U^2 and V^2 are the m_3-dimensional dual vectors associated with direct input constraints and variables to the follower, respectively, U^3 and V^3 are the m_4-dimensional dual vectors correspond to the intermediate input constraints and variables of the follower, respectively, U^4 and V^4 are the n-dimensional dual vectors, and U^5 is an m_1-dimensional dual vectors correspond to the constrained resources. Correspondingly, W_i^T, $i = 1, \ldots, 9$, are the zero-one vectors. e and M are the vector of ones and the large positive constant, respectively.

By solving the bi-level programming DEA model, the optimal solution of $\left(\overline{X}^{1*}, \overline{X}^{2*}, \overline{X}^{D1*}, \overline{X}^{D2*}, \overline{Y}^{I1*}, \lambda_j^*, \pi_j^*\right)$ are obtained. Based on optimal solution, the cost efficiency of the leader of DMU$_0$ is defined as:

$$CE_0^L = \frac{C^{1^T}\overline{X}^{1*} + C^{2^T}\overline{X}^{D1*}}{C^{1^T}X_0^1 + C^{2^T}X_0^{D1}} \tag{8}$$

and the cost efficiency of the follower of DMU$_0$ is defined as:

$$CE_0^F = \frac{C^{1^T}\overline{X}^{2*} + D^{1^T}\overline{X}^{D2*} + D^{2^T}\overline{Y}^{I1*}}{C^{1^T}X_0^2 + D^{1^T}X_0^{D2} + D^{2^T}Y_0^{I1}} \tag{9}$$

and the total cost efficiency of DMU$_0$ is defined as:

$$CE_0^S = \frac{\left(C^{1^T}\overline{X}^{1*} + C^{2^T}\overline{X}^{D1*}\right) + \left(C^{1^T}\overline{X}^{2*} + D^{1^T}\overline{X}^{D2*} + D^{2^T}\overline{Y}^{I1*}\right)}{\left(C^{1^T}X_0^1 + C^{2^T}X_0^{D1}\right) + \left(C^{1^T}X_0^2 + D^{1^T}X_0^{D2} + D^{2^T}Y_0^{I1}\right)}. \tag{10}$$

The framework of the efficiency evaluation of the banks

After analyzing the previous literature review, the efficiency of the banks which was proposed throughout this research was shown in Fig. 2. The analytical processes were divided and carried out in six steps: (1) in the first step the efficiency measurements of the bank were determined by reviewing literature and expert ideals; (2) in the second step the hierarchical structure of the bank was determined and the measurements were divided into two levels namely leader and follower levels; (3) in the third step the relationships between the leader and the follower were determined; (4) in the fourth step the main bi-level DEA model of the bank was created; (5) in the fifth step the created bi-level model was converted into a single linear model using mixed integer programming; (6) and finally in the sixth step the efficiency of the bank was obtained and the its ranking was recognized.

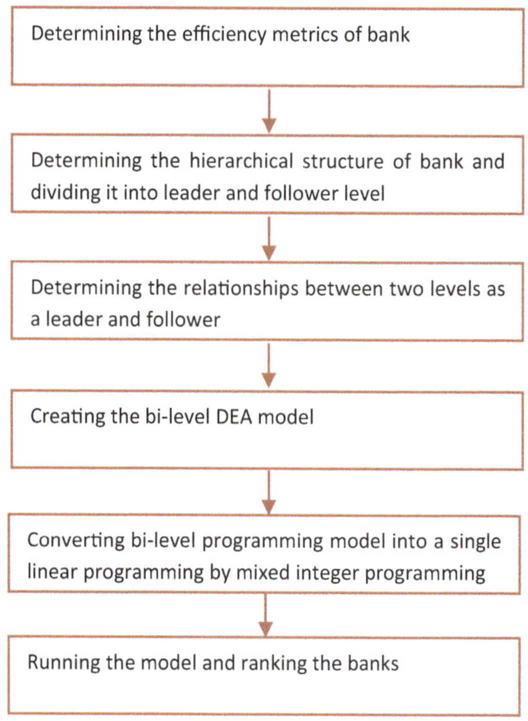

Fig. 2 Proposed performance evaluation model of bank

Empirical study

Banks are financial institutions that gather their assets from different resources and they made available for the sections that need liquidity. Therefore, banks are considered critical currents for each nation. Along with the emergence of private banks in the financial markets, Demand has dramatically increased for variety of banking services. Banks are looking for different procedures of functional improvement to attract customers, since they overtake one another to increase their contribution in market and Profitability; performance evaluation of banks is significantly considered among them and it is become one of the main bank manager activities.

Growth and economic development of every country requires moving additional funds of saver to the investors in a proper way. Availability of an extensive and efficient financial market which in financial resources is directed toward the best investment opportunities is critical. On the other hand, the maximum turnover in Iran is achieved through banking system. In addition, desired functioning of the banking system plays an important role to improve the economic actions.

However, it can be used the most traditional factors of an enterprise performance to evaluate a bank performance with modern procedures. One of the institutions investigating function of banking system is Iran Banking education Institute. Most performance evaluation criteria of it is quantitative, in addition it is considered financial standards. It investigates banks based on Liabilities, Assets, Number of bank branches, international and exchange Activities, The combination of human resources, profits and losses, facilities, overdue demands and the benefit of electronic banking technology. In this study, first it is determined the indicators through bank experts interview and library studies and criteria are extracted. These include: fixed assets, space, non-invest deposits, IT cost, profit, Deposit, Marketability employees and profitability employees. Usually the bank performance evaluation process involves the measurement of the performance of bank through its profitability levels. There are some other factors, in addition to profitable activities, that play an important role in bank performance evaluation process, including: bank location (ratio of residential to commercial region of the banks), cultural context of the region (whatever people tend to deposit their money as long term or current deposit), the services bank provides to compete with other banks and attract customers. These factors affect bank's performance indirectly, and usually are ignored in performance evaluation. So, in addition to the bank's profitability, intangible factors that indirectly affect bank's performance should also be taken into account while measuring the performance of banks. Thus, to measure all aspects of bank performance, the banks were evaluated in two levels so that all activities that affect the performance of banks would be taken into consideration.

Each bank studied in this research considers certain strategies according to the needs of people and market traction due to the geographic location of each region in terms of area, culture, population density, the ratio of commercial to residential area, etc. The banks offer different services to their clients by hiring the required number of employers, and with the deployment of fixed assets such as office supplies, computers, desks, chairs, banking software, IT costs, storage systems, and data protection. Banks collect their resources by attracting deposits from clients, and they begin profiting by investing those resources on various projects and giving loans.

As mentioned, banks can be viewed as an entity in which two decision makers in a hierarchical structure make decisions in turn so as to optimize their performance. Based on the proposal model in the previous section, the bi-level programming DEA model was applied to evaluate the performance of 15 branches of Iranian banks in 2011. Each branch had a certain amount of marketability and profitability level, where the marketability played as a leader and the profitability level served as a follower. Marketability level indicates the ability of the branch in marketing to collect funds from costumer in form of deposit by consuming the bank's resources. Profitability level indicates the ability of the branch to make profit by investing the deposits in other activities. The evaluation index system of bank branch performance evaluation problem is shown in Table 1.

The performance of each branch is characterized by seven variables where fixed assets (X^{D11}), space (X^{D12}), and non-invest deposits (Y^1) are associated with leader level, while IT cost (X^{D2}) and profit (Y^2) are associated with the follower level. Deposit (Y^{I1}) from leader to follower level is intermediate variable. Marketability employees (X^1) and profitability employees (X^2) are resource shared variables. The data related to these variables are presented in Table 2. Shared employees and the space costs are also given in Table 2 and due to cost nature of fixed assets, deposits, IT costs, and the correspond costs are assumed to be unit.

The first column of Table 3 depicts the results of performance evaluation using the CCR model, where each unit is as a black box with only a few input and output while the relationships between the elements are not considered. To calculate the performance of the banks in this way, four

Table 1 Evaluation index system

Factors	Name of index	Unit of index
Leader		
Shared input	Employees	Person
Direct inputs	Fixed assets	1,000,000,000 Riyal
	Space	m^2
Intermediate output	Deposit	10,000,000,000 Riyal
Output	Non-invest deposit	100,000,000 Riyal
Input costs	Employees	1,000,000 Riyal
	Space	1,000,000 Riyal
Follower		
Shared input	Employees	Person
Direct input	IT cost	100,000,000 Riyal
Intermediate input	Deposit	10,000,000,000 Riyal
Output	Profit	10,000,000,000 Riyal
Input cost	Employees	1,000,000 Riyal

Table 2 Data

	Leader employee	Fixed asset	Space	Non-invest deposit	Deposit	Follower employee	IT cost	Profit	Employees cost	Space cost
DMU 1	23	4.93	110	10.57	15.78	14	4.93	3.15	5.97	5.97
DMU 2	34	3.64	167.50	14.82	43.90	28	3.64	3.31	6.25	4.41
DMU 3	14	2.87	150	7.60	7.74	25	2.87	3.53	7.86	6.02
DMU 4	28	1.78	366.6	14.80	45.95	31	1.78	1.98	5.70	10.64
DMU 5	33	3.61	555	13.39	38.70	35	3.61	2.69	9.65	11.78
DMU 6	34	2.24	690	14.15	36.81	28	2.24	1.93	7.59	8.25
DMU 7	25	1.41	750.92	13.08	61.11	28	1.41	2.12	8.53	16.02
DMU 8	35	2.64	368.3	20.31	26.33	18	2.64	2.55	12.37	9.92
DMU 9	27	2.97	205	16.16	11.74	40	2.97	1.62	6.87	12.96
DMU 10	35	2.65	195	16.21	70.42	11	2.65	2.00	7.90	7.21
DMU 11	33	3.76	210	38.24	34.25	15	3.76	2.90	11.24	12.97
DMU 12	28	2.47	481	17.94	37.03	26	2.47	2.25	5.61	7.33
DMU 13	20	2.29	200	13.73	11.66	24	2.29	2.45	8.51	12.05
DMU 14	28	3.55	506	14.17	10.31	36	3.55	4.19	100.03	11.75
DMU 15	35	1.97	814	14.41	31.98	30	1.97	1.96	7.53	10.45

Table 3 Cost efficiency and reference set unit

	Classic DEA efficiency	Bank cost efficiency	Leader cost efficiency	Follower cost efficiency	Reference set for the leader	Reference set for the follower
DMU 1	1	0.66	0.61	1.00	10, 11	1
DMU 2	0.82	0.50	0.90	0.70	10, 11	3
DMU 3	1	0.48	0.37	1.00	10, 11	3
DMU 4	0.84	0.14	0.41	0.28	10, 11	1
DMU 5	0.58	0.08	0.24	0.35	10, 11	1
DMU 6	0.63	0.10	0.20	0.31	10, 11	1
DMU 7	1	0.05	0.24	0.31	10	1
DMU 8	0.75	0.14	0.41	0.92	10, 11	3
DMU 9	0.44	0.19	0.54	0.29	11	3
DMU 10	0.61	0.32	1.00	0.52	10	1
DMU 11	0.67	0.18	1.00	0.79	11	1
DMU 12	0.69	0.15	0.33	0.52	10, 11	3
DMU 13	0.86	0.21	0.44	0.71	11	3
DMU 14	0.92	0.09	0.20	0.83	11	3
DMU 15	0.69	0.07	0.15	0.30	10, 11	1

inputs (Employee, Fixed asset, Space, and IT Cost) are used to obtain one output (Profit). (these inputs and outputs belong to leader and follower levels.)

At this stage we ignored the intermediate variables that exist between the leader and follower. The results showed that three out of the 15 banks were efficient, and the other 12 banks had high performance compared to the performance of the bi-level models. This experiment shows how weak the DEA's classic model is in regards to separation ability. Compared to classic DEA models, bi-level DEA model has higher separation ability, mainly because it provides a tool to reveal the internal activities and relationships between activities within the black box and provides a detailed assessment of the existing subsystem performance.

Based on mixed integer single-level linear programming DEA model (7), cost efficiency scores for the bank branches and the followers and the leaders are obtained. Cost efficiency scores and the reference units correspond to the leader and the follower are given in following Table 3.

According to Table 3, there are no cost efficient banks, because banks are not performing efficiently at both decision levels. The tenth and eleventh banks are cost efficient at the leader, and the first and third banks are cost efficient at the follower level, and because the other player in these banks is inefficient, these banks systematically are termed as inefficient. Table 3 also indicates that the second bank is more efficient than the third, tenth, and eleventh banks which are only efficient in one level, so that, it can be the advantages of coordination between players. In addition, the classic DEA efficiency score and reference set for the leader and follower are shown in the Table 3. DMU 1, 3 and 7 are efficient in classic DEA model, DMU 10 and 11 are reference for the leader and DMU 1 and 3 are reference for the follower.

Conclusion

In the real world, banks have a decentralized structure in which multi decision makers in a hierarchical structure makes decision in turn or at the same time to optimize their objective function. In this rapidly changing world, responding to change needs the ability of management to identify the location of inefficiency. Thus, efficiency analysis can be a source of competitive advantages (Avkiran 2009). DEA is a better way to measure efficiency since it requires no prior assumption on the specification of the efficient frontier. In this paper, we applied bi-level programming DEA model with two inter-related decision makers in a decentralized decision structure to evaluate the performance of 15 Iranian bank branches with one level correspond to a leader while the other a follower. Bi-level programming DEA model proposed by (Wu 2010) can provide insight and detailed information to bank managers when measuring the efficiency of a bank with Stackelberg-game relationships. The results obtained from bi-level programming DEA model have a strong discriminating power due to considering internal operations in the banking chain.

Further researchers can develop our model in the three-level or multi-level. Since the multi-level DEA models are NP-hard problem, we proposed to use heuristic model to solve them. Finally, we propose to further researcher to create a benchmark unit in bi-level DEA model and determine VRS in efficiency model.

References

Abtahi AR, Khalili-Damghani K (2011) Fuzzy data envelopment analysis for measuring agility performance of supply chains. Int J Model Oper Manag 1(3):263–288

Alimardani M, Jolai F, Rafiei H (2013) Bi-product inventory planning in a three-echelon supply chain with backordering, Poisson demand, and Limited warehouse apace. J Ind Eng Int 9:22

AriaNezhad MG, Makuie A, Khayatmoghadam S (2013) Developing and solving two-echelon inventory system for perishable items in a supply chain: case study (Mashhad Behrouz Company). J Ind Eng Int 9:39

Arora SR, Gupta R (2009) Interactive fuzzy goal programming approach for bi-level programming problem. Eur J Oper Res 194:368–376

Avkiran NK (2009) Opening the black-box of efficiency analysis: an illustration with UAE banks. Omega 37:930–941

Babazadeh R, Razmi J, Ghodsi R (2012) Supply chain network design problem for a new market opportunity in an agile manufacturing system. J Ind Eng Int 8:19

Bahri M, Tarokh MJ (2012) A seller-buyer supply chain model with exponential distribution lead time. J Ind Eng Int 8:13

Bhattacharya A, Lovell CAK, Sahay P (1997) The impact of liberalization on the productive efficiency of indian commercial banks. Eur J Oper Res 98(2):332–345

Camanho AS, Dyson RG (1999) Efficiency, size, benchmarks and targets for bank branches: an application of data envelopment analysis. J Oper Res Soc 50:903–915

Camanho AS, Dyson RG (2005) Cost efficiency measurement with price uncertainty: a DEA application to bank branch assessments. Eur J Oper Res 161(2):432–446

Charnes A, Cooper WW, Rhodes E (1978) Measuring the efficiency of decision making units. Eur J Oper Res 2:429–444

Cook WD, Hababou M (2001) Sales performance measurement in bank branches. Omega. Int J Manag Sci 29(4):299–307

Cooper WW, Seiford LM, Tone K (2007) Data envelopment analysis: a comprehensive text with models, applications, references and DEA-solver software. Springer, New York

Elyasiani E, Mehdian SM (1990) A nonparametric approach to measurement of efficiency and technological change: the case of large US banks. J Financ Serv Res 4(2):157–168

Fukuyama H, Weber WL (2010) A slacks-based inefficiency measure for a two-stage system with bad outputs. Omega 38:398–409

Hongjie L, Ruxian L, Zhigao L, Ruijiang W (2011) Study on the inventory control of deteriorating items under VMI model based on bi-level programming. Expert Syst Appl 38:9287–9295

Huijun S, Ziyou G, Jianjum W (2008) A bi-level programming model and solution algorithm for the location of logistics distribution centre. Appl Math Model 32:610–616

Kao C, Hwang SN (2008) Efficiency decomposition in two-stage data envelopment analysis: an application to non-life insurance companies in Taiwan. Eur J Oper Res 185(1):418–429

Kao C, Liu ST (2004) Predicting bank performance with financial forecasts: a case of Taiwan commercial banks. J Bank Finance 28(10):2353–2368

Khalili-Damghani K, Hosseinzadeh Lotfi F (2012) Performance measurement of police traffic centers using fuzzy DEA-based Malmquist productivity index. Int J Multicrit Decis Mak 2(1):94–110

Khalili-Damghani K, Sadi-Nezhad S, Hosseinzadeh-Lotfi F (2014) Imprecise DEA models to assess the agility of supply chains. In: Kahraman C, Öztayşi B (eds) Supply chain management under fuzziness, Studies in fuzziness and soft computing. Springer, Berlin, Heidelberg, vol 313, pp 167–198

Khalili-Damghani K, Taghavifard M (2012) A three-stage fuzzy DEA approach to measure performance of a serial process including JIT practices, agility indices, and goals in supply chains. Int J Serv Oper 13(2):147–188

Khalili-Damghani K, Taghavifard B (2013) Sensitivity and stability in two-stage models with fuzzy data. Int J Oper Res 17(1):1–37

Khalili-Damghani K, Tavana M (2013) A new fuzzy network data envelopment analysis model for measuring the performance of agility in supply chains. Int J Adv Manuf Technol 69(1–4):291–318

Khalili-Damghani K, Taghavifard M, Olfat L, Feizi K (2011) A hybrid approach based on fuzzy DEA and simulation to measure the efficiency of agility in supply chain: real case of dairy industry. Int J Manag Sci Eng Manag 6(3):163–172

Khalili-Damghani K, Taghavi-Fard M, Abtahi AR (2012) A fuzzy two-stage DEA approach for performance measurement: real case of agility performance in dairy supply chains. J Appl Decis Sci 5(4):293–317

Khalili-Damghani K, Tavana M, Haji-Saami E (2015) A data envelopment analysis model with interval data and undesirable output for combined cycle power plant performance assessment. Expert Syst Appl 42(2):760–773

Lachhwani K, Poonia MP (2012) Mathematical solution of multilevel fractional programming problem with fuzzy goal programming approach. J Ind Eng Int 8:16

Li Y, Chen Y, Liang L, Xie J (2012) DEA models for extended two-stage network structure. Omega 40:611–618

Liang L, Cook WD, Zhu J (2008) DEA models for two-stage processes: game approach and efficiency decomposition. Nav Res Logist 55:643–653

Momen O, Kimiagari A, Noorbakhsh E (2012) Modeling the operational risk in Iranian commercial banks: case study of a private bank. J Ind Eng Int 8:15

Paradi CJ, Rouatt S, Zhu H (2011) two stage evaluation of bank branch efficiency using data envelopment analysis. Omega 39:99–109

Roghanian E, Sajjadi SJ, Aryanezhad MB (2007) A probabilistic bi-level linear multi-objective programming problem to supply chain planning. Appl Math Comput 187:786–800

Ryu J-H, Dua V, Pistikopoulos EN (2004) A bi-level programming framework for enterprise-wide process networks under uncertainty. Comput Chem Eng 28:1121–1129

Sakawa M, Nishizaki I (2009) Cooperative and non-cooperative multi-level programming, concepts, system, algorithms & case studies. Springer, New York

Sakawa M, Nishizaki I, Uemura Y (2002) A decentralized two-level transportation problem in a housing material manufacturer: interactive fuzzy programming approach. Eur J Oper Res 141:167–185

SanieeMonfared MA, Safi M (2013) Network DEA: an application to analysis of academic performance. J Ind Eng Int 9:15

Seiford LM, Zhu J (1999) Profitability and marketability of the top 55 US commercial banks. Manage Sci 45(9):1270–1288

Sexton TR, Lewis HF (2003) Two-stage DEA: an application to major league baseball. J Prod Anal 19:227–249

Stackelberg HV (1952) The theory of the market economy. Oxford University Press, Oxford

Tavana M, Khalili-Damghani K (2014) A new two-stage Stackelberg fuzzy data envelopment analysis model. Measurement 53:277–296

Tavana M, Khalili-Damghani K, Sadi-Nezhad S (2013) A fuzzy group data envelopment analysis model for high-technology project selection: a case study at NASA. Comput Ind Eng 66(1):10–23

Tavana M, Khalili-Damghani K, Rahmatian R (2014) A hybrid fuzzy MCDM method for measuring the performance of publicly held pharmaceutical companies. Ann Oper Res 226(1):589–621

Tohidi GH, Khodadadi M (2013) Allocating models for DMUs with negative data. J Ind Eng Int 9:16

Tone K, Tsutsui M (2009) Network DEA: a slacks-based measure approach. Eur J Oper Res 197(1):243–252

Wang H (1997) Contingent valuation of environmental resources: a stochastic perspective, Ph.D. Dissertation. Department of Environmental Sciences and Engineering, School of Public Health, University of North Carolina at Chapel Hill, Chapel Hill, North Carolina, USA

Wu D (2010) Bi-level programming data envelopment analysis with constrained resource. Eur J Oper Res 207:856–864

11

Prioritizing critical success factors for reverse logistics implementation using fuzzy-TOPSIS methodology

Saurabh Agrawal[1] · Rajesh K. Singh[1] · Qasim Murtaza[1]

Abstract Electronics industry is one of the fastest growing industries in the world. In India also, there are high turnovers and growing demand of electronics product especially after post liberalization in early nineties. These products generate e-waste which has become big environmental issue. Industries can handle these e-waste and product returns efficiently by developing reverse logistics (RL) system. A thorough study of critical success factors (CSFs) and their ordered implementation is essential for successful RL implementation. The aim of the study is to review the CSFs, and to prioritize them for RL implementation in Indian electronics industry. Twelve CSFs were identified through literature review, and discussion with the experts from the Indian electronics industry. Fuzzy-Technique for Order Preference by Similarity to Ideal Solution (TOPSIS) approach is proposed for prioritizing these CSFs. Perusal of literature indicates that fuzzy-TOPSIS has not been applied earlier for prioritization of CSFs in Indian electronics industry. Five Indian electronics companies were selected for evaluation of this methodology. Results indicate that most of the identified factors are crucial for the RL implementation. Top management awareness, resource management, economic factors, and contracts terms and conditions are top four prioritized factor, and process capabilities and skilled workers is the least prioritized factor. The findings will be useful for successful RL implementation in Indian electronics industry.

✉ Saurabh Agrawal
agrawals.iit@gmail.com

Rajesh K. Singh
rksdce@yahoo.com

Qasim Murtaza
qasimmurtaza@gmail.com

[1] Mechanical, Production and Industrial Engineering Department, Delhi Technological University, Delhi 110042, India

Keywords Reverse logistics · Critical success factors · Indian electronics industry · Environment · Fuzzy TOPSIS

Introduction

The last decade has seen remarkable growth in the Indian economy due to economic liberalization started in early nineties. Also, during the post liberalization era, electronics industry is growing very fast in India. Due to the presence of global electronics companies and tremendous expansion of telecommunication, and information technology, Indian markets are flooded with electronics goods. This results in a new kind of waste known as electronics waste or e-waste. E-waste contains both toxic and valuable materials. The fraction including iron, copper, aluminium, gold and other metals in e-waste is over 60 %, while plastics account for about 30 % and the hazardous pollutants comprise only about 2.70 % (Widmer et al. 2005). This large percentage of valuable materials offers a business opportunity to recover value from the e-waste. It may be even profitable to collect and process e-waste. The hazardous pollutants though small in quantity but contain highly toxic materials like lead, mercury, arsenic, cadmium, chromium and many more. When these materials enter into the environment through land filling, causes damage to human health and the environment (Lee et al. 2000). E-waste can be well managed through development of RL system. Cheng and Lee (2010) found that effective RL focuses on the backward flow of materials to maximize value from returned items and guarantee their proper disposal. However, many companies are not yet ready to implement RL including Indian companies. Indian electronics industry has been selected for the study because it has a high consumption volume, and major source of e-waste generation. Also, this is one of the few sectors which come under e-waste regulations. A thorough study of CSFs and their ordered implementation is essential for successful RL implementation. Most of the researches on the CSFs of RL implementation are concentrated on developed countries, with relatively little attention being given to developing countries (Abdulrahman et al. 2014). The major intention of this study is to understand various CSFs for RL implementation in Indian electronics industry. The identification and prioritization of these factors will help the researchers and the managers in strategic decision making for RL implementation. After review of literature on RL and the opinion of experts from Indian electronics industry, 12 CSFs factors of RL implementation were identified. The experts were asked to rate each of these 12 factors in terms of their importance. A decision matrix was developed from these responses which are used in the application of fuzzy-TOPSIS methodology for prioritizing CSFs.

The main objectives of this paper are:

1. to study the literature available on CSFs for RL implementation;
2. to identify the CSFs which are important in Indian electronics industry;
3. to find out the relative importance of these factors; and
4. to discuss the managerial implications of this research.

The remaining of this paper is structured as follows: Section "Literature review" comprises a literature review on RL. CSFs for RL implementation are identified in Section "Identification of CSFs for RL implementation". Step by step fuzzy-TOPSIS approach with case example of Indian electronics industry is discussed in Section "Fuzzy-TOPSIS methodology". Section "Results and discussion", summarizes all the findings and analyzed them in context of Indian electronics industry. Finally, Section "Conclusion" concludes the study along with future outlook and limitations of this research.

Literature review

Growing concern for the environment and government regulations in many countries has increased the interest in reverse flows, which has become the subject of attention over the last decade (Fleischmann et al. 1997). RL is the process of planning, implementing, and controlling the efficient, cost effective flow of raw materials, in-process inventory, finished goods, and related information from the point of consumption to the point of origin for the purpose of recapturing or creating value or proper disposal (Rogers and Tibben-Lembke 1999). Srivastava (2008) explained the flow of RL showing all the basic RL activities. The RL activities include collection, grading, reprocessing and redistribution (Fleischmann 2003). A well-managed RL can provide important cost savings in procurement, disposal, inventory carrying and transportation (Kannan et al. 2009).

Most of the organizations in the world are presently exploring the RL to make it profitable business. RL has been found to be beneficial to some of the industries in terms of the improvement of profits. Some of the organizations like HP, Dell have implemented RL for competitive advantage. Jayaraman and Luo (2007) mentioned that Kodak has successfully implemented recycling facilities and is able to reuse up to 86 % of a camera's parts. Similarly other leading manufacturers such as Canon and Xerox have also attained remanufacturing rates of nearly 90 %. In fact, implementing RL programs to reduce, reuse, and recycle wastes produces tangible and intangible value and may lead to better corporate image (Carter and Ellram 1998). Lau and Wang (2009) studied the electronics industry in China, explored the problems encountered in the implementation of RL. Janes et al. (2010) worked on implementation of reserve logistics in consumer electronics industry of USA. Jayaraman et al. (2003) discussed RL systems for recycling and reusing of beverage containers. Biehl et al. (2007) worked on carpet industry, Bernon et al. (2011) worked on retail industry, Gonzalez-Torre et al. (2004) worked on bottling, and Gonzalez-Torre and Adenso-Diaz (2006) worked on packaging firms, Rahman and Subramanian (2012) worked on manufacturing, Clottey et al. (2013) worked on remanufacturing, Vishkaei et al. (2014) worked on returnable defective items, Khalili-Damghani and Najmodin (2014) worked on automobile industry; are some of the examples of previous research on RL implementation.

RL implementation involves many financial and operational issues which determine the productivity and performance of the RL. A critical analysis of the variables affecting RL and their mutual interactions can be a valuable source of information for the RL implementation (Ravi et al. 2005). Rogers and Tibben-Lembke (1999) suggested that there are a number of factors affecting RL practices. The presence or absence of these factors can become drivers or barriers to RL implementation in an industry. Several conceptual models have been developed for assisting in design and implementation of RL system. Brito and Dekker (2002) differentiate two types of factors, internal and external factors for existence of reverse flows. Carter and Ellram (1998) identified internal and external factors to examine whether a company is reactive, proactive, or value-seeking in RL implementation. They considered the internal factor of policy entrepreneur and two external factors of government regulations, and customer demands as the main factors of RL systems. Holt and Ghobadian (2009) suggested that the internal factors such as culture of organization, internal resources, and operation management control practices have positive correlation to environmental thoughts in the green supply chain. Stock (1998) mentioned that factors related to management and control, measurement, and finance are crucial for the successful RL implementation. Dowlatshahi (2000) focused on internal strategic and operational issues that may require consideration in RL implementation. Strategic factors include legislative concerns, environmental concerns, customer service, quality, and strategic costs while operational factors include cost benefit analysis, transportation, warehousing, supply management, remanufacturing or recycling and packaging. Later on Dowlatshahi (2005) suggested a five-factor strategic framework for successful RL implementation. He proposed these five factors as strategic costs, strategic quality, customer service, environmental concerns, and legal concerns. In a survey of consumer electronics, Janes et al. (2010) identified main facilitators of RL as top management awareness, strategic partnerships with supply chain partners, detailed insight into cost and performance, reclaiming value from returned products, and capability to put products rapidly back into the market. Rahman and Subramanian (2012) worked on end of life computers and found customer demand as one of the major factors. These factors have great impact on availability of resource, coordination and integration of recycling tasks, volume and quality of recyclable materials. Most of these factors are described from different perspectives, i.e. for particular sector or industry of a country or region. Khalili-Damghani et al. (2015) considered the factors such as green image, flexibility, quality, responsiveness, expenses, and value recovered for studying the relationship between supply chain performance and RL. The literature review of RL implementation across the industry is summarized in Table 1. It is evident from Table 1 that CSFs vary from industry to industry and country to country but many of them are common to all of these studies. A lot of work has been done in electronics industry in many countries and different methodologies have been used for prioritizing CSFs. Indian electronics industry is more dependent on the import of components and products for fulfilling growing demand of electronics goods. So CSFs may be different from other developing country like china which has own manufacturing facilities. Remanufacturing of products may not be economical (because of transportation cost) and technically feasible within the country. Therefore, it is important to identify the CSFs addressing these issues. No study was found on identification and prioritization of CSFs for RL implementation in Indian electronics industry. Also, Fuzzy-TOPSIS methodology is first time being utilized for prioritizing CSFs for Indian electronics industry.

Identification of CSFs for RL implementation

Several useful factors for RL implementation are pointed out in the literature review discussed in last section. Many CSFs are common to all of these studies and these factors can be utilized as base for discussion with expert from Indian electronics industry. Twelve CSFs were identified after pertinent literature review including studies discussed in "Introduction" and discussion with the number of experts from the Indian electronics industry. These factors are shown in Fig. 1 and are explained as follows.

Legislation

Legislation refers to regulations or acts passed by the government authorities to minimize the effect of end of life products on environment. Ravi et al. (2005) define legislation as one of the determinants of RL. In fact, focusing on environmental concerns is partly enforced by government legislation (Prendergast and Pitt 1996).

Recently, Government of India has instituted e-waste (Management and Handling) Rules, 2011 which have come into effect from May 2012. Experts mentioned that sooner or later Indian electronics manufacturer will have to comply with these e-waste management regulation.

Economic factors

Economics is seen as one of the driving forces to RL relating all the recovery options, where the company receives both direct as well as indirect economic benefits (Ravi et al. 2005). In a survey of mobile manufacturing company in Hong Kong, Chan and Chan (2008) found that majority of the returned products add extra value to the company. The recovery of the products for remanufacturing, repair, reconfiguration, and recycling can lead to profitable business opportunities (Andel 1997). Guide and Wassenhove (2003) discussed an example of the US firm ReCellular, which had gained economic advantage by refurbishing the cell phones.

Experts comment that RL practices are assumed to be cost driven activity and the companies in India are waiting for the response from each other for adopting these practices. There is also need of analyzing indirect benefits like green image, tax benefits, preference for projects in government, and image of environmentally conscious organization in Indian electronics industry.

Table 1 Literature review of reverse logistics implementation

References	Factors	Methodology	Sector	Country
Carter and Ellram (1998)	Customer demand, regulations, resource constraints, policy entrepreneur		General	
Knemeyer et al. (2002)	Costs, quality, customer service, environment, legislation, transportation, warehousing, recycling, remanufacturing	Qualitative techniques	End of life computers	USA
Ravi et al. (2005)	Economic factors, environmental factors, corporate citizenship	ANP	End of life computers	USA
Dowlatshahi (2005)	Strategic costs, Strategic quality, Customer service, Environmental concerns, Legal concerns	Grounded theory approach	Electronics	USA
Lau and Wang (2009)	Economic policies, environment protection, customer service, publicity and knowledge of reverse logistics	Case study approach	Electronics	China
Janes et al. (2010)	Top management awareness, strategic partnerships, performance visibility, reclaiming value from returns, products rapidly back into the market	Empirical study	Electronics	Europe
Sharma et al. (2011)	Lack of awareness about reverse logistics, management inattention, financial constraints, legal issues	Interpretive Structural Modeling	Indian industry	India
Rahman and Subramanian (2012)	Legislation, customers, strategic cost, environmental concerns, volume and quality, incentives, resources, and integration and coordination	DEMATEL	EOL-computers	Australia
Chiou et al. (2012)	Economic needs, environmental needs, social needs, recycled volumes, recycling costs, increase of sales volume for new product	FAHP	Electronics	Taiwan
Tyagi et al. (2012)	Facilities, handling, ease of access, information	Empirical study	Hospitals	Canada
Gonzalez-Torre et al. (2012)	Customer, reluctance on the part of social actors, lack of know-how, top management commitment, information systems, financial and human resources	Structural Equation Modelling	Automotive	Spain
Jindal and Sangwan (2013)	Economical, environmental, and social drivers	Interpretive Structural Modelling	Indian industry	India
Kannan et al.(2014)	Extended producer, responsibility, codes of conduct, and resource scarcity	Interpretive Structural Modelling	EOL tire Manufac.	India
Mittal and Sangwan (2013)	Legislation, public pressure, competitiveness, customer demand, top management commitment, technology, organizational resources	Interpretive Structural Modelling	Indian industry	India

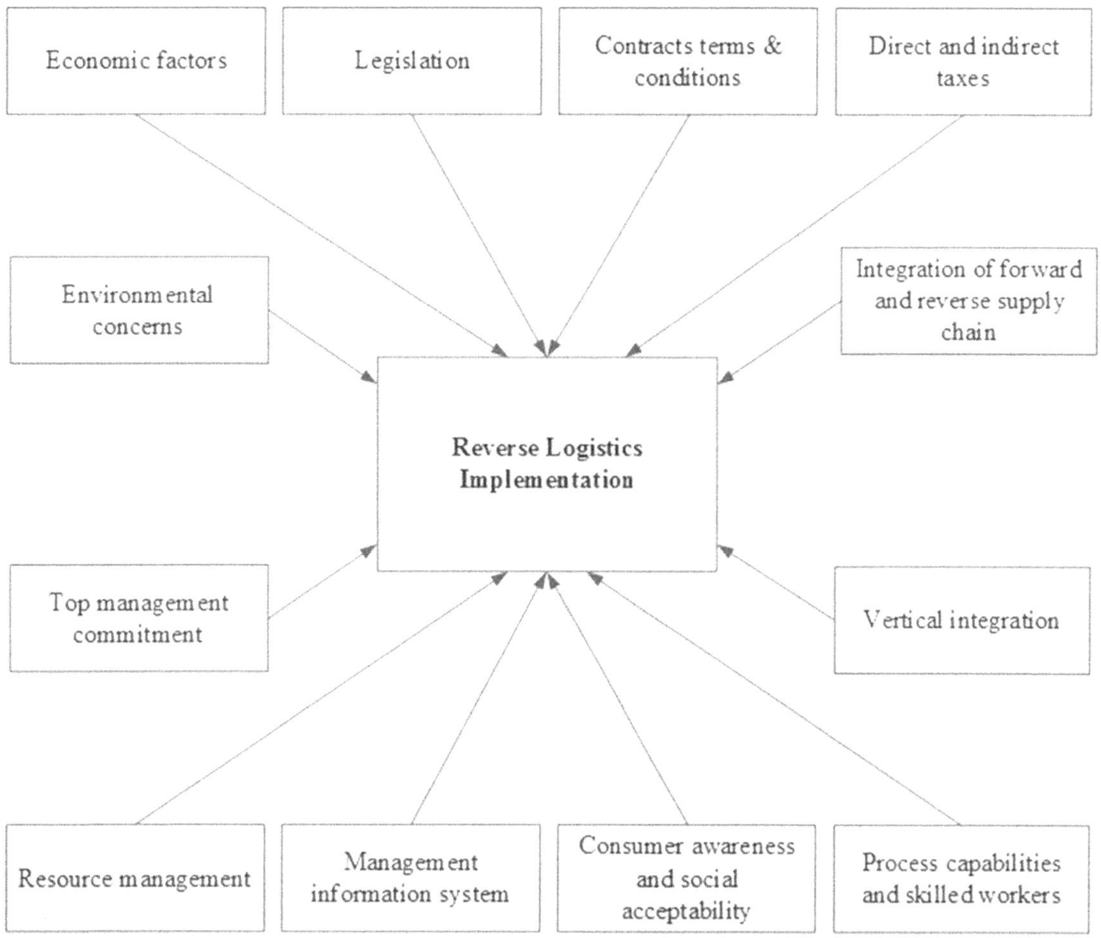

Fig. 1 Identified CSFs for Indian electronics industry

Fig. 2 Linguistics scales and triangular fuzzy numbers

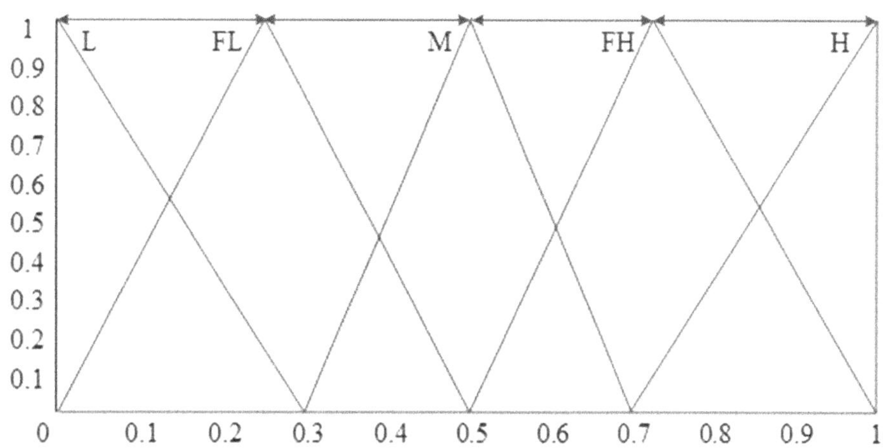

Environmental concerns

Environmental concerns are significant force shaping the economy, as well as one of the most important issues faced by businesses (Murphy and Poist 2003). Many companies have focused on RL operations because of environmental reasons (Rogers and Tibben-Lembke 1999).

Many power projects and real state projects have been delayed because of environmental clearances from Government of India. In last few years, Government of

India has taken several initiatives because of environmental concerns. This is an indication for other industries to focus on environmental concerns.

Top management awareness

Top management awareness is very crucial for the success of RL implementation. A sincere and committed effort from the top management is essential for successful deployment of RL programs (Carter and Ellram 1998). Mintzberg (1973) stated that top management awareness is the dominant factor of corporate endeavors.

Top management awareness is needed to provide clear vision and value to RL programs. Top management awareness motivates employees and ensures full support from seniors.

Resource management

Miller and Shamsie (1996) categorized the resources into property-based resources and knowledge-based resources. Property-based resources including the physical facility, automated machines and equipment, financial and human resource are regarded as critical indicators of the competitiveness (Das and Teng 2000). Knowledge-based resources including managerial resources and technology are also critical for the success of RL. Availability and effective utilization of both types of resources is essential for exploring the true value potential of RL.

Human resource is crucial for RL implementation. Companies encounter challenges while implementing RL because of lack of knowledge of RL among their employees. In India, there are very few RL experts available in the market. Companies, willing to adopt RL will have to develop their own expertise through various education and training programs for promoting the environmental awareness in their organization.

Management information system

Information support is one of the important factors for developing linkages to achieve efficient RL operations (Daugherty et al. 2005). Efficient information systems are needed for individually tracking and tracing the information on reverse flows. Information and communication technologies assume tremendous importance in RL, which are needed to process and transmit information (Brito and Dekker 2002). IT enablement is necessary and one of the important factors for effective communication (Kumar et al. 2013).

Availability of prompt and accurate information may help managers in achieving operational efficiency in managing their RL. This is an important tool but cost is a concern. Integration with the current management information system is also crucial for successful implementation.

Contract terms and conditions

Contract terms and conditions with suppliers are one of the most important factors for RL implementation. Most of the parts and components are outsourced by the electronics companies in India. Legal terms and conditions with the contractors are important. Companies may enforce regulatory requirements in the contacts to meet the criteria for parts and components from environmental perspectives. Contractor's ability to meet regulatory criteria and corresponding costs are still to be analyzed.

Direct and indirect taxes

According to Sharma et al. (2011), complex flows of goods as well as the diverse bought-in services engrained in the reverse chain create a high degree of tax complexity and lead to unexpected tax exposures and costs.

Direct and indirect taxes are very important factor for the financial consideration. Tax structure is very complex because of involvement of import–export, and no special consideration is given to the remanufactured products in India. In fact, direct and indirect taxes for remanufactured products need to be relooked for the promotion of environmental friendly practices.

Integration of forward and reverse supply chain

Integration of forward and reverse supply chain implies simultaneous management of material, information, and monetary flows as suggested by Fleischmann (2001) and by Tibben-Lembke and Rogers (2002). According to Mehrbod et al. (2014) the integration of forward and RL has attracted growing attention with the stringent pressures of customer expectations, environmental concerns, and economic factors. Greater resource utilization can be achieved through integration of forward and reverse supply chain.

In general, responsibility of RL is assigned to the supply chain department rather than having separate department. Therefore integration of forward and reverse supply chain plays an important role because same people work on both forward and reverse supply chains. One of the big concerns is the impact of reverse supply chain on forward supply chain. Experts fear that integration may disturb the whole of the forward supply chain. Employee's awareness and motivation for the change is essential for successful integration.

Joint consortium

Experts' opinion that joint consortium may be one of the options for handling returns just like telecommunication tower sharing in India. Earlier companies had their own telecommunication towers. Later on companies started tower sharing and now one tower in a particular region is being used by many companies reducing their investments and operational costs. There is need of exploring such kind of business model, for example common collection centre for all used cellular phones may reduce collection cost and also, economies of scale can be achieved. Joint consortium may be helpful, particularly in case of recycling where high product volume is important.

Process capabilities and skilled workers

Experts state that process capabilities and skilled workers are very important for successful implementation. Workers must be skilled enough to work simultaneously both on manufacturing and re-manufacturing efficiently. Machines, equipments and tools must also be developed to perform both the operations simultaneously as much as possible. This is important for effective utilization of resources because of uncertainty of product returns.

Consumer awareness and social acceptability

Sharma et al. (2011) suggested that the awareness of RL could bring economic benefits by recovery of the returned product for use. Research suggests that there is an increasing customer demand for green products and for organizations to engage in environmental supply chain practices (New et al. 2000).

Social acceptability of remanufactured products among the Indian consumers and society is crucial for the success of RL. Most of the remanufactured products in India are sold in secondary market at lower prices because these products are purchased by lower income group. Consumer awareness and social acceptability will not only increase the product returns but also will motivate them to buy refurbished or remanufactured products at reasonable price.

Fuzzy-TOPSIS methodology

There are various methods used to prioritize CSFs. Multiple criteria decision making (MCDM) is one of the powerful tools widely used for dealing with unstructured problems containing multiple and potentially conflicting objectives (Lee and Eom 1990). A number of approaches have been developed for solving MCDM problems such as analytical hierarchy process (AHP), data envelopment analysis (DEA), and TOPSIS. These traditional MCDM approaches measure the alternative ratings and weights of the criteria in crisp or precise numbers which depends upon decision makers preferences (Wang and Lee 2009). The TOPSIS method was developed by Hwang and Yoon (1981) to provide solutions of the MCDM the problems. Kim et al. (1997) stated the advantages of TOPSIS as follows:

- A sound logic that represents the rationale of human choice;
- a scalar value that accounts for both the best and worst alternatives simultaneously; and
- a simple computation process that can be easily programmed into a spreadsheet.

TOPSIS is useful particularly when there are a large number of alternatives and criteria. In such cases, methods like AHP which require pair wise comparison are avoided. Also, TOPSIS has the fewest rank changes reversals when an alternative is added or removed in comparison to other MCDM methods (Zanakis et al. 1998). These advantages make TOPSIS a major MCDM technique as compared with other related techniques such as analytical hierarchical process (AHP) and ELECTRE. The traditional TOPSIS method considers ratings and weights of criteria's in crisp numbers. However, crisp data are inadequate to represent the real life situation since human judgements are vague and cannot be estimated with exact numeric values. In such situations, the fuzzy set theory is useful to capture the uncertainty of human judgments. Zadeh (1965) first introduced fuzzy set theory into MCDM including TOPSIS as an approach for effectively working with the vagueness and ambiguity of the human judgements. In fuzzy TOPSIS, all the ratings and weights are defined by means of linguistic variables. There are following two main characteristics of fuzzy systems given by Kahraman et al. (2007):

- Fuzzy systems are suitable for uncertain or approximate reasoning, especially for the system with a mathematical model that is difficult to derive; and
- Fuzzy logic allows decision-making with estimated values under incomplete or uncertain information.

Because of all these advantages, fuzzy logic has been combined and used along with TOPSIS known as fuzzy-TOPSIS methodology. The fuzzy-TOPSIS methodology has been used to solve many problems ranging from facility location selection (Chu 2002), robot selection (Chu and Lin 2003), selection and ranking of the most suitable third party logistics service provider (Bottani and Rizzi 2006) to service quality in airline service (Nejati et al. 2009), competitive advantage of shopping web-sites (Sun and Lin 2009), e-sourcing problem (Singh and

Table 2 Linguistics terms and corresponding fuzzy number

Linguistic term	Fuzzy number
Low	(0.0,0.1,0.3)
Fairly low	(0.1,0.3,0.5)
Medium	(0.3,0.5,0.7)
Fairly high	(0.5,0.7,0.9)
High	(0.7,0.9,1.0)

Benyoucef 2011), maintenance problem (Ding and Kamaruddin 2014), traffic police center performance (Sadi-Nezhad and Khalili-Damghani 2014), and sustainable project selection (Khalili-Damghani and Sadi-Nezhad 2014). Fuzzy-TOPSIS methodology based on the technique introduced by Chen (1997) is selected for this study. The technique given by Chen (1997) is selected for prioritizing CSFs because this technique gives better result in comparison to other techniques. The following steps of fuzzy TOPSIS are used for the proposed research.

Step 1 Collect the required data containing linguistics terms. A proper scale must be chosen to represent the data accurately and more precisely. Respondents must be asked to choose the best alternative among the linguistics terms for a given question. Linguistics terms must be converted into the fuzzy number. For example, triangular fuzzy numbers are used for the study and a 5-point scale having the linguistic terms low (L), fairly low (FL), medium (M), fairly high (FH), and high (H); are selected as shown in Fig. 2. Triangular fuzzy numbers are used because it is intuitively easy for the respondents to use and calculate.

The fuzzy number of each linguistic term is determined with the help of Fig. 2. Fuzzy numbers for the selected linguistics terms are presented in Table 2.

Step 2 The TOPSIS method evaluates the following fuzzy decision matrix

$$D = \begin{bmatrix} x_{11} & x_{12} & \cdots & x_{1j} & \cdots & x_{1n} \\ x_{21} & x_{22} & \cdots & x_{2j} & \cdots & x_{2n} \\ \cdots & \cdots & \cdots & \cdots & \cdots & \cdots \\ x_{i1} & x_{i2} & \cdots & x_{ij} & \cdots & x_{in} \\ \cdots & \cdots & \cdots & \cdots & \cdots & \cdots \\ x_{m1} & x_{m2} & \cdots & x_{mj} & \cdots & x_{mn} \end{bmatrix}, \quad (1)$$

where x_{ij} $(= (a_{ij}, b_{ij}, c_{ij}))$ is a fuzzy number corresponding to the linguistic term assigned by the *i*th Decision Maker (DM) to the *j*th factor. $i = 1, 2, \ldots, m$ are the number of DMs and $j = 1, 2, \ldots, n$ are the number of factors (CSFs).

Step 3 This step includes neutralizing the weight of decision matrix and generating fuzzy un-weighted matrix (R).

To generate R, following relationship can be applied.

$$R = [r_{ij}]_{m\times n}, \quad r_{ij} = \left(\frac{a_{ij}}{c_j^*}, \frac{b_{ij}}{c_j^*}, \frac{c_{ij}}{c_j^*}\right), \quad (2)$$

where $c_j^* = \max_i c$

Step 4 Calculate the weighted normalized decision matrix

$$V = [v_{ij}]_{m\times n}; \quad i = 1, 2, \ldots, m; j = 1, 2, n \quad (3)$$

The weighted normalized value v_{ij} is calculated as

$$v_{ij} = r_{ij} * w_j; \quad (4)$$

where w_j is the weight given to each decision maker. $w_j = (1, 1, 1, 1, 1)\forall j \in n$, because all the DMs are considered to have same weight for this study.

Step 5 Determine the ideal and negative-ideal solution for the CSFs

$$A^* = \{v_1^*, v_2^*, \ldots v_n^*\} \quad (5)$$

$$A^- = \{v_1^-, v_2^-, \ldots v_n^-\} \quad (6)$$

Since the positive and negative ideas introduced by Chen (1997) are used for the research. The following terms are used for ideal and negative ideal solution.

$$v_j^* = (1, 1, 1) \quad (7)$$

$$v_j^- = (0, 0, 0) \quad (8)$$

Step 6 Calculate the sum of distances from positive and negative ideal solution for each factor.

$$D_j^* = \frac{\sum_{i=1}^{m} d(v_{ij} - v_i^*)}{m}, \quad j = 1, 2, \ldots, n \quad (9)$$

$d(v_{ij} - v_i^*)$ is the distance between two fuzzy numbers which can be calculated using the vector algebra. For example distance between two numbers $A1(a_1, b_1, c_1)$ and $A2(a_2, b_2, c_2)$ can be calculated as

$$d(A1 - A2) = \sqrt{\frac{1}{3}\left[(a_2 - a_1)^2 + (b_2 - b_1)^2 + (c_2 - c_1)^2\right]}.$$

Similarly, the separation from the negative ideal solution is given as

$$D_j^- = \frac{\sum_{i=1}^{m} d(v_{ij} - v_i^-)}{m}, \quad j = 1, 2, \ldots, n \quad (10)$$

Step 7 Calculate the relative closeness to the ideal solution. The relative closeness with respect to A^* is defined as

$$C_j = D_j^- / (D_j^* + D_j^-) \quad (11)$$

Step 8 Prioritize the preference order based on the order of the values of C_j.

Application of the fuzzy-TOPSIS methodology

The fuzzy-TOPSIS methodology, presented in this research paper has been evaluated in context of Indian electronics industry. Five experts from electronic companies participated in this study. Name of the companies are not mentioned because of confidentiality of data. Profile of the decision makers and their respective organization is given as follows:

First decision maker (DM1) is a supply chain manager in a mobile manufacturing company, ABC-1 limited which is interested in RL implementation. DM1 has responsibility of developing a RL system for the company. The company is a pioneer in the manufacturing of mobile phones. The company has received many awards for best quality and management practices. The company has annual turnover of approximately USD 200 million from its business in India. In India, the company has a mobile handset manufacturing facility in Chennai. At present the company has approximately 110,000 outlets including 50,000 stores selling company's product exclusively. The company has outsourced its forward logistics to other computer manufacturing companies for distribution to city warehouses and city warehouse company's own employee for distributing products to the retailers. Recently, company has decided to develop its own forward logistics system along with development of RL system.

Second decision maker (DM2) is a logistics manager in an electronics manufacturing company, ABC-2 limited. The company manufactures, assembles, and distributes a comprehensive range of electronic hardware including computer peripherals in India. The company has annual turnover of approximately USD 50 million. The company has manufacturing facilities in Chennai, Pondicherry, and Uttaranchal. It has strong chain of distributors and dealers with 92,500 outlets in 8700 towns in India. The company has not given much attention to the EOL computers. Green awareness and implementation of e-waste management rules prompted them to think about implementing RL system for handling product returns and EOL computers. This company is also interested in working towards sustainability.

Third decision maker (DM3) is a logistics manager in an electronics manufacturing company, ABC-3 limited. The company assembles and distributes consumer electronics products in India including refrigerators, LCD, CTV, mobiles, washing machines, and microwave ovens. The company has annual turnover of approximately USD 120 million. The company has manufacturing facility in NCR Delhi having more than 1200 employees. The company has mother warehouse in the NCR Delhi and four child warehouses in the cities Chennai, Ahmadabad, Kolkata, and Bangalore in India. The company has its own well-established distribution system and logistics facilities. The major challenge for the company is to implement RL without effecting the current operations. This company has already taken several green initiatives including take back program for used products.

Fourth decision maker (DM4) is a marketing executive looking after north India region of an electronics manufacturing company, ABC-4 limited. The company manufactures, assembles, and distributes colour television sets in India. The company has annual turnover of approximately USD 30 million. The company has manufacturing facility in NCR Delhi having more than 350 employees. The company has strong chain of distributors and dealers. The company manufactures CTV mainly in rural markets in India. Growing demand for the LCDs and LEDs may hamper the demand of CTVs in future for the company. The company is willing to introduce new electronics product in the market for sustaining their business.

Fifth decision maker (DM5) is vice president of operations management of an electronics company, ABC-5 limited engaged in manufacturing of consumer electronics products. The company has annual turnover of approximately USD 130 million. The company has manufacturing facilities in NCR Delhi and in Bangalore. The company has more than 1500 employees. The company has strong supply chain for forward operations and willing to integrate it with its reverse supply chain.

Results and discussion

To prioritize the CSFs for RL implementation in Indian electronics industry, 12 factor legislation, economic factors, environmental concerns, top management awareness, resource management, management information system, contracts terms and conditions, direct and indirect taxes, integration of forward and reverse supply chain, joint consortium, process capabilities and skilled workers, and consumer awareness and social acceptability, identified in section "Identification of CSFs for RL implementation" are considered for the prioritization. Five decision makers DM1, DM2, DM3, DM4, and DM5 were asked to rate the importance of the above mentioned each CSF on a 5-point scale having the linguistic terms low (L), fairly low (FL), medium (M), fairly high (FH), and high (H). The decision-makers used the linguistic variables shown in Table 2 to assess the importance of the CSFs. A decision matrix was prepared based on the responses received from the DMs shown in Table 3.

Table 3 Decision matrix using linguistic variable

S. No.	CSFs for RL implementation	DM1	DM2	DM3	DM4	DM5
1	Legislation	FH	FH	M	H	M
2	Economic factors	H	H	H	M	M
3	Environmental concerns	FH	M	M	M	M
4	Top management awareness	H	FH	H	H	H
5	Resource management	FH	FH	H	H	H
6	Management information system	M	L	M	M	M
7	Contracts terms and conditions	FH	FH	FH	FH	H
8	Direct and indirect taxes	M	M	L	L	L
9	Integration of forward and reverse supply chain	FL	FL	M	M	M
10	Joint consortium	FH	FH	M	M	FH
11	Process capabilities and skilled workers	FL	FL	L	L	L
12	Consumer awareness and social acceptability	FH	H	M	M	M

L Low, *FL* fairly low, *M* medium, *FH* fairly high, *H* high

Table 4 Closeness coefficient matrix and priority

S. No.	CSFs for RL implementation	D^*	D^-	C	Priority
1	Legislation	0.384	0.673	0.637	5
2	Economic factors	0.32	0.736	0.697	3
3	Environmental concerns	0.489	0.565	0.536	8
4	Top management awareness	0.214	0.844	0.797	1
5	Resource management	0.246	0.813	0.768	2
6	Management information system	0.562	0.541	0.491	9
7	Contracts terms and conditions	0.285	0.633	0.689	4
8	Direct and indirect taxes	0.736	0.32	0.303	11
9	Integration of forward and reverse supply chain	0.603	0.452	0.429	10
10	Joint consortium	0.415	0.642	0.607	6
11	Process capabilities and skilled workers	0.813	0.246	0.232	12
12	Consumer awareness and social acceptability	0.42	0.634	0.601	7

As mentioned in the fuzzy-TOPSIS methodology step 1, triangular fuzzy numbers were used to convert linguistics variable into the fuzzy numbers. By converting the fuzzy linguistic variables into triangular fuzzy numbers using Table 2, the fuzzy decision matrix D was obtained. In the next step un-weighted fuzzy decision matrix R was enumerated. Further steps were followed to obtain the weighted fuzzy normalized decision matrix, to find the ideal and negative-ideal solutions for the CSFs. The distance D^- and D^* of each CSF is derived, respectively, by using Eqs. (7), (8), (9), and (10). The closeness coefficient C for each CSF is obtained by using Eq. (11). Values of D^-, D^* and closeness coefficient C for each CSF are shown in Table 4. The prioritization of CSFs was obtained and is shown in Table 4. The most important CSF among the 12 CSFs is top management awareness and the least important CSF is process capabilities and skilled workers.

The overall prioritization of CSFs is

$$CSF4 > CSF5 > CSF2 > CSF7 > CSF1 > CSF10 > CSF12 > CSF3 > CSF6 > CSF9 > CSF8 > CSF11$$

Top management awareness has the highest value and is prioritized as top most factors. Top management initiates, guides, and motivates the organization for adoption and implementation of RL implementation. Resource management is prioritized as 2nd most important CSF. Previous studies also support this result like Richey et al. (2005) showed that resource commitment makes RL more efficient and more effective. Economic factors are prioritized 3rd and are very crucial for RL implementation. Recaptured value is the major source of direct revenue generation from RL implementation. Higher recapturing value motivates the management for RL implementation. Ravi et al. (2005) also found economic factor as important factor for

RL implementation. Contracts term and conditions is prioritized fourth and are very important to meet environmental objectives. Most of the outsourced parts/components can be reprocessed through legal terms and conditions for reprocessing returned products. Contract terms and conditions are particularly important in cases where component/products are outsourced from other countries. Legislation is prioritized 5th and has become very important for Indian electronics industry after enforcement of e-waste management rules and regulations in 2012. This finding is also supported by earlier research (Walker et al. 2008). Joint consortium is ranked 6th and has great influence on the success of RL implementation in Indian electronics industry. Joint consortium is important to have co-operation with other companies to minimize the e-waste and to recapture maximum value while satisfying regulatory requirements. Joint consortium may help in achieving economies of scale and also, may help in reduced investment for joint reprocessing/recycling facilities. Consumer awareness and social acceptability is prioritized 7th. This factor is very important to achieve a good volume and quality of returned products, and to gain profit through resale of remanufactured products. Ravi and Shankar (2004) also found lack of awareness as a chief barrier of RL benefits in Indian automobile supply chain. Environmental concerns are ranked 8th and are mentioned by most of previous research on RL implementation. Management information system is ranked 9th. This factor is important for operational efficiency but ranked lower because managers of the opinion that it is not easy to integrate reverse logistics in current management information system. Daugherty et al. (2005) also suggested that development of information technology capabilities may give better performance in reverse supply chain organizations. Integration of forward and reverse supply chain is ranked lower because it is more dependent on the characteristics of the organization. Direct and indirect tax is ranked 11th and does not make much difference to implementation because of being part of government policy, and industry does not have any control on this factor. Process capabilities and skilled workers are ranked 12th. This factor is important for the operational performance of RL. Managers stated that it may require considerable investment in education and training of employees.

Conclusion

RL is in focus worldwide because of its inherent advantages of reducing the impact of hazard materials on human life and environment. Reuse/recycle of materials is important because of rising costs of materials, limited resources and growing environmental concerns. RL is relatively new for Indian industry and limited studies are available for RL practices. This research paper provides the valuable information on RL implementation for Indian electronics industry. The research identified 12 CSFs for RL implementation in Indian electronics industry. The identified factors are somewhat similar to those identified by various researchers all over the world. Still, factors like contracts terms and conditions, direct and indirect taxes, joint consortium, process capabilities, and skilled workers are rarely included in other studies. Analysis of the findings shows that top four prioritized factors top management awareness, resource management, economic factors, and contracts terms and conditions are the most important among all 12 factors. Briefly, the contributions of this study are summarized as follows:

(a) The study provides the insight into previous research on RL implementation.
(b) Identifies the CSFs based on past literature review and experts opinion for successful reverse logistics implementation.
(c) The research work proposes a framework for evaluating and prioritizing the CSFs by using Fuzzy-TOPSIS methodology for RL implementation.
(d) The study will help the managers and practitioners implementation of RL. It will enable the managers in identifying the factors which they need to work out for successful implementation.

The findings of the research will help the managers and academicians in the development of RL strategies and practices in Indian electronics industry. These CSFs can also be used for RL implementation in other sectors of Indian industry. Like other studies; this study also has some limitations. This study is conducted using five experts from the Indian electronics industry. Future studies may consider larger sample size to assess the methodology and the effectiveness of the proposed solution to enable generalization. Furthermore, the wider rating of the 7 or 11-point linguistic scale could be used instead of using a 5-point linguistics scale. Researchers may utilise other methodologies including other MCDM methodologies and may compare the results. Future studies may be carried out to identify company-specific or product-specific identification of CSFs for RL implementation.

Acknowledgments We would like to express our special gratitude and thanks to electronics industry experts for giving us such attention and time.

References

Abdulrahman MD, Gunasekaran A, Subramanian N (2014) Critical barriers in implementing reverse logistics in the Chinese manufacturing sectors. Int J Prod Econ 147(B):460–471

Andel T (1997) Reverse logistics: a second chance to profit. Transp Distrib 38(7):61–64

Bernon M, Rossi S, Cullen J (2011) Retail reverse logistics: a call and grounding framework for research. Int J Phys Distrib Logist Manag 41(5):484–510

Biehl M, Prater E, Realff MJ (2007) Assessing performance and uncertainty in developing carpet reverse logistics systems. Comput Oper Res 34:443–463

Bottani E, Rizzi A (2006) A fuzzy TOPSIS methodology to support outsourcing of logistics services. Supply Chain Manag Int J 11(4):294–308

Carter C, Ellram L (1998) Reverse logistics: a review of literature and framework for future investigation. J Bus Logist 19(1):85–102

Chan FTS, Chan HK (2008) A survey on reverse logistics system of mobile phone industry in Hong Kong. Manag Decis 46(5):702–708

Chen CT (1997) Extensions of the TOPSIS for group decision-making under fuzzy environment. Fuzzy Sets Syst 114:1–9

Cheng YH, Lee F (2010) Outsourcing reverse logistics of high-tech manufacturing firms by using a systematic decision-making approach: TFT-LCD sector in Taiwan. Ind Mark Manag 39(7):1111–1119

Chiou YC et al (2012) Consideration factors of reverse logistics implementation—a case study of Taiwan's electronics industry. Procedia Soc Behav Sci 40:375–381

Chu TC (2002) Selecting plant location via a fuzzy TOPSIS approach. Int J Adv Manuf Technol 20:859–864

Chu TC, Lin YC (2003) A fuzzy TOPSIS method for robot selection. Int J Adv Manuf Technol 21:284–290

Clottey T, Benton JWC, Srivastava R (2013) Forecasting product returns for remanufacturing operations. Decis Sci 43(4):589–614

Das TK, Teng BS (2000) A resource-based theory of strategic alliances. J Manag 26(1):31–61

Daugherty PJ, Richey RG, Genchev SE, Chen H (2005) Reverse logistics: superior performance through focused resource commitments to information technology. Trans Res: Part E 41(2):77–92

Brito MP De, Dekker R (2002) Reverse logistics: a framework. Econometric Institute report EI 38

Ding SH, Kamaruddin S (2014) Assessment of distance-based multi-attribute group decision making methods from a maintenance strategy perspective. J Ind Eng Int 10:78. doi:10.1007/s40092-014-0078-2

Dowlatshahi S (2000) Developing a theory of reverse logistics. Interfaces 30(3):43–55

Dowlatshahi S (2005) A strategic framework for the design and implementation of remanufacturing operations in reverse logistics. Int J Prod Res 43(16):3455–3480

Fleischmann M (2001) Reverse logistics network structures and design. In: Guide VDR Jr, Van Wassenhove LN (eds) Business aspects of closed-loop supply chains. Carnegie Bosch Institute, Pittsburgh, pp 117–148

Fleischmann M (2003) Reverse logistics network structures and design. In: Guide VDR, Van Wassenhove LN (eds) Business aspects of closed-loop supply chains. Carnegie Mellon University Press, Pittsburgh, pp 117–148

Fleischmann M, Bloemhof-Ruwaard JM, Dekker R, van der Laan E, Van Nunen JAEE, van Wassenhove LN (1997) Quantitative models for reverse logistics: a review. Eur J Oper Res 103(1):1–17

Gonzalez-Torre PL, Adenso-Diaz B (2006) Reverse logistics practices in the glass sector in Spain and Belgium. Int Bus Rev 15(5):527–546

Gonzalez-Torre PL, Adenso-Diaz B, Artiba H (2004) Environmental and reverse logistics policies in European bottling and packaging firms. Int J Prod Econ 88(1):95–104

Gonzalez-Torre PL, Alvarez M, Sarkis J, Adenso-Diaz B (2012) Barriers to the implementation of environmentally oriented reverse logistics: evidence from the automotive industry sector. Br J Manag 21:889–904

Guide VDR, Wassenhove LN (2003) Full cycle supply chains. Carnegie Mellon Press, Pittsburgh

Holt D, Ghobadian A (2009) An empirical study of green supply chain management practices among UK manufacturers. J Manuf Technol Manag 20(7):933–956

Hwang CL, Yoon KP (1981) Multiple attribute decision making: methods and applications. Business and Economics. Springer, New York

Janes B, Schuur P, Brito DMP (2010) A reverse logistics diagnostic tool: the case of the consumer electronics industry. Int J Adv Manuf Technol 47(5–8):495–513

Jayaraman V, Luo Y (2007) Creating competitive advantages through new value creation: a reverse logistics perspective. Acad Manag Perspect 21(2):56–73

Jayaraman V, Patterson RA, Rolland E (2003) The design of reverse distribution networks: models and solution procedures. Eur J Oper Res 150:128–149

Jindal A, Sangwan KS (2013) Development of an interpretive structural model of drivers for reverse logistics implementation in Indian industry. Int J Bus Perform Supply Chain Modell 5(4):325–342

Kahraman C, Ates NY, Cevik S (2007) Hierarchical fuzzy TOPSIS model for selection among logistics information technologies. J Enterp Inf Manag 20(2):143–168

Kannan G, Haq AN, Devika M (2009) Analysis of closed loop supply chain using genetic algorithm and particle swarm optimization. Int J Prod Res 47(5):1175–1200

Kannan D, Diabat A, Shankar KM (2014) Analyzing the drivers of end-of-life tire management using interpretive structural modeling (ISM). Int J Adv Manuf Technol 72:1603–1614

Khalili-Damghani K, Najmodin M (2014) A conceptual model for measuring reverse logistics performance in automobile industry. Int J Strat Decis Sci 5(2):21–29. doi:10.4018/ijsds.2014040102

Khalili-Damghani K, Sadi-Nezhad S (2014) A decision support system for fuzzy multi-objective multi-period sustainable project selection. Comput Ind Eng 64:1045–1060

Khalili-Damghani K, Tavana M, Najmodin M (2015) Reverse logistics and supply chains: a structural equation modeling investigation. J Ind Eng Int 22(3):354–368

Kim G, Park CS, Yoon KP (1997) Identifying investment opportunities for advanced manufacturing systems with comparative-integrated performance measurement. Int J Prod Econ 50(1):23–33

Knemeyer AM, Ponzurick TG, Logar CM (2002) A qualitative examination of factors affecting reverse logistics systems for end of life computers. Int J Phys Distrib Logist Manag 32(6):455–479

Kumar S, Luthra S, Haleem A (2013) Customer involvement in greening the supply chain: an interpretive structural modeling methodology. J Ind Eng Int 9:6

Lau KH, Wang Y (2009) Reverse logistics in the electronic industry of China: a case study. Supply Chain Manag Int J 14(6):447–465

Lee SM, Eom HB (1990) Multiple-criteria decision support systems: the powerful tool for attacking complex, unstructured decisions. Syst Pract 3(1):51–65

Lee CH, Chang SL, Wang KM, Wen LC (2000) Management of scrap computer recycling in Taiwan. J Hazard Mater 73:209–220
Mehrbod M, Tu N, Miao L (2014) A hybrid solution approach for a multi-objective closed-loop logistics network under uncertainty. J Ind Eng Int. doi:10.1007/s40092-014-0089-z
Miller D, Shamsie J (1996) The resource-based view of the firm in two environments: the Hollywood film studios from 1936 to 1965. Acad Manag J 39(3):519–543
Mintzberg H (1973) Strategy making in three modes. Calif Manag Rev 16(3):44–58
Mittal VK, Sangwan KS (2013) Assessment of hierarchy and inter-relationships of barriers to environmentally conscious manufacturing adoption. World J Sci Technol Sustain Dev 10(4):297–307
Murphy P, Poist R (2003) Green perspectives and practices: a comparative logistics study. Supply Chain Manag Int J 8(2):122–131
Nejati M, Nejati M, Shafaei A (2009) Ranking airlines' service quality factors using a fuzzy approach: study of the Iranian society. Int J Qual Reliab Manag 26(3):247–260
New S, Green K, Morton B (2000) Buying the environment: the multiple meanings of green supply. In: Fineman S (ed) The business of greening. Routledge, London, pp 35–54
Prendergast G, Pitt L (1996) Packaging, marketing, logistics and the environment: are there trade-offs? Int J Phys Distrib Logist Manag 26(6):60–72
Rahman S, Subramanian N (2012) Factors for implementing end-of-life computer recycling operations in reverse supply chains. Int J Prod Econ 140(1):239–248
Ravi V, Shankar R (2004) Analysis of interactions among the barriers of reverse logistics. Technol Forecast Soc Change 72(8):1011–1029
Ravi V, Shankar R, Tiwari MK (2005) Productivity improvement of a computer hardware supply chain. Int J Prod Perform Manag 54(4):239–255
Richey RG, Genchev SE, Daugherty PJ (2005) The role of resource commitment and innovation in reverse logistics performance. Int J Phys Distrib Logist Manag 35(4):233–257
Rogers DS, Tibben-Lembke R (1999) Going backwards: reverse logistics trends and practices. RLEC Press, Pittsburgh
Sadi-Nezhad S, Khalili-Damghani K (2014) Application of a fuzzy TOPSIS method base on modified preference ratio and fuzzy distance measurement in assessment of traffic police centers performance. Appl Soft Comput 10:1028–1039
Sharma SK, Panda BN, Mahapatra SS, Sahu S (2011) Analysis of barriers for reverse logistics: an Indian perspective. Int J Model Optim 1(2):101–106
Singh RK, Benyoucef L (2011) A fuzzy TOPSIS based approach for e-sourcing. Eng Appl Artif Intell 24:437–448
Srivastava S (2008) Network design for reverse logistics. Omega 36(4):535–548
Stock JR (1998) Development and implementation of reverse logistics program. Council of Logistics Management, Oak Brook
Sun C, Lin GTR (2009) Using fuzzy TOPSIS method for evaluating the competitive advantages of shopping websites. Exp Syst Appl 36(9):11764–11771
Tibben-Lembke RS, Rogers DS (2002) Differences between forward and reverse logistics in a retail environment. Supply Chain Manag Int J 7(5):271–282
Tyagi RK, Dhanda KK, Young S (2012) An operational framework for reverse supply chains. Int J Manag Inf Syst 16(2):137–150
Vishkaei BM, Niaki STA, Farhangi M, Rashti MEM (2014) Optimal lot sizing in screening processes with returnable defective items. J Ind Eng Int 10:70. doi:10.1007/s40092-014-0070-x
Walker H, Sisto LD, McBain D (2008) Drivers and barriers to environmental supply chain management practices: lessons from public and private sectors. J Purch Supply Manag 14:69–85
Wang TC, Lee HD (2009) Developing a fuzzy TOPSIS approach based on subjective weights and objective weights. Expert Syst Appl 36(5):8980–8985
Widmer R, Oswald-Krapf H, Sinha-Khetriwal D, Schnellmann M, Boeni H (2005) Global perspectives on e-waste. Environ Impact Assess Rev 25:436–458
Zadeh LA (1965) Fuzzy sets. Inf Control 8:338–353
Zanakis SH, Solomon A, Wishart N, Dublish S (1998) Multi-attribute decision making: a simulation comparison of selection methods. Eur J Oper Res 107:507–529

Saurabh Agrawal is an Assistant Professor in Mechanical, Production and Industrial Engineering Department at Delhi Technological University, Delhi, India. He has vast experience of academics, research and the industry both in India and in USA. His research focus is in the areas of supply chain management, reverse logistics, sustainability, and e-waste management. He received his Master of Business Administration (MBA) from Oregon State University, USA, and Master in Technology (M.Tech) from Indian Institute of Technology (IIT), Delhi, India.

Dr. Rajesh K. Singh is an Associate Professor in Mechanical, and Production and Industrial Engineering Department at Delhi Technological University, Delhi, India. He has published about 85 research papers in reputed international/national journals and conferences. His areas of interest include competitiveness, small business management, Quality Management and Supply Chain Management. He has published papers in journals such as Industrial management and Data Systems, Singapore Management Review, International Journals of Productivity and Performance Management, International Journal of Automotive Industry and Management, Competitiveness Review: An International Journal, International Journals of Services and Operations Management, Global Journal of Flexible Systems and Management, International Journals of Productivity and Quality Management, South Asian Journal of Management, Productivity, IIMB Management Review and Productivity Promotion. He is also on editorial board of some reputed journals.

Dr. Qasim Murtaza is an Associate Professor in Mechanical, and Production and Industrial Engineering Department at Delhi Technological University, Delhi, India. He has vast experience of work and research both in India and abroad. He has received his doctorate in Manufacturing Engineering from Dublin City University, Dublin, Ireland. He has published his research work in reputed journals and attended several international conferences.

12

An alternative transformation in ranking using l_1-norm in data envelopment analysis

S. Ziari[1]

Abstract Jahanshahloo et al. (Appl Math Comput 153:215–224, 2004) propose a method for ranking extremely efficient decision making units (DMUs) in data envelopment analysis (DEA) using super-efficiency technique and l1-norm and they show that the presented method is able to eliminate the existing difficulties in some methods. This paper suggests an alternative transformation to convert the nonlinear model proposed by Jahanshahloo et al. (Appl Math Comput 153:215–224, 2004) into a linear programming form. The present paper shows that model with this transformation is equivalent to the above-mentioned nonlinear model. The motivation of this work is to linearize the proposed nonlinear model by Jahanshahloo et al. (Appl Math Comput 153:215–224, 2004) which has the higher order of complexity.

Keywords Data envelopment analysis (DEA) · Ranking · Efficiency · Extremely efficient · Chinese cities

Introduction

For many applications, ranking DMUs is an important and essential procedure to decision makers in DEA, especially when there are extremely efficient DMUs. In these regards, several methods have been proposed for ranking of the extreme efficient DMUs, see Andersen and Petersen (1993), Mehrabian et al. (1999) and Jahanshahloo et al. (2004). A DMU is called extremely efficient if it cannot be represented as a linear combination (with nonnegative coefficients) of the remaining DMUs (Charnes et al. 1991). Andersen and Petersen (AP) (Andersen and Petersen 1993) proposed a new procedure to rank efficient DMUs. The AP model determines the rank of a given DMU by removing it from the reference set and by computing its super-efficiency score. However, the AP model may be infeasible in some cases. It is proved that super-efficient DEA models are infeasible (see Thrall 1996; Seiford and Zhu 1999). Mehrabian et al. (1999) suggested the MAJ model for complete ranking of efficient DMUs, but their approach lacks feasibility in some cases, too. To overcome the drawbacks of the AP (Andersen and Petersen 1993) and MAJ (Mehrabian et al. 1999) models, Jahanshahloo et al. (2004) presented a method to rank the extremely efficient DMUs in DEA models with constant and variable returns to scale using L_1-norm. Their proposed model is a nonlinear programming form which has the higher order of complexity in solving. In addition, a complex procedure was applied in Jahanshahloo et al. (2004) to provide the nonlinear model into a linear one which obtains an approximately optimal solution. Wu and Yan (2010) have also suggested an effective transformation to convert the nonlinear model in Jahanshahloo et al. (2004) into a linear model. Recently, Ziari and Raissi (2016) proposed an approach to rank the efficient DMUs in DEA based on minimizing distance of the under evaluation DMU to the frontier of efficiency. Following this trend, the present paper is an attempt to provide an alternative transformation to convert the nonlinear model in Jahanshahloo et al. (2004) into a linear model. The proposed model in this paper linearizes model in Jahanshahloo et al. (2004) in a way which is different from the presented method in Wu and Yan (2010). To linearize model in Jahanshahloo et al. (2004), the presented model uses the number of fewer

✉ S. Ziari
shok_ziari@yahoo.com

[1] Department of Mathematics, Firoozkooh Branch, Islamic Azad University, Firoozkooh, Iran

auxiliary variables with respect to proposed model in Wu and Yan (2010). Furthermore, it is shown that the model with this transformation is equivalent to the nonlinear model and is easier to be solved. The rest of the paper is organized as follows. Section 2 reviews some ranking methods, especially the model by Jahanshahloo et al. (2004). The paper proposes an alternative transformation in Sect. 3. Section 4 re-executes the empirical example by Jahanshahloo et al. (2004) for illustration. The last Section concludes the study.

Review of some ranking models

In this section, we review some ranking models in data envelopment analysis. In the following subsections, it is assumed that there are n DMUs and for each DMU_j ($j = 1,\ldots,n$) a vector of inputs (X_j) is considered to produce a vector of outputs (Y_j), where $X_j = (x_{1j}, x_{2j}, \ldots, x_{mj})$ and $Y_j = (y_{1j}, y_{2j}, \ldots, y_{sj})$. It is also assumed that $X_j \geq 0, Y_j \geq 0, X_j \neq 0$, and $Y_j \neq 0$ for every $j = 1,\ldots,n$.

AP model

The ranking method proposed by Andersen and Petersen (1993) is a supper efficiency model. In the AP model, DMU under evaluation is excluded from reference set and using other units, the rank of given DMU is obtained.

The input-oriented AP model using the CRS super-efficiency model is as follows:

$$\begin{aligned} &\min \theta \\ &\text{s.t.} \sum_{j=1, j\neq k}^{n} \lambda_j x_{ij} \leq \theta x_{ik}, \quad i = 1,\ldots,m \\ &\quad \sum_{j=1, j\neq k}^{n} \lambda_j y_{rj} \geq y_{rk}, \quad r = 1,\ldots,s \\ &\quad \lambda_j \geq 0, \quad j = 1,\ldots,n,\ j \neq k, \end{aligned} \tag{1}$$

where λ_j, $j = j = 1,\ldots,n, j \neq k$ and θ are the variables of model.

In addition, the output-oriented AP model using the CRS super-efficiency model is as follows:

$$\begin{aligned} &\max \phi \\ &\text{s.t.} \sum_{j=1, j\neq k}^{n} \lambda_j x_{ij} \leq x_{ik}, \quad i = 1,\ldots,m \\ &\quad \sum_{j=1, j\neq k}^{n} \lambda_j y_{rj} \geq \phi y_{rk}, \quad r = 1,\ldots,s \\ &\quad \lambda_j \geq 0, \quad j = 1,\ldots,n, \quad j \neq k, \end{aligned} \tag{2}$$

where $\lambda_j, j = j = 1,\ldots,n, \quad j \neq k$ and ϕ are the variables of model.

The main drawbacks of this model are infeasibility and instability for some DMUs. It can be said that a model is stable if the DMU under evaluation which is efficient remains efficient after perturbation on data.

MAJ model

This ranking model is proposed by Mehrabian et al. (1999) to solve infeasibility of AP models in some cases. The MAJ model can be expressed as follows:

$$\begin{aligned} &\min 1 + w \\ &\text{s.t.} \sum_{j=1, j\neq k}^{n} \lambda_j x_{ij} \leq x_{ik} + w, \quad i = 1,\ldots,m \\ &\quad \sum_{j=1, j\neq k}^{n} \lambda_j y_{rj} \geq y_{rk}, \quad r = 1,\ldots,s \\ &\quad \lambda_j \geq 0, \quad j = 1,\ldots,n, j \neq k, \end{aligned} \tag{3}$$

where $\lambda_j, j = j = 1,\ldots,n, \quad j \neq k$ and w are the variables of model.

L_1-norm model

In this subsection, the model of Jahanshahloo et al. (2004) is explained as follows. Then the production possibility set (PPS) with constant returns to scale T_c and the PPS with variable returns to scale T_v are defined as:

$$T_c = \left\{ (X,Y) | X \geq \sum_{j=1}^{n} \lambda_j X_j, Y \leq \sum_{j=1}^{n} \lambda_j Y_j, \lambda_j \geq 0, j = 1,\ldots,n \right\}, \tag{4}$$

and

$$T_v = \left\{ (X,Y) | X \geq \sum_{j=1}^{n} \lambda_j X_j, Y \leq \sum_{j=1}^{n} \lambda_j Y_j, \sum_{j=1}^{n} \lambda_j = 1 \lambda_j \geq 0, j = 1,\ldots,n \right\}, \tag{5}$$

respectively.

DMU_k is assumed to be extremely efficient. By removing (X_k, Y_k) from T_c, the production possibility set T_c' is defined as:

$$T_c' = \left\{ (X,Y) | X \geq \sum_{j=1, j\neq k}^{n} \lambda_j X_j, Y \leq \sum_{j=1, j\neq k}^{n} \lambda_j Y_j, \lambda_j \geq 0, j = 1,\ldots,n \right\}. \tag{6}$$

To obtain the ranking score of DMU_k the model in Jahanshahloo et al. (2004) is considered as follows:

$$\min \Gamma_k^c(X,Y) = \sum_{i=1}^{m} |x_i - x_{ik}| + \sum_{r=1}^{s} |y_r - y_{rk}|$$
$$\text{s.t.} \sum_{j=1, j\neq k}^{n} \lambda_j x_{ij} \leq x_i, \quad i = 1,\ldots,m$$
$$\sum_{j=1, j\neq k}^{n} \lambda_j y_{rj} \geq y_r, \quad r = 1,\ldots,s \tag{7}$$
$$x_i \geq 0, \quad y_r \geq 0 \quad i = 1,\ldots,m, \ r = 1,\ldots,s$$
$$\lambda_j \geq 0, \quad j = 1,\ldots,n, \quad j \neq k,$$

where $X = (x_1, \ldots, x_m)$, $Y = (y_1, \ldots, y_s)$ and $\lambda = (\lambda_1, \ldots, \lambda_{k-1}, \lambda_{k+1}, \ldots, \lambda_n)$ are the variables of the model (7) and $\Gamma_k^c(X, Y)$ is the distance between (X_k, Y_k) and (X, Y) by l_1-norm.

To convert model (4) into a linear model, [1] defines the set T_c'' as follows:

$$T_c'' = T_c' \cap \left\{(X, Y)|X \geq X_k \quad \text{and} \quad Y \leq Y_k\right\} \tag{8}$$

and they apply the scaling input and output data by normalization. After these changes, an approximately optimal solution of model (7) is obtained by solving a linear programming model related to it.

In next section, an alternative transformation to model (7) is considered and to obtain the optimal solution, an equivalent linear programming model is solved.

An alternative transformation

For converting model (7) into a linear model, the following transformation is utilized: $|x_i - xik| \leq a_i \quad i = 1,\ldots,m$ and $|y_r - yrk| \leq b_r \quad r = 1,\ldots,s$. Thus, we have:

$$\begin{cases} x_i - x_{ik} \leq a_i, & i = 1,\ldots,m, \\ x_i - x_{ik} \geq -a_i, & i = 1,\ldots,m, \end{cases} \quad \text{and} \quad \begin{cases} y_r - y_{rk} \leq b_r, & r = 1,\ldots,s, \\ y_r - y_{rk} \geq -b_r, & r = 1,\ldots,s. \end{cases}$$

Then Model (7) can be converted into the following linear programming problem:

$$\min \Gamma_k^c(X,Y) = \sum_{i=1}^{m} a_i + \sum_{r=1}^{s} b_r$$
$$\text{s.t.} \sum_{j=1, j\neq k}^{n} \lambda_j x_{ij} \leq x_i, \quad i = 1,\ldots,m$$
$$\sum_{j=1 j\neq k}^{n} \lambda_j y_{rj} \geq y_r, \quad r = 1,\ldots,s$$
$$x_i - x_{ik} \leq a_i, \quad i = 1,\ldots,m,$$
$$-x_i + x_{ik} \leq a_i, \quad i = 1,\ldots,m,$$
$$y_r - y_{rk} \leq b_r, \quad r = 1,\ldots,s,$$
$$-y_r + y_{rk} \leq b_r, \quad r = 1,\ldots,s,$$
$$x_i \geq 0, \quad y_r \geq 0, \quad a_i \geq 0, \quad b_r \geq 0, \quad i = 1,\ldots,m, r = 1,\ldots,s,$$
$$\lambda_j \geq 0, \quad j = 1,\ldots,n, \quad j \neq k, \tag{9}$$

where $X = (x_1, \ldots, x_m), Y = (y_1, \ldots, y_s), a = (a_1, \ldots, a_m)$, $b = (b_1, \ldots, b_s)$ and $\lambda = (\lambda_1, \ldots, \lambda_{k-1}, \lambda_{k+1}, \ldots, \lambda_n)$ are the variables of model (9).

It is obvious that model (9) is equivalent with model (7). Furthermore, model (9) is a linear programming problem which can easily provide the optimal solution of model (7). The above-proposed model includes $3(m+n)$ constraints whereas Wu and Yan (2010) model has $2(m+n)$ constraints that obviously the Wu and Yan (2010) model is more efficient than the above model. To overcome this problem, we proposed the following model which is more efficient with respect to Wu and Yan (2010) model. The efficient proposed model is as follows:

$$\min \Gamma_k^c(X,Y) = \sum_{i=1}^{m} (x_i - x_{ik}) + \sum_{r=1}^{s} (-y_r + y_{rk})$$
$$\text{s.t.} \sum_{j=1, j\neq k}^{n} \lambda_j x_{ij} \leq x_i, \quad i = 1,\ldots,m$$
$$\sum_{j=1, j\neq k}^{n} \lambda_j y_{rj} \geq y_r, \quad r = 1,\ldots,s \tag{10}$$
$$x_i \geq x_{ik}, \quad i = 1,\ldots,m$$
$$y_r \leq y_{rk}, \quad r = 1,\ldots,s$$
$$x_i \geq 0, \quad y_r \geq 0 \quad i = 1,\ldots,m, r = 1,\ldots,s$$
$$\lambda_j \geq 0, \quad j = 1,\ldots,n, \ j \neq k,$$

where $X = (x_1, \ldots, x_m), Y = (y_1, \ldots, y_s)$ and $\lambda = (\lambda_1, \ldots, \lambda_{k-1}, \lambda_{k+1}, \ldots, \lambda_n)$ are the variables of model (10).

Next, model (10) is reformulated by imposing the convexity constraint $\sum_{j=1, j\neq k}^{n} \lambda_j = 1$ on (10), to extend the model (10) from the constant returns to scale to the variable returns to scale case. Then, to obtain the ranking score of DMU under evaluation, the following linear programming problem is solved:

$$\min \Gamma_k^c(X,Y) = \sum_{i=1}^{m} (x_i - x_{ik}) + \sum_{r=1}^{s} (-y_r + y_{rk})$$
$$\text{s.t.} \sum_{j=1, j\neq k}^{n} \lambda_j x_{ij} \leq x_i, \quad i = 1,\ldots,m$$
$$\sum_{j=1, j\neq k}^{n} \lambda_j y_{rj} \geq y_r, \quad r = 1,\ldots,s$$
$$\sum_{j=1, j\neq k}^{n} \lambda_j = 1 \tag{11}$$
$$x_i \geq x_{ik}, \quad i = 1,\ldots,m$$
$$y_r \leq y_{rk}, \quad r = 1,\ldots,s$$
$$\lambda_j \geq 0, \quad j = 1,\ldots,n, \ j \neq k$$
$$x_i \geq 0, \quad y_r \geq 0 \quad i = 1,\ldots,m, \ r = 1,\ldots,s$$
$$\lambda_j \geq 0, \quad j = 1,\ldots,n, \quad j \neq k,$$

Table 1 The data for 28 Chinese cities

DMU1	Input 1	Input 2	Input 3	Output 1	Output 2	Output 3
1	483.01	1,397,736	616,961	6,785,798	1,594,957	1,088,699
2	371.95	855,509	385,453	2,505,984	545,140	835,745
3	268.23	685,584	341,941	2,292,025	406,947	473,600
4	202.02	452,713	117,424	1,158,016	135,939	336,165
5	197.93	471,650	112,634	1,244,124	204,909	317,709
6	178.96	423,124	189,743	1,187,130	190,178	605,037
7	148.04	367,012	97,004	658,910	86,514	239,760
8	184.93	408,311	111,904	993,238	1,411,954	353,896
9	123.33	245,542	91,861	854,188	135,327	239,360
10	116.91	305,316	91,710	606,743	78,357	208,188
11	129.62	295,812	92,409	736,545	114,365	298,112
12	106.26	198,703	53,499	454,684	67,154	233,733
13	89.70	210,891	95,642	494,196	78,992	118,553
14	109.26	282,209	84,202	842,854	149,186	243,361
15	85.50	184,992	49,357	776,285	116,974	234,875
16	72.17	222,327	73,907	490,998	117,854	118,924
17	76.18	161,159	47,977	482,448	67,857	158,250
18	73.21	144,163	43,312	515,237	114,883	101,231
19	86.72	190,043	55,326	625,514	173,099	130,423
20	69.09	158,436	66,640	382,880	74,126	123,968
21	77.69	135,046	46,198	867,467	65,229	262,876
22	97.42	206,926	66,120	830,142	128,279	242,773
23	54.96	79,563	43,192	521,684	37,245	184,055
24	67.00	144,092	43,350	869,973	86,859	194,416
25	46.30	100,431	31,428	604,715	55,989	127,586
26	65.12	96,873	28,112	601,299	37,088	224,855
27	20.09	50,717	54,650	145,792	11,816	24,442
28	69.81	117,790	30,976	319,218	31,726	169,051

Table 2 The results of ranking by applying model (11)

DMU	1	2	6	8	21	23	24	25	26	27
Ranking result	1	8	3	2	7	10	9	5	6	4
Value of obj. function	1.521	0.015	0.128	0.700	0.024	0.010	0.013	0.038	0.037	0.090

Model (11) is similar to model (10) which can be solved by any classical mathematical tool such as GAMS.

Empirical example

In this section, we apply DEA model (11) on the data set used by Jahanshahloo et al. (2004), with the assumption of variable returns to scale. The data set consists of 28 DMUs with 3 inputs and 3 outputs. Data are originally reported by Charnes et al. (1989) which comprised 28 Chinese cities (DMUs) in 1983. The inputs are labor, working funds, and investment. The outputs are gross industrial output value, profit and taxes, and retail sales. The data in Table 1 should be normalized before applying model (11). Table 2 includes the ranking results for 10 extremely efficient DMUs in model (11) ($DMU_1, DMU_2, DMU_6, DMU_8, DMU_{21}, DMU_{23}, DMU_{24}, DMU_{25}, DMU_{26}, DMU_{27}$). Note that in the empirical study these results are the same as the studies by Jahanshahloo et al. (2004) and Wu and Yan (2010). In this study, we have presented a method for converting the nonlinear programming model by Jahanshahloo et al. (2004) into linear programming problem using an alternative transformation, and it can easily be solved. It is worth nothing that some of the input data of Table 3 in Jahanshahloo et al. (2004) were mistakenly recorded and in this Section we have corrected them.

Conclusion

Jahanshahloo et al. (2004) propose a new ranking method using super-efficiency technique and l_1-norm. It was shown that the proposed method is able to eliminate the existing problems in some methods. In this regard, this paper provides an alternative transformation for converting the nonlinear programming model by Jahanshahloo et al. (2004) into the linear programming model which is equivalent to the original nonlinear model. In addition, we proposed the new efficient model which gives the same solutions of original model proposed by Jahanshahloo et al. (2004). Considering the higher order of complexity of nonlinear model (7), the proposed treatment in this article is easier to be utilized.

References

Andersen P, Petersen NC (1993) A procedure for ranking efficient units in data envelopment analysis. Manage Sci 39:1261–1264

Charnes A, Cooper WW, Li S (1989) Using DEA to evaluate relative efficiencies in the economic performance of Chinese-key cities. Soc Econ Plan Sci 23:325–344

Charnes A, Cooper WW, Thrall RM (1991) A structure for classifying and characterizing efficiency and inefficiency in data envelopment analysis. J Prod Anal 2:197–237

Jahanshahloo GR, Hosseinzadeh Lotfi F, Shoja N, Tohidi G, Razavian S (2004) Ranking by using L_1-norm in data envelopment analysis. Appl Math Comput 153:215–224

Liang L, Wu J, Cook WD, Zhu J (2008) Alternative secondary goals in DEA cross efficiency evaluation. Int J Prod Econ 113:1025–1030

Mehrabian S, Alirezaee MR, Jahanshahloo GR (1999) A compelete efficiency ranking of decision making units in data envelopment analysis. Comput Optimiz Appl 14:261–266

Seiford LM, Zhu J (1999) Infeasibility of super-efficiency data envelopment analysis models. INFOR 37(2):174–187

Thrall RM (1996) Duality, classification and slacks in DEA. Ann Oper Res 66:109–138

Wu J, Yan H (2010) An effective transformation in ranking using l1-norm in data envelopment analysis. Appl Math Comput 217:4061–4064

Ziari S, Raissi S (2016) Ranking efficient DMUs using minimizing distance in DEA. J Ind Eng Int. doi:10.1007/s40092-016-0141-2

13

A time-shared machine repair problem with mixed spares under N-policy

Madhu Jain[1] · **Chandra Shekhar**[2] · **Shalini Shukla**[3]

Abstract The present investigation deals with a machine repair problem consisting of cold and warm standby machines. The machines are subject to breakdown and are repaired by the permanent repairman operating under N-policy. There is provision of one additional removable repairman who is called upon when the work load of failed machines crosses a certain threshold level and is removed as soon as the work load again ceases to that level. Both repairmen recover the failed machines by following the time sharing concept which means that the repairmen share their repair job simultaneously among all the failed machines that have joined the system for repair. Markovian model has been developed by considering the queue dependent rates and solved analytically using the recursive technique. Various performance indices are derived which are further used to obtain the cost function. By taking illustration, numerical simulation and sensitivity analysis have been provided.

✉ Shalini Shukla
sshukla.dav@gmail.com

Madhu Jain
drmadhujain.iitr@gmail.com

Chandra Shekhar
chandrashekhar@pilani.bits-pilani.ac.in

[1] Department of Mathematics, Indian Institute of Technology Roorkee, Roorkee 247 667, Uttarakhand, India

[2] Department of Mathematics, Birla Institute of Technology and Science, Pilani 333 031, Rajasthan, India

[3] Department of Mathematics, D.A.V. (PG) College, Dehra Dun 248 001, Uttarakhand, India

Keywords Markov chain · Machine repair · Mixed spares · Time sharing · Threshold policy · Additional repairman · Queue length

Introduction

In the present era of modernization, machines have become part and partial of our day to day life. Due to automation of many systems, we are now completely dependent on the machines as it has become very difficult to even imagine our life without machines. But unfortunately, we cannot rely on the machines completely as they are always prone to failure. The failures of the machines affect the system adversely by reducing the efficiency and thereby increasing the overall cost of the system. Thus, it has become a very difficult task for the system developer to design a completely reliable machining system which can operate without any hindrance in spite of component failures. The provision of having spare machines in the system is one of the key approaches to cope up with the failure of operating machines and carrying out the machining operation smoothly without any interruptions. Multi-component systems with the provision of redundancy and maintainability are commonly seen in industrial scenarios, namely production systems, computer networks, transportation system, etc.

The spare machines are those machines which are put in place of the failed machines in the main system just like the main operating machines to carry out the functions properly and continuously. In the present paper, we study a Markovian machine repair problem with mixed standbys under the care of a repair facility having one permanent repairman and another additional removable repairman which turn on according to N-policy and threshold policy, respectively. The repairs to the failed machines are rendered on the time sharing basis. The time-shared policy is used for allocating the repairman capacity of the repair job to be utilized simultaneously by all failed machines by slicing the unit repair time among all failed machines as per round robin discipline. Each failed machine receives time slice of permanent as well as additional removable repairmen and recovers from the faults after receiving several quantum of repair time depending on the severity of the faults/damage. Taylor and Jackson (1954) introduced the standby provisioning in machine repair system by incorporating the use of cold standbys in a machine repair system. In the past years, a lot of researches have been done on the repairable machining system with standbys by many renowned queue theoreticians (cf. Albright 1980; Wang and Sivazlian 1989; Wang 1995; Jain and Baghel 2001). Haque and Armstrong (2007), Jain et al. (2010) and Jain and Gupta (2011) presented a brief review on the machine interference problem (MRP) with spares in their review articles. In recent years, many researchers also contributed to study on the performance prediction of the machine repair problems with standbys by incorporating some other distinct features like vacation, heterogeneous repairmen, N-policy, etc. (see Ke et al. 2009; Maheshwari et al. 2010; Jain et al. 2012). By including the F-policy Kumar and Jain (2013a) developed a queueing model for the performance prediction of the machine repair problems with standbys. A machine repair problem with standbys, having unreliable repairman and working vacation for the repairman, was investigated by Jain (2014).

In many machine repair systems, it has been often seen that the service rendered by the permanent repairman becomes too expensive as it may be idle most of the times. To reduce the cost for such system, the repairman can initiate the service only when a certain workload is build up. Yadin and Naor (1963) introduced the N-policy which can be used to control the service rendered by the repairman in optimal manner. Optimal N-policy ensures that the repairman will be activated only when N failed machines are accumulated in the system. This will be helpful in reduction of the expenditures such as set up cost to initiate the busy period after each idle period. According to N-policy, there will be no service provided by the repairman until the queue length is build up to N and once the service is started, the repairman stops service only when the queue becomes empty. There are some important research works available in the literature on MRP operating under N-policy (cf. Shawky 2000; Jain et al. 2004; Jain and Bhargava 2009). A few papers on N-policy for the MRP have been reported in the survey article by Sharma (2012). Machine repair problems with spares were investigated by Yue et al. (2012) and Kumar and Jain (2013a, b) by incorporating the N-policy. The performance modeling of multi-component machining systems under the care of unreliable repairman which operates according to N-policy was done by Jain et al. (2014b, c). Jayachitra and Albert (2014) presented an elaborated survey on various queueing models under N-policy which also includes the works on MRP under N-policy.

To tackle with the congestion problem, the provision of additional repairmen to the system may be helpful in reducing the workload of the permanent repairman and up gradation of the repair facility provided to the system. The feature of varying number of repairmen in queueing system has been studied by many researchers in the recent past (Shawky 1997; Jain 1998; Jain and Singh 2003). Later some works were also done on MRP with additional repairman by some renowned queue theorists (Al-seedy and Al-Ibraheem 2001; Sharma et al. 2005; Jain et al. 2007a, b). In recent years, some remarkable works have been done by the researchers by deploying the additional repairmen in machine repair system based on workload level. The important Markovian studies done on the same line in the last couple of years have been reviewed here. Markovian machine repair problems with additional repairman were investigated by Huang et al. (2011) and Liou et al. (2013) by including threshold control policies. Maheshwari and Ali (2013) studied an M/M/C/K/N machine repair problem with additional repairman by incorporating the concept of discouragement. A MRP with mixed standbys having the provision of permanent as well as additional repairmen was also investigated by Jain and Preeti (2014) by evaluating the probabilities for the transient states of the system which are further used to carry out the performance prediction of system characteristics.

The time sharing concept emerges in the 1960s to make the computer systems an object of public utility, i.e., to make it useful for more and more people (cf. Kleinrock 1967). In the context of machine repair system, the time-shared systems are the repair facility wherein the technical staff repairs the failed machines on time slicing basic, i.e., by dividing its unit time among all those failed machines which are presently waiting in the queue to be repaired by the repairman. The repairman will attend a failed machine for a pre-specified fixed quantum of time only and then it will move to the next machine in the queue. If the service of the first machine is not completed in that time interval then it will be put in the end in the queue to be served again. The repairman will return to the first machine after rendering its service to each of all other failed machines present in the queue for the fixed (pre-specified) small duration of time. This way, the whole cycle will go on by following the round-robin discipline. As soon as the repair of individual failed machine is completed during the repair cycle, it is removed from the queue. The newly failed machines will join the queue at the last position. The fraction of the total service time offered to any failed machine will depend upon the number of failed machines waiting in the queue for their service at the repair facility. In a specific case, we can assume the sharing factor of servers' time according to harmonic variation of individual capacity among. The number of users present in the systems (cf. Kleinrock 1967; Coffman and Kleinrock 1968; Adiri and Avi-Itzhak 1969). For the early notable contribution on time-shared systems, we refer Klienrock (1967). In the past, some other important contributions on time-shared computer systems are due to Rasch (1970), Yashkov (1992), Wang and Tai (2000). Yashkov and Yashkova (2007) have presented a survey article on processor-shared queueing systems which presents an overview of the work done so far on the concerned topic. In recent years, other related research works on the time-shared queueing systems by incorporating various distinct features have been done by Zhean and Knessl (2009), Altman et al. (2010), Tahar and Jean-Marie (2012), and others.

Sometimes, it is also observed that due to fewer failed machines in the system there may be less workload of the repairing as such the repairman may remain idle most of the time which is the wastage of resources and time. So in order to avoid this situation and to utilize the repair facility optimally, the concept of additional removable repairman is better option and can be employed in time sharing machine repair systems also. In the timesharing system, all the failed machines are served by the repairman at the same time through various repair positions. Such scenario of time sharing in machine repair problem can be seen in automobile repair shop of travel agency where a limited registered vehicles are repaired and the permanent repairman starts the concurrent repair jobs on the vehicles by slicing unit time only when some vehicles have joined the repair shop. In case of high workload when a certain number of failed vehicles have already joined the repair shop, the secondary repairman is called upon to render the repair. A lot of research works on the time-shared systems have appeared in the queueing literature (cf. Jain and Lata 1995; Jain et al. 2005; Kim and Kim 2007). In recent years, Chandrasekaran et al. (2013) and Jeong et al. (2014) investigated the optimization issues of machining system used for cloud computing. More recently, time-shared machining systems have been studied by Flapper et al. (2014) in manufacturing–remanufacturing system. Jain et al. (2014a) analyzed the sensitivity of a machine repair problem with two types of spares and controlled rates by incorporating the concept of time sharing.

In the present investigation, a time-shared machine repair problem with mixed (cold and warm) standbys has been studied. There is provision of permanent as well as one additional repairman; the permanent repairman follows the N-policy whereas additional removable repairman is introduced when the workload of failed machines crosses a certain threshold level. The noble feature of the present model over other existing models lies in the incorporation of many key realistic factors such as N-policy, time sharing concept, provision of mixed standbys, and facility of additional removable server in case of heavy workload in a combined and collaborated manner for the performance modeling of machine repair system. It is to be worth mentioning that the permanent repairman follows N-policy, i.e., starts working only when N failed machines are accumulated in the system. The secondary additional repairman is called upon as and when the workload of failed machines crosses a critical threshold level. Both repairmen work on the time sharing basis which means that both of them repair the failed machines present in the queue by sharing their time with all failed machines accumulated in the system. By constructing Chapman–Kolmogorov equations, the steady-state probabilities have been evaluated using the recursive solution approach. The rest of the paper is organized in different sections as follows. The description of the model and the differential equations, which governs the model, is given in "The model" and "Governing equations", respectively. In "Queue size distribution", the queue size distribution is obtained using recursive method. In "Special cases", some particular cases are deduced by setting appropriate parameter values. The queue size distribution is used to derive various performance measures and cost function which has been explained in "Performance indices" and "Cost function", respectively. To validate the tractability of the analytical results, the numerical simulation has been provided in "Numerical analysis". To summarize the findings and highlight the noble features of the work done, the concluding remarks have been given in the last section on "Discussion".

The model

Consider a time-shared machining system with mixed standby support and under the care of repair facility having permanent and additional removable servers. The permanent repairman operates under the N-policy whereas the additional repairman is called upon according to a threshold policy to reduce the workload of permanent repairman. For developing Markov model, we have made the underlying assumptions:

- The machining system is composed of Y cold and S warm standbys machines along with M operating machines. The system operates under the (m, M) policy, i.e., the system can work with at least m $(<M)$ machines in short mode whereas M operating machines are required for the normal functioning of the machining system.
- The life times of the operating and standby machines follow the exponential distribution. The operating machines may fail with rate of λ and the failure rate

of the cold standby is zero whereas the warm standbys fail with a rate of α.

- After the use of all spares, the system starts to fail in a degraded fashion with a failure rate λ_d.
- There is provision of two repairmen for the repair of the failed machines in the maintenance facility; the first one is appointed on the permanent basis and the second one is secondary removable repairmen which can be called upon to reduce the burden of loaded permanent repairman. The permanent and additional repairmen provide the repair following the exponential distribution at the rate of μ and μ_a, respectively. The permanent and additional repairmen turn on according to N-policy and a threshold policy respectively, on time sharing basis. The first permanent repairman follows the N-policy according to which it starts the repair work only when there are N failed machines accumulated in the system. Once the permanent repairman initiates the repair, it continues its job in time sharing manner till all the failed machines are repaired.
- The additional removable repairman gets activated when all spare machines are exhausted and the system will go in the short mode with the occurrence of failure of next machine. Thus, to prevent the system to work in degraded mode, the additional server will be called upon at a threshold level $N_1 = Y + S$. Furthermore, it becomes deactivated as soon as the workload of failed machines drops below N_1.
- The failed machines are repaired by the repairmen following the FCFS rule, i.e., the failed machines are queued up in the order in which they failed and join the system. Both the repairmen provide the repair on the time sharing basis. They take care of all the failed machines in the queue for a small interval of time as the time has been shared equally by all available failed machines in the queue. The machine which has been attended by the repairman will join the queue to be served again if its repair has not been completed otherwise it will leave the system. The rate of sharing time by both repairmen is ϕ (n) which can be considered as the reciprocal of the available numbers of failed machines in the queue.
- In case of failure of any machine, the switchover time of the standby machine (if available) from standby state to operating state of the machines is considered to be instantaneous. It is to be mentioned that the cold standbys are used to switch over the failed machines before warm standbys (cf. Gross et al. 2009; Jain et al. 2012; Maheshwari and Ali 2013).

Let λ_n and μ_n denote the down and up transition rates corresponding to exponentially distributed life and repair processes of the machines, respectively; here, suffix 'n' denote the number of failed machines in the system. The state transition diagram, showing in-flows and out-flows of system states, is depicted in Fig. 1. The state-dependent failure and repair rates are defined as follows:

$$\lambda_n = \begin{cases} M\lambda + S\alpha; & 0 \le n \le Y \\ M\lambda + (Y + S - n)\alpha; & Y < n \le Y + S \\ (M + Y + S - n)\lambda_d; & Y + S < n \le L = M + Y + S - m + 1 \end{cases}$$

and

$$\mu_n = \begin{cases} \mu\phi(n); & 1 \le n < Y + S \\ (\mu + \mu_a)\phi(n); & Y + S \le n \le L = M + Y + S - m + 1 \end{cases}$$

We denote the steady-state probabilities of the system states when there are 'n' failed machines in the system, as follows:

$P_{0,n}$ The steady-state probability that there is n failed machine in the system which is in accumulation state.

$P_{1,n}$ The steady-state probability that the first or both repairmen are activated and there are n numbers of failed machines present in the system at any instant.

$P_{Y+S}(1)$ Probability that first permanent repairman is performing the repair work of the failed machine at the threshold level $Y + S$ when all standby machines are used.

$P_{Y+S}(2)$ Probability that second additional repairman is performing the repair work of the failed machine at the threshold level $Y + S$ when all standby machines are used.

Governing equations

In this section, Chapman–Kolmogorov equations for all the states of the system using the appropriate transition rates for three different situations ($N = Y, N < Y$ and $N > Y$) have been constructed.

Case I: The first repairman starts repair when all cold standby machines (Y) are exhausted, i.e. when $N = Y$

$$-(M\lambda + S\alpha)P_{0,0} + \mu\phi(1)P_{1,1} = 0 \tag{1}$$

$$-(M\lambda + S\alpha)P_{0,n} + (M\lambda + S\alpha)P_{0,n-1} = 0; \quad 1 \le n \le N - 1 \tag{2}$$

$$-[M\lambda + S\alpha + \mu\phi(1)]P_{1,1} + \mu\phi(2)P_{1,2} = 0 \tag{3}$$

$$-[M\lambda + S\alpha + \mu\phi(n)]P_{1,n} + (M\lambda + S\alpha)P_{1,n-1} + \mu\phi(n+1)P_{1,n+1} = 0; \quad 2 \le n \le N - 1 \tag{4}$$

$$-[M\lambda + S\alpha + \mu\phi(N)]P_{1,N} + (M\lambda + S\alpha)P_{1,N-1} + \mu\phi(N+1)P_{1,N+1} + (M\lambda + S\alpha)P_{0,N-1} = 0 \tag{5}$$

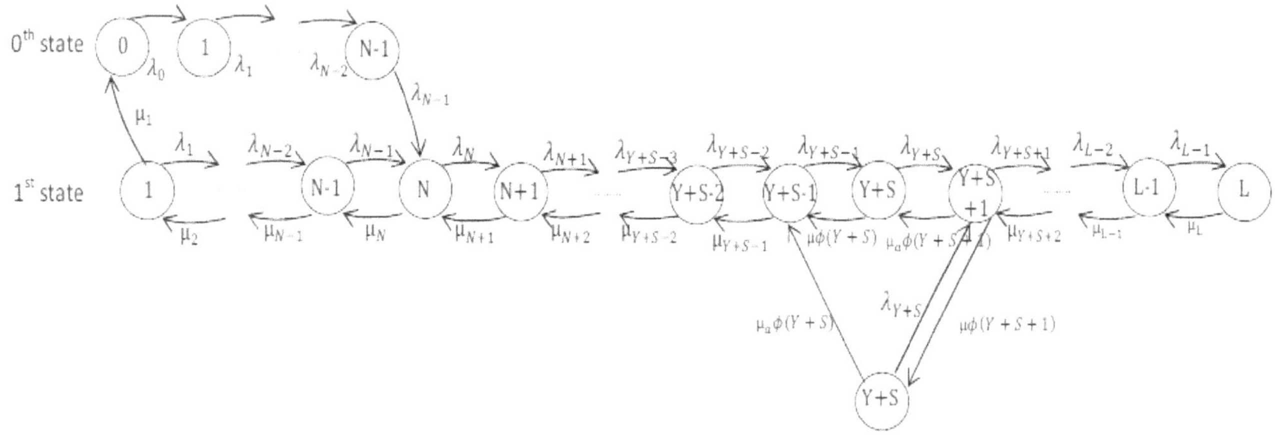

Fig. 1 State transition diagram

$$-[M\lambda + (Y+S-n)\alpha + \mu\phi(n)]P_{1,n} + [M\lambda + (Y+S-n+1)\alpha]P_{1,n-1} + \mu\phi(n+1)P_{1,n+1} = 0; \quad N+1 \le n \le Y+S-2 \tag{6}$$

$$-[M\lambda + \alpha + \mu\phi(Y+S-1)]P_{1,(Y+S-1)} + (M\lambda + 2\alpha)P_{1,Y+S-2} + \mu\phi(Y+S)P_{Y+S}(1) + \mu_a\phi(Y+S)P_{Y+S}(2) = 0 \tag{7}$$

$$-[M\lambda + \mu\phi(Y+S)]P_{(Y+S)}(1) + (M\lambda + \alpha)P_{1,Y+S-1} + \mu_a\phi(Y+S+1)P_{1,Y+S+1} = 0 \tag{8}$$

$$-[M\lambda + \mu_a\phi(Y+S)]P_{Y+S}(2) + \mu\phi(Y+S+1)P_{1,Y+S+1} = 0 \tag{9}$$

$$-[(M-1)\lambda_d + (\mu+\mu_a)\phi(Y+S+1)]P_{1,Y+S+1} + M\lambda P_{Y+S}(1) + M\lambda P_{Y+S}(2) \; (\mu+\mu_a)\phi(Y+S+2)P_{1,Y+S+2} = 0 \tag{10}$$

$$-[(M+Y+S-n)\lambda_d + (\mu+\mu_a)\phi(n)]P_{1,n} + (M+Y+S-n+1)\lambda_d P_{1,n-1} + (\mu+\mu_a)\phi(n+1)P_{1,n+1} = 0; \quad Y+S+2 \le n < L \tag{11}$$

$$-(\mu+\mu_a)\phi(L)P_{1,L} + m\lambda_d P_{1,L-1} = 0 \tag{12}$$

Case II: The number of cold standby machines (*Y*) is less than the threshold value (*N*) at which the repair starts, i.e. when *N* < *Y*

The steady-state probabilities of the states (0, n) for $0 \le n \le N-1$ are governed by Eqs. (1)–(2). Also for the states (1, n) when $1 \le n \le N$, we can refer Eqs. (3)–(5).

Now, we construct the equations for the states (1, $N+1$) to (1, $Y+S-2$) as follows:

$$-[M\lambda + S\alpha + \mu\phi(n)]P_{1,n} + (M\lambda + S\alpha)P_{1,n-1} + \mu\phi(n+1)P_{1,n+1} = 0; N+1 \le n \le Y \tag{13}$$

$$-[M\lambda + (S-1)\alpha + \mu\phi(Y+1)]P_{1,Y+1} + (M\lambda + S\alpha)P_{1,Y} + \mu\phi(Y+2)P_{1,Y+2} = 0 \tag{14}$$

$$-[M\lambda + (Y+S-n)\alpha + \mu\phi(n)]P_{1,n} + [M\lambda + (Y+S-n+1)\alpha]P_{1,n-1} + \mu\phi(n+1)P_{1,n+1} = 0; \quad Y+2 \le n \le Y+S-2 \tag{15}$$

For the range (1, $Y+S-1$) to (1, L), Eqs. (7)–(12) hold.

Case III: The threshold parameter (*N*) is more than the number of cold standby machines (*Y*) and is less than the total number (*Y* + *S*) of standbys machines, i.e. when *Y* < *N* < *Y* + *S*

In this case, Eqs. (1) and (3), will hold for the states (0, 0) and (1, 1), respectively. For the states (0, 1) to (0, $N-1$), we have the following equations:

$$-(M\lambda + S\alpha)P_{0,n} + (M\lambda + S\alpha)P_{0,n-1} = 0; 1 \le n \le Y \tag{16}$$

$$-[M\lambda + (S-1)\alpha]P_{0,Y+1} + (M\lambda + S\alpha)P_{0,Y} = 0 \tag{17}$$

$$-[M\lambda + (Y+S-n)\alpha]P_{0,n} + [M\lambda + (Y+S-n+1)\alpha]P_{0,n-1} = 0; \quad Y+2 \le n \le N-1 \tag{18}$$

For the states (1, 2) to (1, N), we construct the following equations:

$$-\,[M\lambda + S\alpha + \mu\phi(n)]P_{1,n} + (M\lambda + S\alpha)P_{1,n-1} + \mu\phi(n+1)P_{1,n+1} = 0; 2 \le n \le Y \quad (19)$$

$$-\,[M\lambda + (S-1)\alpha + \mu\phi(Y+1)]P_{1,Y+1} + (M\lambda + S\alpha)P_{1,Y} + \mu\phi(Y+2)P_{1,Y+2} = 0 \quad (20)$$

$$-[M\lambda + (Y+S-n)\alpha + \mu\phi(n)]P_{1,n} + [M\lambda + (Y+S-n+1)\alpha]P_{1,n-1} + \mu\phi(n+1)P_{1,n+1} = 0; \; Y+2 \le n \le N-1 \quad (21)$$

$$-\,[M\lambda + (Y+S-N)\alpha + \mu\phi(N)]\,P_{1,N} + [M\lambda + (Y+S-N+1)\alpha]P_{1,N-1} + \mu\phi(N+1)P_{1,N+1} + [M\lambda + (Y+S-N+1)\alpha]P_{0,N-1} = 0 \quad (22)$$

For the states in the range $(1, N+1)$ to $(1, L)$, Eqs. (6)–(12) also hold.

Queue size distribution

The queue size for the steady-state probabilities $P_{0,n}$ and $P_{1,n}$ can be obtained by solving the governing equations, which can be further used for the evaluation of the performance measures of interest.

Case I: When $N = Y$

In this case, when the number of cold standbys and the threshold level N are equal, the queue size for the steady-state probabilities $P_{0,n}$, $P_{1,n}$, $P_{Y+S}(1)$ and $P_{Y+S}(2)$ can be obtained by solving the Eqs. (1)–(12) recursively in the following manner.

Equation (1) can be written as:

$$P_{1,1} = \frac{\Lambda}{a_1}P_{0.0} \quad (23)$$

where $\Lambda = M\lambda + S\alpha; \quad a_n = \mu\phi(n)$.

From Eq. (2), we can get

$$P_{0,n} = P_{0,0}; 1 \le n \le N-1 \quad (24)$$

Solving Eqs. (3) and (4), we obtain

$$P_{1,n} = \frac{\Lambda^n + \sum_{k=1}^{n-1}\Lambda^{n-k}\gamma^{(k)}}{\gamma^{(n)}}P_{0.0}; \quad 1 \le n \le N \quad (25)$$

where $\gamma^{(n)} = \prod_{j=1}^{n} a_j$.

On solving Eq. (5) for $n = N$, we obtain

$$P_{1,N+1} = \frac{B_1}{\gamma^{(N+1)}}P_{0.0} \quad (26)$$

where, $B_1 = \Lambda^{Y+1} + \sum_{k=1}^{Y-1}\Lambda^{Y+1-k}\gamma^{(k)}$ (as we already know that we are considering the case when $N = Y$).

Again, solving recursively for $n = N+1$ to $Y+S-2$, we obtain

$$P_{1,n} = \frac{B_1}{\gamma^{(n)}}\prod_{i=1}^{n-Y-1}(\lambda_{Y+i})P_{0.0}; \quad N+2 \le n \le Y+S-1 \quad (27)$$

where $\lambda_n = M\lambda + (Y+S-n)\alpha$.

To obtain the steady-state probabilities for the nodes $Y+S$, $Y+S+1$ and $Y+S+2$, i.e., $P_{1,Y+S}$, $P_{1,Y+S+1}$, $P_{1,Y+S+2}$, we solve the Eqs. (7), (8), (9) and (10), recursively. Solving Eq. (7) and (8), we get

$$P_{Y+S}(1) = \frac{B_1}{\gamma^{(Y+S)}}\prod_{i=1}^{S-1}(\lambda_{Y+i})P_{0.0} - \frac{\mu_a}{\mu}P_{Y+S}(2) \quad (28)$$

$$P_{1,Y+S+1} = \frac{B_1\prod_{i=1}^{S}(\lambda_{Y+i})}{\mu_a\phi(Y+S+1)\gamma^{(Y+S)}}P_{0,0} - \left[\frac{\lambda_{Y+S} + a_{Y+S}}{a_{Y+S+1}}\right]P_{Y+S}(2) \quad (29)$$

Multiplying the above equation by $\mu\phi(Y+S+1)$ on both sides and then solving it simultaneously with Eq. (9), we get

$$P_{Y+S}(2) = \frac{\mu}{\mu_a}C_1\xi_1P_{0,0} \quad (30)$$

where $\xi_1 = \frac{B_1\prod_{i=1}^{S}(\lambda_{Y+i})}{\gamma^{(Y+S)}}$ and $C_1 = \frac{1}{2\lambda_{Y+S} + (\mu+\mu_a)\phi(Y+S)}$. On substituting the value of $P_{Y+S}(2)$ in Eqs. (28) and (29), we get

$$P_{Y+S}(1) = \frac{\lambda_{Y+S} + (\mu + \mu_a)\phi(Y+S)}{\lambda_{Y+S}}C_1\xi_1P_{0,0} \quad (31)$$

and

$$P_{1,Y+S+1} = \frac{\lambda_{Y+S} + \mu_a\phi(Y+S)}{\mu_a\phi(Y+S+1)}C_1\xi_1P_{0,0} \quad (32)$$

$$P_{1,Y+S} = P_{Y+S}(1) + P_{Y+S}(2) \quad (33)$$

From Eqs. (10), (11) and (12), we can obtain the results for the remaining states as:

$$P_{1,n} = \frac{\lambda_{Y+S} + \mu_a\phi(Y+S)}{\mu_a\phi(Y+S+1)} \times \prod_{i=1}^{n-(Y+S)-1}\left[\frac{\lambda'_{Y+S+i}}{(\mu+\mu_a)\phi(Y+S+i-1)}\right] \times C_1\xi_1P_{0,0}; \quad Y+S+2 \le n \le L \quad (34)$$

where $\lambda'_n = (M + Y + S - n)\lambda_d$.

Thus, the queue size distribution for case (I) is given in the Eqs. (24)–(27) and (30)–(34). To obtain the probability $P_{0,0}$, the following normalizing condition is used:

$$\sum_{i=0}^{N-1} P_{0,i} + \sum_{j=1}^{L} P_{1,j} = 1 \tag{35}$$

Case II: When $N < Y$

In the similar manner as in case I, on solving Eqs. (1)–(5), (7)–(15) recursively and using the notations:

$$B = \Lambda^n + \sum_{k=1}^{n-2} \Lambda^{n-k}\gamma^{(k)}, \xi = \frac{B\prod_{i=1}^{S}\lambda_{Y+i}}{\gamma^{(Y+S)}},$$

$$C = \frac{1}{2\lambda_{Y+S} + (\mu + \mu_a)\phi(Y+S)},$$

$$Z = \sum_{k=1}^{n-Y-1}\left(\prod_{i=k+1}^{n-Y-1}\lambda_{Y+i}\right)\gamma^{(Y+k)},$$

$$P_{Y+S}(1) = \frac{\lambda_{Y+S} + (\mu+\mu_a)\phi(Y+S)}{\lambda_{Y+S}} \times C\xi P_{0,0},$$

and $P_{Y+S}(2) = \frac{\mu}{\mu_a} C\xi P_{0,0}$

and other notations being same as used in case I, we obtain the queue size distribution as follows:

$$P_{0,n} = P_{0,0}; 1 \le n \le N-1 \tag{36}$$

$$P_{1,n} = \begin{cases} \left(\frac{\Lambda^n + \sum_{k=1}^{n-1}[\Lambda^{n-k}\gamma^{(k)}]}{\gamma^{(n)}}\right)P_{0,0}; 1 \le n \le N \\ \left(\frac{\Lambda^n + \sum_{k=1}^{n-2}[\Lambda^{n-k}\gamma^{(k)}]}{\gamma^{(n)}}\right)P_{0,0}; \quad N+1 \le n \le Y+1 \\ \frac{B\prod_{i=1}^{n-Y-1}(\lambda_{Y+i}) + \frac{\Lambda^2}{a_Y}Z}{\gamma^{(n)}}P_{0,0}; \quad Y+2 \le n \le Y+S-1 \\ P_{Y+S}(1) + P_{Y+S}(2); n = Y+S \\ \frac{\lambda_{Y+S} + \mu_a\phi(Y+S)}{\mu_a\phi(Y+S+1)}C\xi P_{0,0}; \quad n = Y+S+1 \\ \frac{\lambda_{Y+S} + \mu_a\phi(Y+S)}{\mu_a\phi(Y+S+1)} \\ \times \prod_{i=1}^{n-(Y+S)-1}\left[\frac{\lambda'_{Y+S+i}}{(\mu+\mu_a)\phi_{Y+S+i+1}}\right] \\ \times C\xi P_{0,0}; \\ Y+S+2 \le n \le L \end{cases} \tag{37}$$

Case III: When $N > Y$

Using the following notations,

$$B_2 = \Lambda^{Y+1} + \sum_{k=1}^{Y}(\Lambda^{Y-k+1}\gamma^{(k)}),$$

$$\xi_2 = \frac{B_2\prod_{i=1}^{Y+S-N}\lambda_{N+i}}{\gamma^{(Y+S)}},$$

$$C_2 = \frac{1}{2\lambda_{Y+S} + (\mu+\mu_a)\phi(Y+S)},$$

$$Z_1 = \sum_{k=1}^{n-Y-1}\left(\prod_{i=k+1}^{n-Y-1}\lambda_{Y+i}\right)\gamma^{(Y+k)},$$

we have

$$P_{Y+S}(1) = \frac{\lambda_{Y+S} + (\mu+\mu_a)\phi(Y+S)}{\lambda_{Y+S}} \times C_2\xi_2 P_{0,0}$$

and

$$P_{Y+S}(2) = \frac{\mu}{\mu_a}C_2\xi_2 P_{0,0}$$

From Eqs. (1), (3), (6)–(12) and (16)–(22), we derive the queue size as follows:

$$P_{0,n} = \begin{cases} P_{0,0}; 1 \le n \le Y \\ \frac{\Lambda}{\gamma^{(n)}}P_{0,0}; \quad Y+1 \le n \le N \end{cases} \tag{38}$$

$$P_{1,n} = \begin{cases} \left(\frac{\Lambda^n + \sum_{k=1}^{n-1}(\Lambda^{n-k}\gamma^{(k)})}{\gamma^{(n)}}\right)P_{0,0}; \quad 1 \le n \le Y+1 \\ \frac{B_2\prod_{i=1}^{n-Y-1}(\lambda_{Y+i}) + \Lambda Z_1}{\gamma^{(n)}}P_{0,0}; \quad Y+2 \le n \le N \\ \frac{B_2}{\gamma^{(n)}}\prod_{i=1}^{n-Y-1}\lambda_{Y+i}P_{0,0}; \quad N+1 \le n \le Y+S-1 \\ P_{Y+S}(1) + P_{Y+S}(2); n = Y+S \\ \frac{\lambda_{Y+S} + \mu_a\phi(Y+S)}{\mu_a\phi(Y+S+1)}C_2\xi_2 P_{0,0}; \quad n = Y+S+1 \\ \frac{\lambda_{Y+S} + \mu_a\phi(Y+S)}{\mu_a\phi(Y+S+1)} \\ \times \prod_{i=1}^{n-(Y+S)-1}\left[\frac{\lambda'_{Y+S+i}}{(\mu+\mu_a)\phi_{Y+S+i+1}}\right] \\ \times C_2\xi_2 P_{0,0}; \quad Y+S+2 \le n \le L \end{cases} \tag{39}$$

Special cases

To establish the validity of the results obtained in "Queue size distribution" (Case I), we explore some special cases by varying the values of $\phi(n)$ and N as follows:

1. If $\phi(n) = \frac{1}{n}$, the model portrays a time sharing and state-dependent MRP problem working under N-policy. For the state (0, n), the state probabilities are same

as given by Eq. (24). For the state (1, n) using

$$D = \Lambda^{N+1} + \sum_{k=1}^{N-1} \frac{(\Lambda^{N+1-k}\mu^k)}{k!}$$

and

$$E = \left[\frac{(Y+S+1)!}{\mu^{Y+S} \times \mu_a}\right]\left[\frac{(Y+S)\lambda_{Y+S} + \mu + \mu_a}{2(Y+S)\lambda_{Y+S} + \mu + \mu_a}\right]$$

we have

$$P_{Y+S}(1) = \left[\frac{(Y+S)\lambda_{Y+S} + \mu + \mu_a}{2(Y+S)\lambda_{Y+S} + \mu + \mu_a}\right] \times \frac{(Y+S)!D}{\mu^{Y+S} \times \lambda_{Y+S}} \times \left[\prod_{i=1}^{s} \lambda_{Y+i}\right] P_{0,0}$$

$$P_{Y+S}(2) = \frac{\mu}{\mu_a} \times \frac{(Y+S)\lambda_{Y+S}}{2(Y+S)\lambda_{Y+S} + \mu + \mu_a} \times \frac{(Y+S)!D}{\mu^{Y+S}} \times \left[\prod_{i=1}^{s} \lambda_{Y+i}\right] P_{0,0}$$

Thus, we get the following probabilities:

$$P_{1,n} = \begin{cases} \frac{n!}{\mu^n}\left[\Lambda^n + \sum_{k=1}^{n-1} \frac{(\Lambda^{n-k}\mu^k)}{k!}\right] P_{0,0};\ 1 \le n \le N \\ \frac{(N+1)!}{\mu^{N+1}}[D]P_{0,0};\ n = N+1 \\ \frac{n!D}{\mu^n}\left[\prod_{i=1}^{n-Y-1} \frac{\lambda_{Y+i}}{\mu}(Y+i)\right] P_{0,0};\ Y+2 \le n \le Y+S-1 \\ P_{Y+S}(1) + P_{Y+S}(2);\ n = Y+S \\ \left[\frac{D(Y+S+1)!}{\mu^{Y+S} \times \mu_a}\right] \times \left[\frac{(Y+S)\lambda_{Y+S} + \mu_a}{(2Y+S)\lambda_{Y+S} + \mu + \mu_a}\right] \times \left[\prod_{i=1}^{s} \lambda_{Y+i}\right] P_{0,0};\ n = Y+S+1 \\ \left[\prod_{i=1}^{n-(Y+S)-1} \frac{\lambda'_{Y+S+i}(Y+S+i+1)}{(\mu+\mu_a)}\right] \times \left[\prod_{i=1}^{s} \lambda_{Y+i}\right] \times \mathrm{DE} \times \mathrm{P}_{0,0};\ Y+S+2 \le n \le L \end{cases} \quad (40)$$

2. If $\phi(n) = 1$, the model reduces to a state-dependent N-policy MRP model with additional repairman. In this case, for brevity the following notations have been used

$$G = \left[\Lambda^{N+1} + \sum_{k=1}^{N-1} (\Lambda^{N+1-k}\mu^k)\right],$$

$$H = \left[\frac{\lambda_{Y+S} + \mu_a}{2\lambda_{Y+S} + \mu + \mu_a}\right]$$

$$P_{Y+S}(1) = \left[\frac{GH}{\mu^{Y+S} \times \mu_a}\right] \times \left[\prod_{i=1}^{s} \lambda_{Y+i}\right] P_{0,0}$$

and $P_{Y+S}(2) = \frac{\mu \times G}{\mu_a \times \mu^{Y+S}} \times \frac{1}{2\lambda_{Y+S} + (\mu + \mu_a)} \times [\prod_{i=1}^{s} \lambda_{Y+i}] P_{0,0}$

The steady-state probabilities $P_{0,n}$ can be obtained by Eq. (24). The steady-state probabilities $P_{1,n}$ become:

$$P_{1,n} = \begin{cases} \frac{1}{\mu^n}\left[\Lambda^n + \sum_{k=1}^{n-1} (\Lambda^{n-k}\mu^k)\right] P_{0,0};\ 1 \le n \le N \\ \frac{1}{\mu^{N+1}}[G]P_{0,0};\ n = N+1 \\ \frac{G}{\mu^n}\left[\prod_{i=1}^{n-Y-1} \lambda_{Y+i}\right] P_{0,0};\ Y+2 \le n \le Y+S-1 \\ P_{Y+S}(1) + P_{Y+S}(2);\ n = Y+S \\ \left[\frac{GH}{\mu^{Y+S} \times \mu_a}\right] \times \left[\prod_{i=1}^{s} \lambda_{Y+i}\right] P_{0,0};\ n = Y+S+1 \\ \left[\prod_{i=1}^{n-(Y+S)-1} \frac{\lambda'_{Y+S+i}}{(\mu+\mu_a)}\right] \times \left[\prod_{i=1}^{s} \lambda_{Y+i}\right] \times \frac{GH}{\mu^{Y+S} \times \mu_a} P_{0,0};\ +S+2 \le n \le L \end{cases} \quad (41)$$

3. If $\phi(n) = \frac{1}{n}$, $N = 1$, the machining system reduces to time sharing state-dependent queueing system. The repairman gets activated as soon as a machine fails, i.e., N-policy is not taken into account. The queue size distribution can be obtained by substituting $N = 1$ and $\phi(n) = \frac{1}{n}$ in the Eqs. (25)–(27), (30) and (32)–(34).
4. If $\phi(n) = 1$, $N = 1$, the model provides results for a machining system with additional repairman but without N-policy and time sharing factor.

Performance indices

The performance indices of the concerned system can help the system engineer to develop an appropriate design for the concerned machining system. Using the probabilities obtained in the case I, some performance indices, viz. expected number of failed machines in the system, probability that the first repairman is busy, probability that both repairmen are busy, probability that the system is in accumulation state, throughput of the system and variance of the number of failed machines in the system have been established as follows:

- The expected number of failed machines in the system is

$$E(n) = \sum_{n=0}^{N-1} nP_{0,n} + \sum_{n=1}^{L} nP_{1,n} \quad (42)$$

- The probability of the system being in accumulation state is

$$P(A) = \sum_{n=0}^{N-1} P_{0,n} \quad (43)$$

- The probability that only first permanent repairman being in busy state is

$$P(\text{FB}) = \sum_{n=0}^{Y+S-1} P_{1,n} \quad (44)$$

- The probability that both the repairmen being in busy state is

$$P(\text{BB}) = \sum_{n=Y+S}^{L} P_{1,n} \quad (45)$$

- The throughput of the time-shared system is

$$\tau = \mu \sum_{n=1}^{Y+S-1} P_{1,n} + (\mu + \mu_a) \sum_{n=Y+S}^{L} P_{1,n} \quad (46)$$

- The variance of the number of failed machines is

$$\text{Var}(n) = \sum_{n=1}^{L} n^2 P_{1,n} + -(E(n))^2 \quad (47)$$

Cost function

The cost function for the time-shared machine repair problem has been constructed to make the system economic by the optimal choice of repair rates. It is desirable to reduce the cost as much as possible by setting the optimal service rate. For the concerned system, we define the cost factors associated with main activities as follows:

C_f Cost per unit time for each failed machine present in the system
C_a Cost per unit time in the accumulation state
C_p Cost per unit time of the permanent repairman
C_b Cost per unit time of the additional removable repairman

To achieve the maximum net profit, total average cost must be minimized. The total average cost is given by

$$E\{\text{TC}\} = C_f E(N) + C_a P(A) + C_p P(\text{FB}) + (C_b + C_p) P(\text{BB}) \quad (48)$$

Numerical analysis

To establish the utility of the performance model of the queueing system, the analytical solution is not enough as such it is important to do numerical simulation. The numerical results of the performance measures will be of great help to the system engineers and decision makers in improving and future designing the system. In this section, the sensitivity analysis is carried out for case I, by setting the default parameters for the numerical results depicted in Figs. 2, 3, 4 and Tables 1, 2, 3 as $Y = 4$, $S = 3$, $\lambda = 0.3$, $\alpha = 0.2$, $\lambda_d = 0.4$, $\mu = 0.5$, $\mu_a = 0.2$. For Tables 1, 2, 3 and Figs. 3, 4, the numerical results are obtained for $M = 5, 7, 9$. To obtain the variation in the cost function $E\{\text{TC}\}$ in Fig. 2, the results are obtained by setting $M = 7$. The effects of the failure rates of operating machines (λ, λ_d), the failure rate of spares (α), service rates (μ, μ_a) of the repairmen and the number of operating machines (M) have been examined on various performance measures such as the expectation $E(n)$ and variance Var(n) of the number of failed machines in the system, throughput (τ) and the expected total cost $E\{\text{TC}\}$ incurred on the system. The long run probability measures of the different states of the system like probability of the system being in accumulation state $P(A)$, probability when the first permanent repairman is in busy state $P(FB)$ and the probability when both repairmen are in busy state $P(BB)$ have also been explored numerically for the variation in different parameters. Now, we discuss the sensitivity of the parameters as follows:

- *Effect of the failure rate of machines* The failure rate of the operating machines (λ) affects the performance of a machining system significantly. It is clear from Table 1 that with the increase in the failure rate (λ) of the operating machines, the probability of both permanent and additional repairmen being busy P(BB) increases whereas the probability of the system being in accumulation state $P(A)$, the probability of only first repairman being busy P(FB) and the Var(n) decrease. It is also observed from Figs. 3a and 4a that the queue length $E(n)$ and the throughput (τ), respectively, increase with the increase in the failure rate of the operating machines. When all the spares have exhausted, the machines start failing with a degraded rate (λ_d) due to overload. The queue length $E(n)$ and the throughput (τ) of the system increase with the increase in the degraded failure rate (λ_d) of the machines. A converging pattern is observed in the graphs shown in the Figs. 3c and 4c.

  The spares are also likely to fail with rate (α) and also affect the system performance considerably. Table 2 depicts a similar variation in the probabilities and variance by increasing the failure rate of the spares as

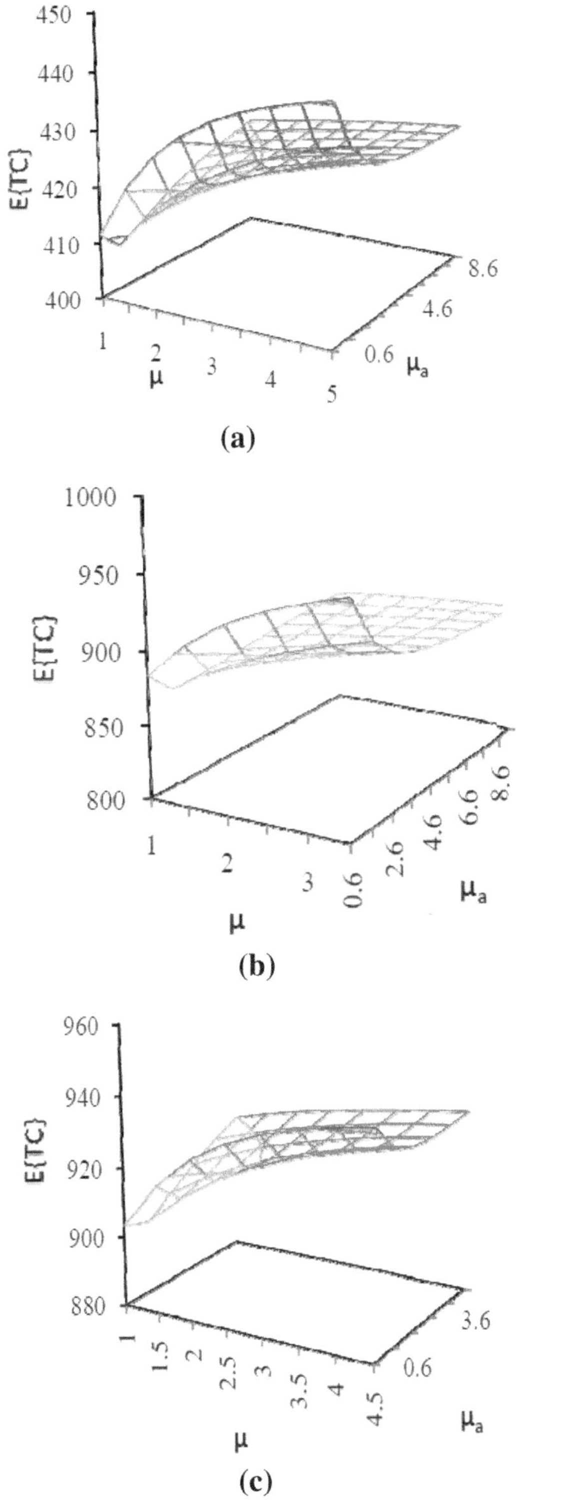

Fig. 2 Total cost of the system by varying μ for different values of μ_a

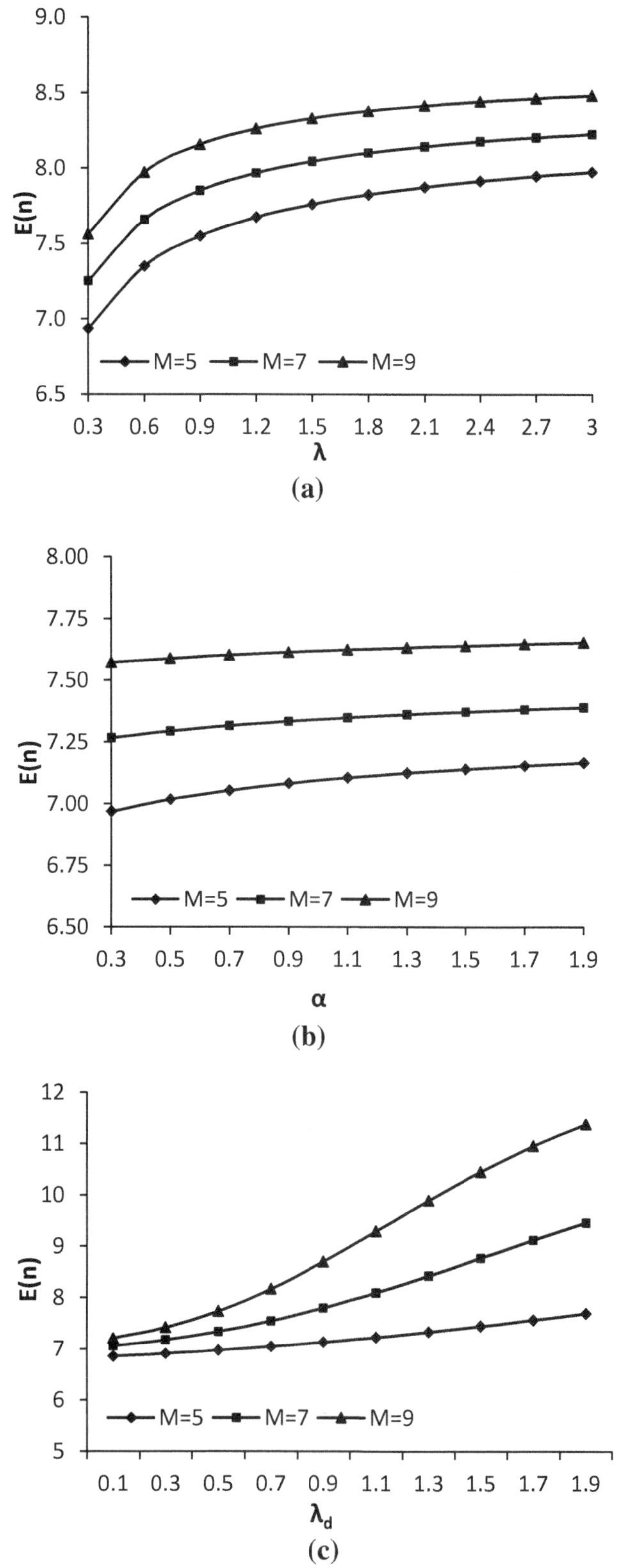

Fig. 3 Expected number of failed machines by varying a λ, b α, c λ_d for different values of M

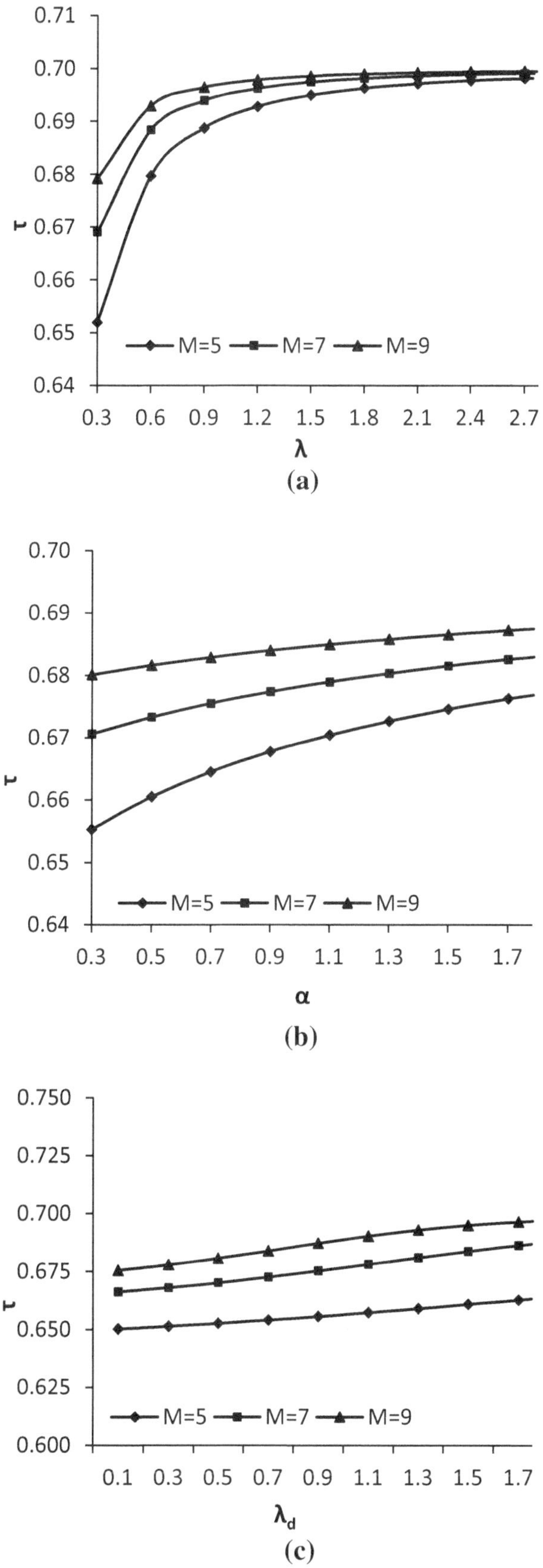

Fig. 4 Throughput of the system by varying **a** λ, **b** α, **c** λ_d for different values of M

Table 1 Performance measures of the system by varying M and λ

M	λ	$P(A)$	P(FB)	P(BB)	Var(n)
5	0.3	0.00070	0.23784	0.76147	0.85857
	1.2	0.00000	0.03607	0.96394	0.66354
	2.1	0.00000	0.01415	0.98586	0.58421
7	0.3	0.00011	0.15464	0.84525	1.01274
	1.2	0.00000	0.01873	0.98127	0.86666
	2.1	0.00000	0.00693	0.99307	0.77330
9	0.3	0.00002	0.10384	0.89614	1.32885
	1.2	0.00000	0.01064	0.98936	1.14656
	2.1	0.00000	0.00379	0.99622	1.04056

Table 2 Performance measures of the system by varying M and α

M	α	$P(A)$	P(FB)	P(BB)	Var(n)
5	0.3	0.00031	0.22237	0.77731	0.80831
	1.1	0.00000	0.14787	0.85213	0.63824
	1.9	0.00000	0.11138	0.88863	0.56696
7	0.3	0.00006	0.14678	0.85316	0.98939
	1.1	0.00000	0.10528	0.89472	0.88215
	1.9	0.00000	0.08256	0.91744	0.82964
9	0.3	0.00001	0.09956	0.90043	1.31454
	1.1	0.00000	0.07539	0.92461	1.24002
	1.9	0.00000	0.06098	0.93902	1.19949

Table 3 Performance measures of the system by varying M and μ

M	μ	$P(A)$	P(FB)	P(BB)	Var(n)
5	0.3	0.00049	0.27418	0.72533	1.1467
	1.2	0.00109	0.15881	0.84010	0.4994
	2.1	0.00130	0.11259	0.88611	0.3518
7	0.3	0.00007	0.16936	0.83057	1.5153
	1.2	0.00018	0.10675	0.89308	0.4701
	2.1	0.00021	0.07541	0.92437	0.2865
9	0.3	0.00001	0.10401	0.89598	2.0977
	1.2	0.00004	0.07709	0.92287	0.5343
	2.1	0.00005	0.05520	0.94476	0.2984

observed by varying the failure rate (λ) of the operating machines in Table 1. The queue length $E(n)$ and the throughput (τ) of the system increase gradually with the increase in the failure rate of the spares which can be seen in the Figs. 3b and 4b, respectively.

- *Effect of the repair rates of the permanent as well as additional repairmen* The repair rates of the permanent as well as additional repairmen affect the total cost of the system remarkably. It is clear from Table 3 that the long run probabilities of the system in accumulation state $P(A)$ and the P(BB) increases whereas the long run probabilities P(FB) and Var(n) decrease with the increase in the repair rate of the permanent repairman (μ).

- *Variation in the cost function* The variation in the expected total cost of the system $E\{TC\}$ has been observed for three different sets of cost parameters which are depicted in Fig. 2a–c. The different cost parameters set for the figures are as follows:

  I. $C_f =$ Rs. 50, $C_a =$ Rs. 10, $C_p =$ Rs. 80, $C_b =$ Rs. 100.
  II. $C_f =$ Rs. 100, $C_a =$ Rs. 50, $C_p =$ Rs. 200, $C_b =$ Rs. 300.
  III. $C_f =$ Rs. 100, $C_a =$ Rs. 50, $C_p =$ Rs. 250, $C_b =$ Rs. 250.

From Fig. 2a–c, it has been observed that the expected total cost of the system $E\{TC\}$increases with the increase in the repair rate of the additional repairman (μ). However, with the increase in the repair rate of the permanent repairman (μ_a), the total cost $E\{TC\}$ first decreases and then increases, i.e., $E\{TC\}$ shows the convexity with respect to repair rate (μ_a). From Fig. 2a, a minimum cost $E\{TC\} =$ Rs. 407.88 is obtained at optimal repair rates of the permanent repairman and additional repairman at $\mu = 1$ and $\mu_a = 1.37$.

Now, we can conclude our results as:

- The failure rates of both operating machines as well as standbys should be kept low to avoid the excessive workload at the repairman.
- The degraded failure rate of the system should also be kept low, otherwise it will result in huge queue at the repairmen.
- The repair rate of the additional repairman should be kept higher as compared to that of the permanent repairmen to minimize the overall cost of the system.

Discussion

In this investigation, threshold-based repair facility for the time-shared Markovian machine repair problem with mixed standbys under the care of one permanent repairman and one additional repairman has been studied. The features of mixed standbys, degraded failure and additional repairman incorporated in the model all together make our study more realistic and can be realized in several real world industrial organizations operating in multi-component machining environment. The repair rate of the failed operating machines and spare machines should be kept higher for smooth functioning of the system. The incorporation of threshold N-policy to turn on the permanent repairman makes our system cost effective and economic. It is realized in many machining systems that the permanent repairman cannot cope up with the increase in work load as such provision of additional repairman may be helpful in faster recovery of the failed machines. The numerical simulation of various performance indices facilitated will definitely provide insight to the system designers and industrial engineers to improve the efficiency and reliability of the concerned machining systems. The cost analysis carried out for the evaluation of minimum value of cost for a given set of other cost parameters signifies the validity and profitability of the model in a very effective manner and will be helpful to the decision makers in minimizing the cost of maintainability and in turn increase in the profit which is a highly desired trait of any organization. This work can be further extended by incorporating some more features, such as bulk failure and the switching failure.

References

Adiri I, Avi-Itzhak B (1969) A time-sharing queue. Manag Sci 15(11):639–657
Albright SC (1980) Optimal maintenance-repair policies for the machine repair problem. Naval Res Logist 27(1):17–27
Al-Seedy RO, Al-Ibraheem FM (2001) An inter arrival hyper-exponential machine interference with balking, reneging, state dependent, spares and an additional server for longer queues. Int J Math Math Sci 27(12):737–749
Altman E, Jimenez T, Kofman D (2010) Discriminatory processor sharing queue with stationary ergodic service times and the performance of TCP in overload. Comput Netw 54(9):1509–1519
Chandrasekaran M, Muralidhar M, Dixit US (2013) Online optimization of multi pass machining based on cloud computing. Int J Adv Manuf Technol 65(1–4):239–250
Coffman EG, Kleinrock L (1968) Feedback queueing models for time-shared systems. J ACM 15(4):549–576
Flapper SD, Gayon JP, Lim LL (2014) On the optimal control of manufacturing and remanufacturing activities with a single shared server. Eur J Oper Res 234(1):86–98
Gross D, Shortle JF, Thompson JM, Harris CM (2009) Fundamentals of Queueing Theory. Wiley, New York
Haque L, Armstrong MJ (2007) A survey of machine interference problem. Eur J Oper Res 179(1):469–482
Huang HI, Hsu PC, Ke JC (2011) Controlling arrival and service of a two-removable server system using genetic algorithm. Expert Syst Appl 38(8):10054–10059
Jain M (1998) M/M/m queue with discouragement and additional servers. Gujarat Stat Rev 25:31–42
Jain M (2014) Cost analysis of a machine repair problem with standby, working vacation and server breakdown. Int J Math Oper Res 6(4):437–451
Jain M, Preeti (2014) Transient analysis of a machine repair system with standby, two modes of failure, discouragement and switching failure. Int J Oper Res 21(3):365–390

Jain M, Baghel KPS (2001) A multi-components repairable problem with spare and state dependent rates. Nepali Math Sci Rep 19:81–92

Jain M, Bhargava C (2009) N-policy machine repair system with mixed standbys and unreliable server. Q Technol Quant Manag 6(2):171–184

Jain M, Gupta R (2011) Redundancy issues in software and hardware systems—an overview. Int J Reliab Qual Saf Eng 18(1):61–98

Jain M, Lata P (1995) Accumulated work process in a time-sharing queue with finite population. Acta Ciencia Indica Math 21(1):33–36

Jain M, Singh P (2003) Performance prediction of loss and delay Markovian queueing model with nopassing and removable additional servers. Comput Oper Res 30:1233–1253

Jain M, Rakhee, Maheshwari S (2004) N-policy for a machine repair system with spares and reneging. Appl Math Model 28(6):513–531

Jain M, Sharma GC, Shekhar C (2005) Processor-shared service systems with queue-dependent processor. Comput Oper Res 32:629–645

Jain M, Sharma GC, Pundhir RS (2007a) Reliability analysis of k-out-of n: G machining systems with mixed spares and multiple modes of failure. Int J Eng Trans A Basics 20(3):243–250

Jain M, Sharma GC, Singh N (2007b) Transient analysis of M/M/R machining system with mixed standbys, switching failures, balking, reneging and additional removable repairmen. Int J Eng Trans A Basics 20(2):169–182

Jain M, Sharma GC, Pundhir RS (2010) Some perspectives of machine repair problem. Int J Eng Trans B Appl 23:253–268

Jain M, Shekhar C, Shukla S (2012) Queueing analysis of a multi-component machining system having unreliable heterogeneous servers and impatient customers. Am J Oper Res 2(3):16–26

Jain M, Sharma GC, Rani V (2014a) M/M/M + r machining system with reneging, spares and interdependent controlled rates. Int J Math Oper Res 6(6):655–679

Jain M, Shekhar C, Rani V (2014b) N-policy for a multi-component machining system with imperfect coverage, reboot and unreliable server. Prod Manuf Res 2(1):457–476

Jain M, Shekhar C, Shukla S (2014c) Vacation queueing model for a machining system with two unreliable repairmen. Int J Oper Res 20(4):469–491

Jayachitra P, Albert AJ (2014) Recent developments in queueing models under N-Policy: a short survey. Int J Math Arch 5(3):227–233

Jeong HY, Park JH, Lee JD (2014) The cloud storage model for manufacturing system in global factory automation. In: The proceedings of 28th international workshop on advanced information networking and applications, Victoria, BC, pp 895–899

Ke JC, Lee SL, Liou CH (2009) Machine repair problem in production systems with spares and server vacations. RAIRO Oper Res 43(1):35–54

Kim J, Kim B (2007) The processor sharing queue with bulk arrivals and phase-type services. Perform Eval 64(4):277–297

Kleinrock L (1967) Time-shared systems: a theoretical treatment. J Assoc Comput Mach 14(2):242–261

Kumar K, Jain M (2013a) Threshold F-policy and N-policy for multi-component machining system with warm standbys. J Ind Eng Int 9(1):1–9

Kumar K, Jain M (2013b) Threshold N-policy for (M, m) degraded machining system with K-heterogeneous servers, standby switching failure and multiple vacations. Int J Math Oper Res 5(4):423–445

Liou CD, Wang KH, Liou MW (2013) Genetic algorithm to the machine repair problem with two removable servers operating under the triadic (0, Q, N, M) policy. Appl Math Model 37(18):8419–8430

Maheshwari S, Ali S (2013) Machine repair problem with mixed spares, balking and reneging. Int J Theor Appl Sci 5(1):75–83

Maheshwari S, Sharma P, Jain M (2010) Machine repair problem with k type warm spares, multiple vacations for repairmen and reneging. Int J Eng Technol 2(4):252–258

Rasch PJ (1970) A queueing theory study of round-robin scheduling of time-shared computer systems. J Assoc Comput Mach 17(1):131–145

Sharma DC (2012) Machine repair problem with spares and N-policy vacation. Res J Recent Sci 1(4):72–78

Sharma GC, Jain M, Pundhir RS (2005) Loss and delay multi-server queuing model with discouragement and additional servers. J Rajasthan Acad Phys Sci 4(2):115–120

Shawky AI (1997) Single server machine interference model with balking, reneging and an additional server for longer queues. Microelectron Reliab 37(2):355–357

Shawky AI (2000) The machine interference model: M/M/C/K/N with balking, reneging and spares. OPSEARCH 37(1):25–35

Tahar AB, Jean-Marie A (2012) The fluid limit of the multi class processor sharing queue. Queueing Syst 71(4):347–404

Taylor J, Jackson RRP (1954) An application of the birth death processes to the provision of spare machines. Oper Res Soc 5(4):95–108

Wang KH (1995) An approach to cost analysis of the machine repair problem with two types of spares and service rates. Microelectron Reliab 35(11):1433–1436

Wang KH, Sivazlian BD (1989) Reliability of a system with warm standbys and repairmen. Microelectron Reliab 29(5):849–860

Wang KH, Tai KY (2000) A queueing system with queue-dependent servers and finite capacity. Appl Math Model 24:807–814

Yadin M, Naor P (1963) Queueing system with a removable service station. Oper Res Q 14(4):393–405

Yashkov SF (1992) Mathematical problems in the theory of shared processor systems. J Soviet Math 58(2):101–147

Yashkov SF, Yashkova AS (2007) Processor sharing: a survey of the mathematical theory. Autom Remote Control 68(9):1662–1731

Yue D, Yue W, Qi H (2012) Performance analysis and optimization of a machine repair problem with warm spares and two heterogeneous repairmen. Optim Eng 13(4):545–562

Zhen Q, Knessel C (2009) On sojourn times in the finite capacity M/M/1 queue with processor sharing. Oper Res Lett 37(6):447–450

14

Pricing in a two-echelon supply chain with different market powers: game theory approaches

Afshin Esmaeilzadeh[1] · **Ata Allah Taleizadeh**[1]

Abstract In this research, the optimal pricing decisions for two complementary products in a two-echelon supply chain under two scenarios are studied. The proposed supply chain in each echelon includes one retailer and two manufacturers and the same complementary products are produced. In the first scenario, we assume the unit manufacturing costs of the complementary products in each echelon are the same, while in the second one the different unit manufacturing costs are supposed and lead to demand leakage from the echelon with the higher unit manufacturing cost to the echelon with the lower unit manufacturing cost. Moreover, under the second scenario, the products with lower price are replaced with the higher price products. The purpose of this study is to analyze the effects of different market powers between the manufacturers and the retailer and the demand leakage on the optimal wholesale and retail prices and also on the profit of the chain. The relationships between the manufacturers and the retailer are modeled by the MS-Stackelberg and MS-Bertrand game-theoretic approach where the manufacturers are leaders and the retailers are followers.

Keywords Pricing · Complementary products · Market power · MS-Stackelberg game · MS-Bertrand game

Introduction and literature review

Market power as the principal companies' success factors is a primitive and important challenge to which companies are faced. The companies, which are competing in the same market, are attempting to increase own market penetrations by using different implements to achieve the more market power than the other rivals. The market power leads to enhance the penetrability of companies so that the market would be handled by the powerful firms (Wei et al. 2013; Zhao et al. 2014). One of the practical and the efficient implements which cause to improve the companies' revenue and also their power market is presenting an optimal price where the same products are launched to the market. So, pricing policy as the useful tool which can solve this imperative problem is recognized by enterprises for decades. In fact, the companies attempt to optimize their selling prices to acquire the more market demand.

Recently, many researchers are focused on the pricing policies. For instance, Starr and Rubinson (1978) proposed a model to survey the relation between the demand of product and its prices. Dada and Srikanth (1987) studied pricing policies under quantity discounts. Kim and lee (1998) employed pricing and ordering strategies for a single item with fixed or variable capacity to maximize the profit of firm faced to price-sensitive and deterministic demand over a planning horizon. Boyaci and Gallego (2002) considered joint pricing and ordering decisions in a supply chain consisting of a wholesaler and one or several retailers. A complete review of dynamic pricing models was presented by Elmaghraby and Keskinocak (2003). Several studies applied pricing policy with coordination mechanisms under different assumptions (Chen and Simchi-Levi 2004a, b, 2006; Chen et al. (2006); Xiao et al. (2010); Wei and Zhao (2011); Yu and Ma (2013); Maihami and Karimi (2014); Taleizadeh and Noori-daryan (2014)). Sinha and Sarmah (2010) studied pricing decisions in a distribution channel under the competition and coordination issues in which two competitive vendors sell products to a common retailer in the same market. A comprehensive

✉ Ata Allah Taleizadeh
Taleizadeh@ut.ac.ir

[1] School of Industrial Engineering, College of Engineering, University of Tehran, Tehran, Iran

review of pricing models for a multi-product system is performed by Soon (2011). Shavandi et al. (2012) presented a new constrained pricing and inventory model for perishable products which those are classified to complementary, substitutable and independent products. Their aim is to optimize the prices, inventory and production decisions such that the total profit is maximized. Mahmoodi and Eshghi (2014) presented three algorithms to obtain the optimal pricing decisions in a duopoly supply chain. Taleizadeh et al. (2014) developed a vendor managed inventory (VMI) model in a two-level supply chain including a vendor and multiple retailers to survey the optimal pricing and inventory policies such that the total profit of the chain is maximized.

The concept of complementary products is suggested when the customer has to purchase more than one product at the same time so that the products could have the required efficiency (Yue et al. 2006). For an instance, software and hardware systems of a computer are two complementary products and should be purchased together to have the required efficiency for the customer. But, if a customer is not satisfied enough with a purchased product and purchases a similar product, then these two products will be substitutable products. For example; different marks of software or hardware systems of a computer may be considered as substitutable products. Several researchers examine the effects of complementary and substitutable products on the profit of inventory systems. For example, the pricing decisions of two complementary products as the bundle policy is studied by Yue et al. (2006) where the products are produced by two separate firms. Mukhopadhyay et al. (2011) considered a duopoly market where two independent firms offer complementary goods under information asymmetry. The Stackelberg game-theoretic model to solve the proposed model is utilized. Yan and Bandyopadhyay (2011) proposed a profit-maximization model and applied a bundle pricing policy for complementary items. Wei et al. (2013) examined the pricing problem under the different market powers structures between members of a supply chain with two manufacturers and one retailer for two complementary products. Wang et al. (2014) employed pricing policy for two complementary products in a fuzzy environment and they survey the changes of the optimal retail prices of two complementary products under two different scenarios. Wei et al. (2015) presented joint optimal pricing and warranty period of two complementary products in a supply chain with two manufacturers and one common retailer under horizontal firm's cooperation/noncooperation strategies.

Tang and Yin (2007) extended the Starr and Rubinson (1978)'s work for two substitutable products under the fixed and variable pricing strategies. The goal of this paper is to jointly determine optimal order quantity and retail price. Hsieh and Wu (2009) and Gurler and Yilmaz (2010) employed coordinating mechanisms for substitutable products under various assumptions. Then two problems are carried out by Zhao et al. (2012a, b) such that in the first one, a pricing problem of substitutable products in a fuzzy environment is discussed. In the second one, a pricing policy in a supply chain including one manufacturer and two competitive retailers for substitutable products where the customers' demand and the manufacturing costs are non-deterministic is employed. Chen et al. (2013) discussed pricing problem for substitutable products under traditional and online channels in a two-stage supply chain including a manufacturer and a retailer where the manufacturer sells a product to a retailer and also sells directly to customers through an online channel. Hsieh et al. (2014) surveyed pricing and ordering decisions of partners of a supply chain including multiple manufacturers and a retailer under demand uncertainty where each manufacturer produces a different substitutable product which is sold through the retailer. Zhao et al. (2014) developed a pricing model for substitutable products under the different market power of firms in a supply chain with two competitive manufacturers and a retailer. Fei et al. (2015) considered a price model for one supplier and multiple retailers under different product substitution degrees. In this article, the authors studied the effect of sub-packaging cost on the retail price.

Panda et al. (2015) studied joint pricing and replenishment policies in a dual-channel supply chain where the manufacturer is the leader of Stackelberg model. Zhang et al. (2014) developed a dynamic pricing model in a competitive supply chain under deterministic demand function to optimize the benefits of supply chain members. Also, they analyzed the profit sensitivity with respect to various factors. Giri and Sharma (2014) developed pricing model under cooperative and non-cooperative advertising in a supply chain with a single manufacturer and two competitive retailers. Consumer demand function depends on price and advertising. They show that cooperative advertising policy is more beneficial.

After reviewing comprehensively pricing problems of complementary and substitutable products, we found although several pricing models are developed to optimize the profit or cost of the inventory systems for complementary and substitutable products, the pricing problem of both complementary and substitutable products in a two-echelon supply chain with market power and demand leakage considerations is not discussed.

In this paper, a pricing model of complementary and substitutable products in a two-echelon supply chain in which each echelon including two manufacturers and one retailer under demand leakage is developed, where the different market powers are assumed for the chain members. Two different game-theoretic approaches including MS-Stackelberg and MS-Bertrand are employed to examine the pricing decisions of the chain members when the market power is different and subsequently demand leaks from one echelon to the second one.

The rest of the paper is organized as follows. The problem is described in Sect. 2. The model is formulated in Sect. 3. Section 4 provides solution methods under MS-Stackelberg and MS-Bertrand game-theoretic approaches. Sections 5 and 6 contain a numerical example, sensitivity analysis and conclusion as a summary of findings and some future researches.

Problem description

Consider a two-echelon supply chain including one retailer and two manufacturers in every echelon where each echelon supplies two complementary products. In the first echelon, manufacturers 1 and 2, respectively, produce two complementary products 1 and 2 and wholesale the products to retailer 1. Then retailer 1 sells the products 1 and 2 to the customers. In the second one, manufacturers 3 and 4 produce two complementary products 3 and 4 and wholesale them to retailer 2. Therefore, retailer 2 sells the products 3 and 4 to the customers. We assume two complementary products produced in each echelon of supply chain are the same such that products 1 and 3 and products 2 and 4 are the same.

In other words, based on Fig. 1 in which the schema of the supply chain is shown, the manufacturer i produces product i at unit manufacturing cost C_i and sells it to retailer j at unit wholesale price W_i. Afterward, the retailer j sells the product i to end users at unit retail price P_i where in the first echelon $i = 1, 2 j = 1$ and in the second echelon $i = 3, 4 j = 2$. Moreover, we assume that if the unit manufacturing cost C_i is different in each echelon of supply chain, the demand leakage from the echelon with the higher unit manufacturing cost to the echelon with lower unit manufacturing cost occurs. Therefore, the products 1 and 3 and also the products 2 and 4 will be transacted in the market as the substitutable products. This scheme can be used for software and hardware systems of a computer as described in previous section. These products are complementary and are produced by manufacturers 3 and 4, as different brands, respectively. So, if a customer is not satisfied enough from the purchased products of manufacturer 1 and 2, then products 1 and 3 and products 2 and 4 will be substitutable products.

The assumptions utilized to model the discussed problem are as follows.

1. Demand is deterministic and price-sensitive.
2. The same complementary products are produced in each echelon.
3. In the first model, the same unit manufacturing costs are considered for each echelon.
4. In the second model, different unit manufacturing costs are assumed for each echelon which is caused demand leakage between two echelons of the chain. So, the product with the higher unit manufacturing cost will be substituted by the products with the lower unit manufacturing cost.
5. The higher market power is assumed for the manufacturers than the retailer in each echelon so that the market is managed by the manufacturers.
6. Shortage is not allowed.
7. All the parameters are deterministic and positive.

The main aim of this paper is to study the optimal pricing policies in a two-echelon supply chain for two complementary products under two scenarios with the different market powers of each echelon partners. Two manufacturers and one retailer are the partners of each echelon and the problem is to determine the optimal values of wholesale prices of the manufacturers and the selling prices of the retailers to maximize the profit of the chain.

The following notations are used to develop the problem.

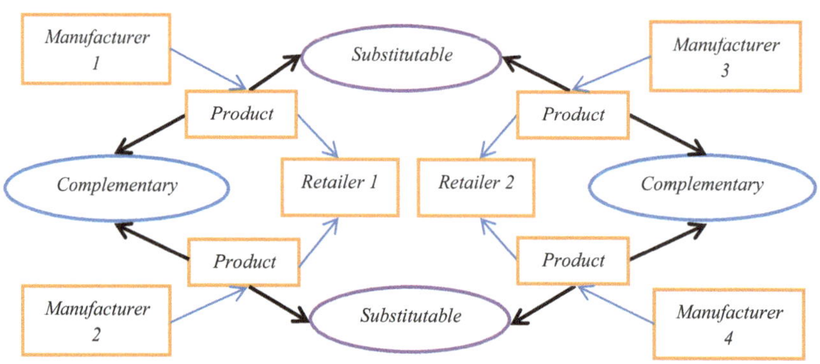

Fig. 1 A two-echelon supply chain

Parameters

C_i The unit manufacturing cost of product i;
A_i The primary demand of customers for product i;
β_{ii} The self-price sensitivity for the demand of ith product respect to its own price;
β_{ij} The cross price sensitivities for the demand of ith product respect to the price of jth product j, $\beta_{ii} > \beta_{ij}$;
L_1 The factor of demand leakage between products 1 and 3;
L_2 The factor of demand leakage between products 2 and 4;
D_i The demand rate of customers for product i under the first scenario;
D'_i The demand rate of customers for product i under the second scenario;
π_{mi} The profit function of manufacturer i under the first scenario;
π'_{mi} The profit function of manufacturer i under the second scenario;
π_{rj} The profit function of retailer j under the first scenario;
π'_{rj} The profit function of retailer j under the second scenario;

Decision variables

W_i The wholesale price of product i per unit, ($);
P_i The retail price of product i per unit, ($)

The optimal values of the decision variables of the models under the both scenarios are shown by sign (*). In addition, some notations utilized to model the first and the second models are defined in Appendices 1 and 2, respectively.

Mathematical model

In this section, two pricing models for the complementary products with and without demand leakage considerations in a two-echelon supply chain are developed where two manufacturers and one retailer are the partners of each echelon.

The first model: without demand leakage

In this model, the same unit manufacturing costs are considered for the manufacturers of each echelon. So, the demand leakage between two echelons is not occurred. Thus, the demand functions of complementary products 1, 2, 3, and 4 are formulated as follows.

$$D_1 = A_1 - \beta_{11}P_1 - \beta_{12}P_2 \tag{1}$$

$$D_2 = A_2 - \beta_{22}P_2 - \beta_{21}P_1 \tag{2}$$

$$D_3 = A_3 - \beta_{33}P_3 - \beta_{34}P_4 \tag{3}$$

$$D_4 = A_4 - \beta_{44}P_4 - \beta_{43}P_3 \tag{4}$$

And the profit functions of the manufacturers and the retailers are represented as follows.

$$\pi_{m1}(W_1) = (W_1 - C_1)[A_1 - \beta_{11}P_1 - \beta_{12}P_2] \tag{5}$$

$$\pi_{m2}(W_2) = (W_2 - C_2)[A_2 - \beta_{22}P_2 - \beta_{21}P_1] \tag{6}$$

$$\pi_{r1}(P_1, P_2) = (P_1 - W_1)[A_1 - \beta_{11}P_1 - \beta_{12}P_2] + (P_2 - W_2)[A_2 - \beta_{22}P_2 - \beta_{21}P_1] \tag{7}$$

$$\pi_{m3}(W_3) = (W_3 - C_3)[A_3 - \beta_{33}P_3 - \beta_{34}P_4] \tag{8}$$

$$\pi_{m4}(W_4) = (W_4 - C_4)[A_4 - \beta_{44}P_4 - \beta_{43}P_3] \tag{9}$$

$$\pi_{r2}(P_3, P_4) = (P_3 - W_3)[A_3 - \beta_{33}P_3 - \beta_{34}P_4] + (P_4 - W_4)[A_4 - \beta_{44}P_4 - \beta_{43}P_3] \tag{10}$$

The second model with demand leakage

In this case, a symmetrical demand leakage between two echelons of supply chain due to the different unit manufacturing costs of two echelons is considered. The demand leakage occurs between products 1 and 3 and also between products 2 and 4. As a result, products 1 and 3 and products 2 and 4 can be traded as the substitutable products. So, the demand functions of products 1, 2, 3, and 4 are obtained as follows:

$$D'_1 = A_1 - \beta_1 P_1 - L_1(P_1 - P_3) \tag{11}$$

$$D'_2 = A_2 - \beta_2 P_2 - L_2(P_2 - P_4) \tag{12}$$

$$D'_3 = A_3 - \beta_3 P_3 + L_1(P_1 - P_3) \tag{13}$$

$$D'_4 = A_4 - \beta_4 P_4 + L_2(P_2 - P_4) \tag{14}$$

Meanwhile, the following relationships are established between β_{ii},β_{ij}, and L_i

$$\beta_{ii} = \beta_i + L_i \tag{15}$$

$$\beta_{ij} = L_i \tag{16}$$

Hence, the profit functions of the manufacturers and retailers are represented as follows:

$$\pi'_{m1}(W_1) = (W_1 - C_1)[A_1 - \beta_1 P_1 - L_1(P_1 - P_3)] \tag{17}$$

$$\pi'_{m2}(W_2) = (W_2 - C_2)[A_2 - \beta_2 P_2 - L_2(P_2 - P_4)] \tag{18}$$

$$\pi'_{r1}(P_1, P_2) = (P_1 - W_1)[A_1 - \beta_1 P_1 - L_1(P_1 - P_3)] + (P_2 - W_2)[A_2 - \beta_2 P_2 - L_2(P_2 - P_4)] \tag{19}$$

$$\pi'_{m3}(W_3) = (W_3 - C_3)[A_3 - \beta_3 P_3 + L_1(P_1 - P_3)] \tag{20}$$

$$\pi'_{m4}(W_4) = (W_4 - C_4)[A_4 - \beta_4 P_4 + L_2(P_2 - P_4)] \tag{21}$$

$$\pi'_{r2}(P_3, P_4) = (P_3 - W_3)[A_3 - \beta_3 P_3 + L_1(P_1 - P_3)] + (P_4 - W_4)[A_4 - \beta_4 P_4 + L_2(P_2 - P_4)] \tag{22}$$

Solution method

For solving the on hand problem, the MS game-theoretic approach is applied, in which the followers first make decision about their decision variables and then the leaders determine the optimal values of own decision variables according to the best reaction of the followers. Here, we consider the manufacturers as the Stackelberg leaders and the retailers as Stackelberg followers where the wholesale prices of the manufacturers and the retail prices of the retailers are the decision variables of the introduced model. So, the manufacturers have more market powers than the retailers and also the market is leaded by the manufacturers. Meanwhile, the theory of MS game consists of two practical approaches which are known as the MS-Bertrand and the MS-Stackelberg models. In this section, we intend to obtain the optimal values of the decision variables by employing the MS-Bertrand and the MS-Stackelberg models under both scenarios.

The MS-Bertrand model

Based on the MS-Bertrand approach, although the manufacturers as the leader have more market power than the retailers as the followers, in each echelon of supply chain the manufacturers have the same power and they move, simultaneously. The solution algorithm of MS-Bertrand model is presented in Fig. 2.

The first model under the MS-Bertrand approach

According to the MS-Bertrand solution algorithm, the optimal values of selling prices of four products versus the wholesale prices are obtained as follows:

$$P_1^*(W_1, W_2) = F_1 + F_2 W_1 + F_3 W_2 \tag{23}$$

$$P_2^*(W_1, W_2) = F_4 + F_5 W_1 + F_6 W_2 \tag{24}$$

$$P_3^*(W_3, W_4) = U_1 + U_2 W_3 + U_3 W_4 \tag{25}$$

$$P_4^*(W_3, W_4) = U_4 + U_5 W_3 + U_6 W_4 \tag{26}$$

Substituting Eqs. (23)–(26) into the manufacturer's profit function, the optimal values of wholesale prices of products are acquired as follows:

$$W_1^* = \frac{E_1 E_6 - E_2 E_3}{E_4 E_6 - E_5 E_3} \tag{27}$$

$$W_2^* = \frac{E_2 E_4 - E_1 E_5}{E_4 E_6 - E_5 E_3} \tag{28}$$

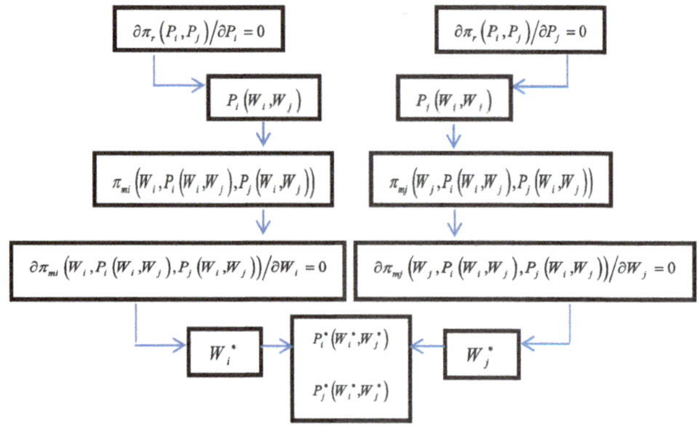

Fig. 2 The MS-Bertrand algorithm

$$W_3^* = \frac{G_1 G_6 - G_2 G_3}{G_4 G_6 - G_5 G_3} \tag{29}$$

$$W_4^* = \frac{G_2 G_4 - G_1 G_5}{G_4 G_6 - G_5 G_3} \tag{30}$$

Then, by substituting Eqs. (27)–(30) into Eqs. (23)–(26), the independent optimal selling prices can be obtained as:

$$P_1^* = F_1 + F_2\left(\frac{E_1 E_6 - E_2 E_3}{E_4 E_6 - E_5 E_3}\right) + F_3\left(\frac{E_1 E_6 - E_2 E_3}{E_4 E_6 - E_5 E_3}\right) \tag{31}$$

$$P_2^* = F_4 + F_5\left(\frac{E_1 E_6 - E_2 E_3}{E_4 E_6 - E_5 E_3}\right) + F_6\left(\frac{E_2 E_4 - E_1 E_5}{E_4 E_6 - E_5 E_3}\right) \tag{32}$$

$$P_3^* = U_1 + U_2\left(\frac{G_1 G_6 - G_2 G_3}{G_4 G_6 - G_5 G_3}\right) + U_3\left(\frac{G_2 G_4 - G_1 G_5}{G_4 G_6 - G_5 G_3}\right) \tag{33}$$

$$P_4^* = U_4 + U_5\left(\frac{G_1 G_6 - G_2 G_3}{G_4 G_6 - G_5 G_3}\right) + U_6\left(\frac{G_2 G_4 - G_1 G_5}{G_4 G_6 - G_5 G_3}\right) \tag{34}$$

The second model under the MS-Bertrand approach

According to the MS-Bertrand solution algorithm, the optimal retail prices of four products versus the wholesale prices of the manufacturers are obtained as follows:

$$P_1^*(W_1, W_3) = K_1 + K_2 W_3 + K_3 W_1 \tag{35}$$

$$P_2^*(W_2, W_4) = K_4 + K_5 W_4 + K_6 W_2 \tag{36}$$

$$P_3^*(W_1, W_3) = K_7 + K_8 W_1 + K_3 W_3 \tag{37}$$

$$P_4^*(W_2, W_4) = K_9 + K_{10} W_2 + K_6 W_4 \tag{38}$$

Substituting Eqs. (35)–(38) into the profit functions of manufacturers, the optimal values of wholesale prices are acquired as follows:

$$W_1^* = \frac{N_1N_6 - N_2N_3}{N_4N_6 - N_5N_3} \quad (39)$$

$$W_2^* = \frac{N_7N_{12} - N_8N_9}{N_{10}N_{12} - N_{11}N_9} \quad (40)$$

$$W_3^* = \frac{N_2N_4 - N_1N_5}{N_4N_6 - N_5N_5} \quad (41)$$

$$W_4^* = \frac{N_8N_{10} - N_7N_{11}}{N_{10}N_{12} - N_{11}N_9} \quad (42)$$

Therefore, by substituting Eqs. (39)–(42) into Eqs. (35)–(38), the optimal retail prices can be obtained independently as:

$$P_1^* = K_1 + K_2\left(\frac{N_2N_4 - N_1N_5}{N_4N_6 - N_5N_3}\right) + K_3\left(\frac{N_1N_6 - N_2N_3}{N_4N_6 - N_5N_3}\right) \quad (43)$$

$$P_2^* = K_4 + K_5\left(\frac{N_8N_{10} - N_7N_{11}}{N_{10}N_{12} - N_{11}N_9}\right) + K_6\left(\frac{N_7N_{12} - N_8N_9}{N_{10}N_{12} - N_{11}N_9}\right) \quad (44)$$

$$P_3^* = K_7 + K_8\left(\frac{N_1N_6 - N_2N_3}{N_4N_6 - N_5N_3}\right) + K_3\left(\frac{N_2N_4 - N_1N_5}{N_4N_6 - N_5N_3}\right) \quad (45)$$

$$P_4^* = K_9 + K_{10}\left(\frac{N_7N_{12} - N_8N_9}{N_{10}N_{12} - N_{11}N_9}\right) + K_6\left(\frac{N_8N_{10} - N_7N_{11}}{N_{10}N_{12} - N_{11}N_9}\right) \quad (46)$$

The MS-Stackelberg model

Under this approach, the manufacturers, because of the more market powers, are considered as the leaders of Stackelberg and the retailers are considered as the followers. Moreover, in each echelon of supply chain, the manufacturers don't have the similar powers and they sequentially make decisions about own decision variables. Also the Stackelberg game is current between them such that one of the manufacturers plays the role of the Stackelberg leader and the other one is the follower of Stackelberg. The figurative MS-Stackelberg solution algorithm is indicated in Fig. 3 in which manufacturer i is the leader and manufacturer j is the follower.

The first model under the MS-Stackelberg approach

Based on the MS-Stackelberg algorithm, the optimal values of selling prices of four products versus the wholesale prices of manufacturers are obtained, similar to the MS-Bertrand model, as follows:

$$P_1^*(W_1, W_2) = F_1 + F_2W_1 + F_3W_2 \quad (47)$$

$$P_2^*(W_1, W_2) = F_4 + F_5W_1 + F_6W_2 \quad (48)$$

$$P_3^*(W_3, W_4) = U_1 + U_2W_3 + U_3W_4 \quad (49)$$

$$P_4^*(W_3, W_4) = U_4 + U_5W_3 + U_6W_4 \quad (50)$$

Since in the first echelon manufacturer 1 is the leader and manufacturer 2 is the follower, so by substituting Eqs. (47) and (48) into the profit functions of manufacturer 1 and 2, the optimal wholesale price of the manufacturer 1 is obtained as:

$$W_2 = \frac{E_2}{E_6} - \frac{E_5}{E_6}W_1 \quad (51)$$

$$W_1^* = \frac{E_7}{E_8} \quad (52)$$

Then by substituting Eq. (52) into Eq. (51), the optimal wholesale price of manufacturer 2 is obtained as follows:

$$W_2^* = \frac{E_2}{E_6} - \frac{E_5}{E_6}\left(\frac{E_7}{E_8}\right) \quad (53)$$

In the second echelon of supply chain, manufacturer 3 is the leader and manufacturer 4 is the follower. Afterward, by substituting Eqs. (49) and (50) into the profit functions of the second echelon manufacturers, the optimal wholesale price of manufacturers 3 is obtained, so we have:

$$W_4 = \frac{G_2}{G_6} - \frac{G_5}{G_6}W_3 \quad (54)$$

$$W_3^* = \frac{G_7}{G_8} \quad (55)$$

Hence, the optimal value of manufacturer 4 is derived by substituting Eq. (55) into Eq. (54) as follows:

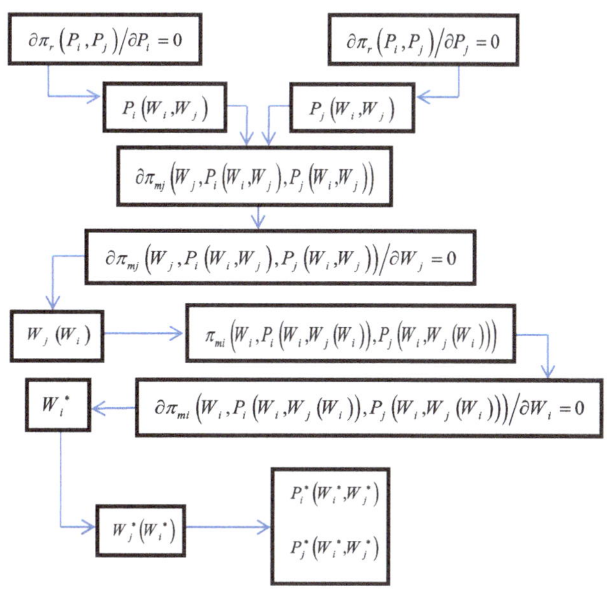

Fig. 3 The MS-Stackelberg algorithm

$$W_4^* = \frac{G_2}{G_6} - \frac{G_5}{G_6}\left(\frac{G_7}{G_8}\right) \quad (56)$$

Therefore, by substituting Eqs. (52)–(56) into Eqs. (47)–(50), the independent retailers' optimal retail prices are obtained which are:

$$P_1^* = F_1 + F_2\left(\frac{E_7}{E_8}\right) + F_3\left(\frac{E_2}{E_4} - \frac{E_5}{E_6}\left(\frac{E_7}{E_8}\right)\right) \quad (57)$$

$$P_2^* = F_4 + F_5\left(\frac{E_7}{E_8}\right) + F_6\left(\frac{E_2}{E_6} - \frac{E_5}{E_6}\left(\frac{E_7}{E_8}\right)\right) \quad (58)$$

$$P_3^* = U_1 + U_2\left(\frac{G_7}{G_8}\right) + U_3\left(\frac{G_2}{G_6} - \frac{G_5}{G_6}\left(\frac{G_7}{G_8}\right)\right) \quad (59)$$

$$P_4^* = U_4 + U_5\left(\frac{G_7}{G_8}\right) + U_6\left(\frac{G_2}{G_6} - \frac{G_5}{G_6}\left(\frac{G_7}{G_8}\right)\right) \quad (60)$$

The second model under MS-Stackelberg approach

Based on the MS-Stackelberg algorithm, the optimal selling prices of four products versus the wholesale prices, which are obtained as the MS-Bertrand model, are as follows.

$$P_1^*(W_1, W_3) = K_1 + K_2W_3 + K_3W_1 \quad (61)$$

$$P_2^*(W_2, W_4) = K_4 + K_5W_4 + K_6W_2 \quad (62)$$

$$P_3^*(W_1, W_3) = K_7 + K_8W_1 + K_3W_3 \quad (63)$$

$$P_4^*(W_2, W_4) = K_9 + K_{10}W_2 + K_6W_4 \quad (64)$$

According to the assumptions, a symmetrical demand leakage occurs between two echelons of supply chain on the same products because of different unit manufacturing costs in the echelons. The demand leakage occurs between products 1 and 3 and also products 2 and 4. Here, we assume that the unit manufacturing costs of manufacturers 1 and 2 are larger than manufacturers 3 and 4. So, the manufacturers 1 and 2 lost their demand and the manufacturers 3 and 4 against earn more demands due to their lower unit manufacturing costs.

Therefore, manufacturers 3 and 4 handle the market owing to having the more powers than the other ones. As a result, manufacturers 3 and 4 are the Stackelberg leaders and manufacturers 1 and 2 are the Stackelberg followers. Thus, by substituting Eqs. (61) and (63) into the profit functions of manufacturers, the optimal wholesale price of manufacturer 1 is derived as follows:

$$W_1 = \frac{N_1}{N_4} - \frac{N_3}{N_4}W_3 \quad (65)$$

$$W_3^* = \frac{N_{13}}{N_{14}} \quad (66)$$

Then, the optimal value of unit wholesale price of manufacturer 1is obtained by substituting Eq. (66) into Eq. (65) which is:

$$W_1^* = \frac{N_1}{N_4} - \frac{N_3}{N_4}\left(\frac{N_{13}}{N_{14}}\right) \quad (67)$$

Furthermore, by substituting Eqs. (62) and (64) into the objective functions of manufacturers 2 and 4, the optimal unit wholesale price of manufacturer 4 is:

$$W_2 = \frac{N_7}{N_{10}} - \frac{N_9}{N_{10}}W_4 \quad (68)$$

$$W_4^* = \frac{N_{15}}{N_{16}} \quad (69)$$

In addition, the optimal unit wholesale price of manufacturer 2 is obtained by substituting Eq. (69) into Eq. (68) which is:

$$W_2^* = \frac{N_7}{N_{10}} - \frac{N_9}{N_{10}}\left(\frac{N_{15}}{N_{16}}\right) \quad (70)$$

Eventually, by substituting Eqs. (66)–(70) into Eqs. (61)–(64), the retailers' optimal unit retail prices can be obtained independently, as follows:

$$P_1^* = K_1 + K_2\left(\frac{N_{13}}{N_{14}}\right) + K_3\left(\frac{N_1}{N_4} - \frac{N_3}{N_4}\left(\frac{N_{13}}{N_{14}}\right)\right) \quad (71)$$

$$P_2^* = K_4 + K_5\left(\frac{N_{15}}{N_{16}}\right) + K_6\left(\frac{N_7}{N_{10}} - \frac{N_9}{N_{10}}\left(\frac{N_{15}}{N_{16}}\right)\right) \quad (72)$$

$$P_3^* = K_7 + K_8\left(\frac{N_1}{N_4} - \frac{N_3}{N_4}\left(\frac{N_{13}}{N_{14}}\right)\right) + K_3\left(\frac{N_{13}}{N_{14}}\right) \quad (73)$$

$$P_4^* = K_9 + K_{10}\left(\frac{N_7}{N_{10}} - \frac{N_9}{N_{10}}\left(\frac{N_{15}}{N_{16}}\right)\right) + K_6\left(\frac{N_{15}}{N_{16}}\right) \quad (74)$$

Numerical example and sensitivity analysis

In this section, a numerical example for a two-echelon supply chain including two manufacturers and one retailer in each echelon is presented. According to the assumption, the model is developed for two complementary products and price-sensitive demand. In addition, the discussed problem is formulated under two different scenarios where the MS-Stackelberg and the MS-Bertrand solution algorithms are employed to solve them. In this example, we consider $A_1 = A_2 = 180$, $A_3 = A_4 = 220$, $C_1 = C_2 = 25$, $C_3 = C_4 = 20$, $\beta_{11} = \beta_{33} = 0.5$, $\beta_{22} = \beta_{44} = 0.6$, $\beta_{12} = \beta_{21} = 0.3$, $\beta_{34} = \beta_{43} = 0.35$, $\beta_{13} = \beta_{31} = 0.3$, $\beta_{24} = \beta_{42} = 0.35$ and the results are shown in Tables 1 and 2.

The findings obtained from Table 1 are summarized as follows.

- According to the obtained results of the first model, retailers 1 and 2 achieve their highest optimal retail prices for products 1 and 3 under the MS-Stackelberg approach and also for products 2 and 4 under the MS-Bertrand approach.

- The highest optimal wholesale prices of products 1 and 3 are acquired under the MS-Stackelberg approach and also for products 2 and 4 under the MS-Bertrand approach in the first model. About the second model, the highest optimal wholesale prices and optimal retail prices of products 1, 2, 3, and 4 are achieved under the MS-Stackelberg approach.

From Table 2, the following results can be obtained too.

- In the first model, manufacturers 1 and 3 achieve their highest profits under the MS-Stackelberg approach and the manufacturers 2 and 4 achieve their highest profits under the MS-Bertrand approach. In the second model, all the manufacturers achieve their highest profits under the MS-Stackelberg approach.
- The retailers 1 and 2 achieve their highest profits using MS-Bertrand game-theoretic approach in the first model, and in the second model retailer 1 achieves his highest profit applying MS-Stackelberg game and retailer 2 achieves his highest profit using MS-Bertrand game.
- The whole supply chain achieves the maximum profit under the MS-Bertrand game-theoretic approach in the first and the second models.

To study the effect of changing the parameter values on the optimal values of the decision variables for this paper, a sensitivity analysis is performed. The sensitivity analysis for the first model is done only at the first echelon of supply chain and for the second model is done only between products 1 and 3. Tables 3, 4, 5 and 6 show the results of the first model under MS-Bertrand and MS-Stackelberg policies, respectively.

The findings obtained from Tables 3 and 4 are summarized as follows.

- W_1^*, W_2^*, P_1^*, P_2^*, D_1, D_2, π_{m1}, π_{m2} and π_{r1} are consumedly sensitive respect to the changes in parameters A_1 and A_2. When A_1 and A_2 are decreased by 25 and 50 %, all of decision variables decrease and vice versa.
- W_1^*, W_2^*, P_1^*, P_2^*, π_{m1} and π_{m2} are consumedly sensitive respect to the changes in parameters β_{11} and β_{22}, while D_1, D_2 and π_{r1} are moderately sensitive respect to the changes in value of β_{11} and β_{22}. When β_{11} and β_{22} are decreased by 25 and 50 %, D_1 and D_2 decrease, while W_1^*, W_2^*, P_1^*, P_2^*, π_{m1}, π_{m2} and π_{r1} increase and vice versa.
- W_1^*, W_2^*, P_1^*, P_2^*, D_1, D_2, π_{m1}, π_{m2} and π_{r1} are moderately sensitive respect to the changes in β_{12} and β_{21}. When β_{12} and β_{21} are decreased by 25 and 50 %, all of the decision variables increase and vice versa.
- W_1^*, W_2^*, P_1^* and P_2^* are slightly sensitive respect to the changes in parameters C_1 and C_2, while D_1, D_2, π_{m1}, π_{m2} and π_{r1} are moderately sensitive respect to the changes in value of C_1 and C_2. When C_1 and C_2 are decreased by 25 and 50 %, W_1^*, W_2^*, P_1^* and P_2^* decrease while D_1, D_2, π_{m1}, π_{m2} and π_{r1} increase and vice versa.

The results of Tables 5 and 6 are similar to the results of Tables 3 and 4, except for sensitivity analysis of β_{11} and β_{22}. We assume manufacture 1 is the leader and manufacturer 2 is the follower. The results show W_1^*, P_1^* and π_{m1} are consumedly sensitive respect to the changes in parameters β_{11} and β_{22}, while W_2^*, P_2^* and π_{m2} are slightly sensitive respect to the changes in value of β_{11} and β_{22}. When β_{11} and β_{22} are decreased by 25 and 50 %, W_1^*, P_1^* and π_{m1} increase while W_2^*, P_2^* and π_{m2} decrease. Also, sensitivity analysis is performed on the second model under MS-Bertrand policy and its results are shown Tables 7 and 8. Moreover the results of sensitivity analysis of the second model under MS-Stackelberg are shown in Tables 9 and 10.

The findings obtained from Tables 7 and 8 are summarized as follows.

- W_1^*, W_3^*, P_1^*, P_3^*, D'_1, D'_3, π'_{m1}, π'_{m3}, π'_{r1} and π'_{r2} are moderately sensitive respect to the changes in parameters A_1 and A_3. When A_1 and A_3 are decreased by 25 and 50 %, all of decision variables decrease and vice versa.
- W_1^*, W_3^*, P_1^*, P_3^*, D'_1, D'_3, π'_{m1}, π'_{m3}, π'_{r1} and π'_{r2} are consumedly sensitive respect to the changes in parameters β_1 and β_3. When β_1 and β_3 are decreased by 25 and 50 %, all of the decision variables increase and vice versa.
- W_1^*, W_3^*, P_1^*, P_3^*, D'_1 and D'_3 are moderately sensitive respect to the changes in parameters β_{13} and β_{31}, while π'_{m1}, π'_{m3}, π'_{r1} and π'_{r2} are slightly sensitive respect to the changes in parameters β_{13} and β_{31}. When β_{13} and β_{31}

Table 1 Optimal decision of retail prices and wholesale prices under different decision scenarios

Decision scenario	Model	P_1^*	P_2^*	P_3^*	P_4^*	W_1^*	W_2^*	W_3^*	W_4^*
MS-Bertrand model	1	186.54	186.54	190.63	190.63	148.08	148.08	149.68	149.68
	2	552.21	449.91	593.26	484.26	388.45	317.33	415.2	339.41
MS-Stackelberg model	1	193.29	184.51	197.27	188.69	161.59	144.02	162.97	145.8
	2	555.97	452.59	601.48	490.3	391.04	319.18	429.38	349.92

Table 2 Maximum profits of the total system and for every firm under different decision scenarios

Decision scenario	Model	π_{m1}	π_{m2}	π_{m3}	π_{m4}	π_{r1}	π_{r2}	Total profit
MS-Bertrand model	1	3786.98	3786.98	5044.87	5044.87	2366.86	3186.23	23,216.79
	2	29,758.21	23,253.79	35,148.92	27,760.29	23,953.38	28,442.13	1,68,352.72
MS-Stackelberg model	1	3824.39	3541.7	5088.86	47,47,071	2092.56	2839.66	22,134.88
	2	30,184.27	23,548.64	35,227.14	27,788.51	24,279.06	26,633.38	1,67,661

Table 3 The sensitivity analysis for the first model in first echelon of supply chain under MS-Bertrand policy

Parameters	% Changes	Optimal values						% Changes in					
		W_1^*	W_2^*	P_1^*	P_2^*	D_1	D_2	W_1^*	W_2^*	P_1^*	P_2^*	D_1	D_2
$A_1 = A_2$	−50	78.85	78.85	95.67	95.67	13.46	13.46	−46.75	−46.75	−48.71	−48.71	−56.25	−56.25
	−25	113.46	113.46	141.11	141.11	22.12	22.12	−23.38	−23.38	−24.36	−24.36	−28.13	−28.13
	+25	182.69	182.69	231.97	231.97	39.42	39.42	23.38	23.38	24.36	24.36	28.13	28.13
	+50	217.31	217.31	277.40	277.40	48.08	48.08	46.75	46.75	48.71	48.71	56.25	56.25
$\beta_{11} = \beta_{22}$	−50	232.81	232.81	280.04	280.04	25.98	25.98	57.22	57.22	50.13	50.13	−15.58	−15.58
	−25	180.36	180.36	223.51	223.51	29.13	29.13	21.80	21.80	19.82	19.82	−5.33	−5.33
	+25	126.21	126.21	160.40	160.40	31.63	31.63	−14.77	−14.77	−14.01	−14.01	2.79	2.79
	+50	110.42	110.42	140.92	140.92	32.03	32.03	−25.43	−25.43	−24.45	−24.45	4.10	4.10
$\beta_{12} = \beta_{21}$	−50	167.39	167.39	222.16	222.16	35.60	35.60	13.04	13.04	19.09	19.09	15.69	15.69
	−25	157.14	157.14	202.71	202.71	33.04	33.04	6.12	6.12	8.67	8.67	7.37	7.37
	+25	140.00	140.00	172.86	172.86	28.75	28.75	−5.45	−5.45	−7.33	−7.33	−6.56	−6.56
	+50	132.76	132.76	161.12	161.12	26.94	26.94	−10.34	−10.34	−13.63	−13.63	−12.45	−12.45
$C_1 = C_2$	−50	143.27	143.27	184.13	184.13	32.69	32.69	−3.25	−3.25	−1.29	−1.29	6.25	6.25
	−25	145.67	145.67	185.34	185.34	31.73	31.73	−1.62	−1.62	−0.64	−0.64	3.13	3.13
	+25	150.48	150.48	187.74	187.74	29.81	29.81	1.62	1.62	0.64	0.64	−3.12	−3.12
	+50	152.88	152.88	188.94	188.94	28.85	28.85	3.25	3.25	1.29	1.29	−6.25	−6.25

are decreased by 25 and 50 %, W_1^*, W_3^*, P_1^* and P_3^* increase, while D_1', D_3', π_{m1}', π_{m3}', π_{r1}' and π_{r2}' decrease and vice versa.

- W_1^*, W_3^*, P_1^*, P_3^*, D_1', D_3', π_{m1}', π_{m3}', π_{r1}' and π_{r2}' are slightly sensitive respect to the changes in value of C_1. When C_1 is decreased by 25 and 50 %, W_1^*, W_3^*, P_1^*, P_3^*, D_3', π_{m3}' and π_{r2}' decrease, while D_1', π_{m1}' and π_{r1}' increase and vice versa.
- W_1^*, W_3^*, P_1^*, P_3^*, D_1', D_3', π_{m1}', π_{m3}', π_{r1}' and π_{r2}' are slightly sensitive respect to the changes in value of C_3. When C_3 is decreased by 25 and 50 %, W_1^*, W_3^*, P_1^*, P_3^*, D_1', π_{m1}' and π_{r1}' decrease, while D_3', π_{m3}' and π_{r2}' increase and vice versa.

The results of Tables 9 and 10 are similar to the results of Tables 7 and 8, except for the sensitivity analysis of β_{13} and β_{31}. W_1^*, W_3^*, P_1^*, P_3^*, D_1' and D_3' are moderately sensitive respect to the changes in parameters β_{13} and β_{31}, while π_{m1}', π_{m3}', π_{r1}' and π_{r2}' are slightly sensitive respect to the changes in parameters β_{13} and β_{31}. When β_{13} and β_{31} are decreased by 25 and 50 %, W_1^*, W_3^*, P_1^*, P_3^* and π_{r2}' increase, while D_1', D_3', π_{m1}', π_{m3}' and π_{r1}' decrease and vice versa.

Some of the sensitivity analyses in Tables 3, 4, 5, 6, 7, 8, 9 and 10 are illustrated by Figs. 4, 5, 6, 7, 8, 9, 10, 11 and 12. Figures 4, 5, 6, 7, 8, 9, 10, 11 and 12 show the effect of some key parameters on optimal wholesale and retail prices and also on the profit of the chain.

Conclusion

We discussed the pricing problem of two complementary and substitutable products in a two-echelon supply chain under two scenarios where two manufacturers and one retailer are the members of each echelon. Under the first scenario, which leads to develop the first model, the same unit manufacturing costs for both echelons are supposed and in the second one we assume that the unit manufacturing costs of echelons are different which causes to leak demand from the echelon with higher unit manufacturing cost to the lower one. Two same complementary products

Table 4 The sensitivity analysis for the first models profit functions in first echelon of supply chain under MS-Bertrand policy

Parameters	% Changes	Optimal values			% Changes in		
		π_{m1}	π_{m2}	π_{r1}	π_{m1}	π_{m2}	π_{r1}
$A_1 = A_2$	−50	724.85	724.85	453.03	−80.86	−80.86	−80.86
	−25	1956.36	1956.36	1222.73	−48.34	−48.34	−48.34
	+25	6216.72	6216.72	3885.45	64.16	64.16	64.16
	+50	9245.56	9245.56	5778.48	144.14	144.14	144.14
$\beta_{11} = \beta_{22}$	−50	5398.25	5398.25	2453.75	42.55	42.55	3.67
	−25	4525.47	4525.47	2514.15	19.50	19.50	6.22
	+25	3201.06	3201.06	2162.88	−15.47	−15.47	−8.62
	+50	2736.00	2736.00	1954.29	−27.75	−27.75	−17.43
$\beta_{12} = \beta_{21}$	−50	5068.82	5068.82	3899.09	33.85	33.85	64.74
	−25	4365.43	4365.43	3010.64	15.27	15.27	27.20
	+25	3306.25	3306.25	1889.29	−12.69	−12.69	−20.18
	+50	2902.98	2902.98	1527.88	−23.34	−23.34	−35.45
$C_1 = C_2$	−50	4275.15	4275.15	2671.97	12.89	12.89	12.89
	−25	4027.37	4027.37	2517.10	6.35	6.35	6.35
	+25	3553.99	3553.99	2221.25	−6.15	−6.15	−6.15
	+50	3328.40	3328.40	2080.25	−12.11	−12.11	−12.11

Table 5 The sensitivity analysis for the first model in first echelon of supply chain under MS-Stackelberg policy

Parameters	% Changes	Optimal values						% Changes in					
		W_1^*	W_2^*	P_1^*	P_2^*	D_1	D_2	W_1^*	W_2^*	P_1^*	P_2^*	D_1	D_2
$A_1 = A_2$	−50	84.76	77.07	98.63	94.79	12.25	13.02	−47.55	−46.49	−48.97	−48.63	−56.25	−56.25
	−25	123.17	110.55	145.96	139.65	20.13	21.39	−23.77	−23.24	−24.49	−24.31	−28.13	−28.13
	+25	200	177.50	240.63	229.38	35.88	38.13	23.77	23.24	24.49	24.31	28.13	28.13
	+50	238.41	210.98	287.96	274.24	43.75	46.49	47.55	46.49	48.97	48.63	56.25	56.25
$\beta_{11} = \beta_{22}$	−50	500.00	72.50	413.64	199.89	16.63	5.94	209.43	−49.66	113.9	8.33	−40.63	−80.05
	−25	216.91	165.74	241.79	216.20	24.47	26.39	34.24	15.07	25.09	17.17	−12.61	−11.32
	+25	132.80	124.63	163.70	159.61	29.81	31.13	−17.82	−13.47	−15.31	−13.50	6.45	4.63
	+50	114.13	109.67	142.78	140.55	30.75	31.75	−29.37	−23.85	−26.13	−23.83	9.82	6.71
$\beta_{12} = \beta_{21}$	−50	170.75	166.89	223.83	221.91	34.80	35.47	5.67	15.87	15.80	20.27	24.27	19.21
	−25	164.59	155.47	206.43	201.87	31.36	32.62	1.86	7.95	6.80	9.41	12.01	9.61
	+25	162.50	131.56	184.11	168.64	24.71	26.64	0.57	−8.65	−4.75	−8.60	−11.76	−10.47
	+50	169.43	116.26	179.45	152.86	21.48	22.81	4.86	−19.28	−7.16	−17.15	−23.27	−23.33
$C_1 = C_2$	−50	157.62	138.96	191.31	181.98	29.75	31.62	−2.45	−3.51	−1.03	−1.37	6.25	6.25
	−25	159.60	141.49	192.30	183.25	28.88	30.69	−1.23	−1.76	−0.51	−0.69	3.12	3.12
	+25	163.57	146.55	194.28	185.78	27.13	28.83	1.226	1.757	0.513	0.686	−3.125	−3.125
	+50	165.55	149.09	195.27	187.04	26.25	27.90	2.45	3.51	1.03	1.37	−6.25	−6.25

are supplied to the market by each echelon of chain to satisfy the customers' demand. The model is developed under price-sensitive and deterministic demand.

The main aim of this research is to analyze the pricing decisions of the members of chain for complementary and substitutable products with the different market powers under two scenarios. In this research, two solution algorithms including MS-Bertrand and MS-Stackelberg game-theoretic approaches are presented to survey the effects of the different market powers on the optimal value of decision variables and also the total profit of the supply chain where the whole sale prices of manufacturers and the retail prices of retailers are the decision variables of the proposed models. Finally, a numerical example to show the applicability of the proposed models is presented and we found that the maximum profit of the whole supply chain is obtained under MS-Bertrand approach in both proposed models. For future works, the model can be extended under

Table 6 The sensitivity analysis for the first models profit functions in first echelon of supply chain under MS-Stackelberg policy

Parameters	% Changes	Optimal values			% Changes in		
		π_{m1}	π_{m2}	π_{r1}	π_{m1}	π_{m2}	π_{r1}
$A_1 = A_2$	−50	732.01	677.90	400.53	−80.86	−80.86	−80.86
	−25	1975.69	1829.65	1081.02	−48.34	−48.34	−48.34
	+25	6278.13	5814.06	3435.16	64.16	64.16	64.16
	+50	9336.89	8646.73	5108.80	144.14	144.14	144.14
$\beta_{11} = \beta_{22}$	−50	7896.88	282.03	−679.44	106.49	−92.04	−132.47
	−25	4695.84	3713.70	1940.40	22.79	4.86	−7.27
	+25	3213.07	3101.82	2010.12	−15.98	−12.42	−3.94
	+50	2740.76	2688.63	1861.40	−28.33	−24.09	−11.05
$\beta_{12} = \beta_{21}$	−50	5071.51	5033.06	3798.90	32.61	42.11	81.54
	−25	4377.88	4255.48	2825.95	14.47	20.15	35.05
	+25	3397.22	2838.89	1521.57	−11.17	−19.84	−27.29
	+50	3103.05	2081.88	1050.48	−18.86	−41.22	−49.80
W	−50	4317.38	3998.25	2362.31	12.89	12.89	12.89
	−25	4067.15	3766.52	2225.39	6.35	6.35	6.35
	+25	3589.10	3323.80	1963.82	−6.152	−6.152	−6.152
	+50	3361.28	3112.82	1839.17	−12.11	−12.11	−12.11

Table 7 The sensitivity analysis for the second model between products 1 and 3 under MS-Bertrand policy

Parameters	% Changes	Optimal values						% Changes in					
		W_1^*	W_3^*	P_1^*	P_3^*	D_1'	D_3'	W_1^*	W_3^*	P_1^*	P_3^*	D_1'	D_3'
A_1	−50	268.66	360.37	378.45	513.72	54.89	76.68	−30.84	−13.21	−31.47	−13.41	−32.96	−13.88
	−25	328.56	387.79	465.33	553.49	68.38	82.85	−15.42	−6.60	−15.73	−6.70	−16.48	−6.94
	+25	448.35	442.62	639.08	633.04	95.37	95.21	15.42	6.60	15.73	6.70	16.48	6.94
	+50	508.24	470.04	725.96	672.81	108.86	101.38	30.84	13.21	31.47	13.41	32.96	13.88
A_3	−50	321.43	268.80	454.98	380.89	66.78	56.05	−17.25	−35.26	−17.61	−35.80	−18.44	−37.05
	−25	354.94	342.00	503.59	487.08	74.33	72.54	−8.63	−17.63	−8.80	−17.90	−9.22	−18.52
	+25	421.96	488.41	600.82	699.45	89.43	105.52	8.63	17.63	8.80	17.90	9.22	18.52
	+50	455.48	561.61	649.43	805.63	96.98	122.01	17.25	35.26	17.61	35.80	18.44	37.05
$\beta_1 = \beta_3$	−50	646.03	678.53	905.73	953.91	103.88	110.15	66.31	63.42	64.02	60.79	26.88	23.73
	−25	484.51	513.82	685.55	729.87	90.47	97.22	24.73	23.75	24.15	23.03	10.49	9.20
	+25	324.78	349.41	462.63	500.88	75.82	83.31	−16.39	−15.85	−16.22	−15.57	−7.40	−6.43
	+50	279.50	302.31	398.26	434.05	71.26	79.05	−28.05	−27.19	−27.88	−26.84	−12.97	−11.21
$\beta_{13} = \beta_{31}$	−50	424.88	466.87	615.20	679.55	66.61	74.44	9.38	12.44	11.41	14.54	−18.64	−16.39
	−25	405.98	438.69	582.12	632.26	74.86	82.27	4.51	5.66	5.42	6.57	−8.57	−7.60
	+25	372.31	394.90	525.30	560.05	87.97	94.96	−4.16	−4.89	−4.87	−5.60	7.44	6.66
	+50	357.45	376.97	501.05	531.16	93.34	100.22	−7.98	−9.21	−9.26	−10.47	14.00	12.57
C_1	−50	381.99	414.02	548.46	591.55	83.24	88.76	−1.66	−0.28	−0.68	−0.29	1.66	−0.30
	−25	385.22	414.61	550.33	592.41	82.56	88.90	−0.83	−0.14	−0.34	−0.14	0.83	−0.15
	+25	391.69	415.80	554.08	594.12	81.20	89.16	0.83	0.14	0.34	0.14	−−0.83	0.15
	+50	394.92	416.39	555.95	594.98	80.52	89.30	1.66	0.28	0.68	0.29	−1.66	0.30
C_3	−50	387.51	410.03	550.83	590.27	81.66	90.12	−0.24	−1.25	−0.25	−0.51	−0.26	1.22
	−25	387.98	412.62	551.52	591.76	81.77	89.57	−0.12	−0.62	−0.12	−0.25	−0.13	0.61
	+25	388.93	417.79	552.89	594.76	81.98	88.49	0.12	0.62	0.12	0.25	0.13	−0.61
	+50	389.40	420.38	553.58	596.26	82.09	87.94	0.24	1.25	0.25	0.51	0.26	−1.22

Table 8 The sensitivity analysis for the second models profit functions between products 1 and 3 under MS-Bertrand policy

Parameters	% Changes	Optimal values				% Changes in			
		π'_{m1}	π'_{m3}	π'_{r1}	π'_{r2}	π'_{m1}	π'_{m3}	π'_{r1}	π'_{r2}
A_1	−50	13,375.09	26,097.77	10,684.74	21,146.28	−55.05	−25.83	−55.39	−25.65
	−25	20,758.53	30,471.97	16,658.45	24,659.28	−30.24	−13.39	−30.45	−13.30
	+25	40,374.13	40,236.60	32,569.53	32,494.82	35.67	14.36	35.97	14.25
	+50	52,606.30	45,627.02	42,506.89	36,817.36	76.78	29.68	77.46	29.45
A_3	−50	19,794.57	13,944.55	15,970.80	11,198.29	−33.48	−60.37	−33.33	−60.63
	−25	24,523.38	23,357.54	19,760.54	18,833.38	−17.59	−33.61	−17.50	−33.78
	+25	35,499.05	49,426.68	28,549.32	40,024.55	19.29	40.48	19.19	40.72
	+50	41,745.91	66,082.84	33,548.37	53,580.65	40.28	87.82	40.06	88.38
$\beta_1 = \beta_3$	−50	64,513.33	72,539.66	37,524.14	42,924.35	116.79	106.17	56.65	50.92
	−25	41,570.01	48,010.27	28,732.71	33,594.07	39.69	36.45	19.95	18.11
	+25	22,728.34	27,442.13	20,996.81	25,208.06	−23.62	−22.01	−12.34	−11.37
	+50	18,135.18	22,315.55	19,008.92	23,003.51	−39.06	−36.58	−20.64	−19.12
$\beta_{13} = \beta_{31}$	−50	26,636.86	33,264.08	23,223.27	28,421.15	−10.49	−5.46	−3.05	−0.07
	−25	28,519.45	34,444.19	23,731.11	28,514.01	−4.16	−2.11	−0.93	−0.25
	+25	30,552.90	35,600.53	24,004.62	28,271.89	2.67	1.18	0.21	0.60
	+50	31,030.40	35,775.83	23,948.88	28,042.33	4.28	1.68	0.02	1.41
C_1	−50	30,754.43	34,974.61	24,843.31	28,258.19	3.35	−0.60	3.72	−0.65
	−25	30,254.27	35,079.69	24,396.29	28,350.08	1.67	−0.30	1.85	−0.32
	+25	29,266.25	35,290.31	23,514.57	28,534.32	−1.65	0.30	−1.83	0.32
	+50	28,778.38	35,395.86	23,079.87	28,626.67	−3.29	0.60	−3.65	0.65
C_3	−50	29,603.45	36,049.64	23,818.34	29,216.22	−0.52	2.46	−0.56	2.72
	−25	29,680.78	35,615.97	23,885.81	28,827.86	−0.26	1.23	−0.28	1.36
	+25	29,835.74	34,756.49	24,021.04	28,059.03	0.26	−1.22	0.28	−1.35
	+50	29,913.37	34,330.69	24,088.80	27,678.55	0.52	−2.43	0.57	−2.68

stochastic demand and also considering competing retailers can develop and enhance our models.

Funding The first author would like to thank the financial support of the University of Tehran for this research under Grant Number 30015-1-02.

Appendix 1: Notations of the first model

The notations employed to solving the first model which is developed under the first scenario are as follows:

$$F_1 = \frac{2\beta_{22}A_1 - A_2(\beta_{12} + \beta_{21})}{4\beta_{11}\beta_{22} - (\beta_{12} + \beta_{21})^2} \tag{75}$$

$$F_2 = \frac{2\beta_{11}\beta_{22} - \beta_{12}(\beta_{12} + \beta_{21})}{4\beta_{11}\beta_{22} - (\beta_{12} + \beta_{21})^2} \tag{76}$$

$$F_3 = \frac{2\beta_{21}\beta_{22} - \beta_{22}(\beta_{12} + \beta_{21})}{4\beta_{11}\beta_{22} - (\beta_{12} + \beta_{21})^2} \tag{77}$$

$$F_4 = \frac{2\beta_{11}A_2 - A_1(\beta_{12} + \beta_{21})}{4\beta_{11}\beta_{22} - (\beta_{12} + \beta_{21})^2} \tag{78}$$

$$F_5 = \frac{2\beta_{11}\beta_{12} - \beta_{11}(\beta_{12} + \beta_{21})}{4\beta_{11}\beta_{22} - (\beta_{12} + \beta_{21})^2} \tag{79}$$

$$F_6 = \frac{2\beta_{11}\beta_{22} - \beta_{21}(\beta_{12} + \beta_{21})}{4\beta_{11}\beta_{22} - (\beta_{12} + \beta_{21})^2} \tag{80}$$

$$U_1 = \frac{2\beta_{44}A_3 - A_4(\beta_{34} + \beta_{43})}{4\beta_{33}\beta_{44} - (\beta_{34} + \beta_{43})^2} \tag{81}$$

$$U_2 = \frac{2\beta_{33}\beta_{44} - \beta_{34}(\beta_{34} + \beta_{43})}{4\beta_{33}\beta_{44} - (\beta_{34} + \beta_{43})^2} \tag{82}$$

$$U_3 = \frac{2\beta_{43}\beta_{44} - \beta_{44}(\beta_{34} + \beta_{43})}{4\beta_{33}\beta_{44} - (\beta_{34} + \beta_{43})^2} \tag{83}$$

Table 9 The sensitivity analysis for the second model between products 1 and 3 under MS-Stackelberg policy

Parameters	% Changes	Optimal values						% Changes in					
		W_1^*	W_3^*	P_1^*	P_3^*	D_1'	D_3'	W_1^*	W_3^*	P_1^*	P_3^*	D_1'	D_3'
A_1	−50	270.90	372.57	381.69	520.79	55.39	74.11	−30.72	−13.23	−31.35	−13.41	−32.82	−13.88
	−25	330.97	400.97	468.83	561.14	68.93	80.08	−15.36	−6.61	−15.67	−6.71	−16.41	−6.94
	+25	451.12	457.78	643.11	641.82	95.99	92.02	30.72	13.23	31.35	13.41	32.82	13.88
	+50	511.19	486.18	730.25	682.17	109.53	97.99	30.72	13.23	31.35	13.41	32.82	13.88
A_3	−50	323.06	277.72	457.35	386.06	67.15	54.17	−17.39	−35.32	−17.74	−35.81	−18.57	−37.05
	−25	357.05	353.55	506.66	493.77	74.80	70.11	−8.69	−17.66	−8.87	−17.91	−9.29	−18.52
	+25	425.04	505.21	605.27	709.19	90.12	101.99	8.69	17.66	8.87	17.91	9.29	18.52
	+50	459.03	581.04	654.58	816.89	97.78	117.93	17.39	35.32	17.74	35.81	18.57	37.05
$\beta_1 = \beta_3$	−50	659.56	730.40	924.92	987.05	106.14	102.66	68.67	70.11	66.36	64.10	28.72	19.30
	−25	489.86	538.79	693.24	744.92	91.52	92.76	25.27	25.48	24.69	23.85	10.99	7.79
	+25	326.21	358.32	464.71	505.90	76.18	81.17	−16.58	−16.55	−16.41	−15.89	−7.62	−5.67
	+50	280.35	308.32	399.52	437.37	71.50	77.43	−28.31	−28.19	−28.14	−27.28	−13.29	−10.01
$\beta_{13} = \beta_{31}$	−50	425.64	473.27	616.32	682.99	66.74	73.40	8.85	10.22	10.86	13.55	−19.07	−14.70
	−25	407.58	449.10	584.46	638.08	75.17	80.32	4.23	4.59	5.13	6.09	−8.84	−6.66
	+25	375.95	412.49	530.55	570.55	88.89	90.89	−3.86	−3.93	−4.57	−5.14	7.80	5.62
	+50	362.15	397.60	507.77	543.80	94.66	95.03	−7.39	−7.40	−8.67	−9.59	14.79	10.43
C_1	−50	384.57	428.15	552.21	599.74	83.82	85.79	−1.66	−0.29	−0.68	−0.29	1.65	−0.30
	−25	387.81	428.76	554.09	600.61	83.14	85.92	−0.83	−0.14	−0.34	−0.14	0.82	−0.15
	+25	394.28	429.99	557.85	602.35	81.78	86.18	0.83	0.14	0.34	0.14	−0.82	0.15
	+50	397.52	430.60	559.73	603.22	81.10	86.31	1.66	0.29	0.68	0.29	−1.65	0.30
C_3	−50	390.13	424.38	554.64	598.58	82.25	87.10	−0.23	−1.16	−0.24	−0.48	−0.25	1.22
	−25	390.59	426.88	555.30	600.03	82.36	86.58	−0.12	−0.58	−0.12	−0.24	−0.12	0.61
	+25	391.50	431.88	556.63	602.93	82.56	85.52	0.12	0.58	0.12	0.24	0.12	−0.61
	+50	391.96	434.38	557.29	604.38	82.67	85.00	0.23	1.16	0.24	0.48	0.25	−1.22

$$U_4 = \frac{2\beta_{33}A_4 - A_3(\beta_{34} + \beta_{43})}{4\beta_{33}\beta_{44} - (\beta_{34} + \beta_{43})^2} \tag{84}$$

$$U_5 = \frac{2\beta_{33}\beta_{34} - \beta_{33}(\beta_{34} + \beta_{43})}{4\beta_{33}\beta_{44} - (\beta_{34} + \beta_{43})^2} \tag{85}$$

$$U_6 = \frac{2\beta_{33}\beta_{44} - \beta_{43}(\beta_{34} + \beta_{43})}{4\beta_{33}\beta_{44} - (\beta_{34} + \beta_{43})^2} \tag{86}$$

$$E_1 = A_1 + \beta_{11}(C_1F_2 - F_1) + \beta_{12}(C_1F_5 - F_4) \tag{87}$$

$$E_2 = A_2 + \beta_{22}(C_2F_6 - F_4) + \beta_{21}(C_2F_3 - F_1) \tag{88}$$

$$E_3 = \beta_{11}F_3 + \beta_{12}F_6 \tag{89}$$

$$E_4 = 2(\beta_{11}F_2 + \beta_{12}F_5) \tag{90}$$

$$E_5 = \beta_{22}F_5 + \beta_{21}F_2 \tag{91}$$

$$E_6 = 2(\beta_{22}F_6 + \beta_{21}F_3) \tag{92}$$

$$E_7 = (A_1 - \beta_{11}F_1 - \beta_{12}F_4)E_6 - E_3E_2 + \left[\frac{E_4E_6}{2} - E_3E_5\right]C_1 \tag{93}$$

$$E_8 = E_4E_6 - 2E_3E_5 \tag{94}$$

$$G_1 = A_3 + \beta_{33}(C_3U_2 - U_1) + \beta_{34}(C_3U_5 - U_4) \tag{95}$$

$$G_2 = A_4 + \beta_{44}(C_4U_6 - U_4) + \beta_{43}(C_4U_3 - U_1) \tag{96}$$

$$G_3 = \beta_{33}U_3 + \beta_{34}U_6 \tag{97}$$

$$G_4 = 2(\beta_{33}U_2 + \beta_{34}U_5) \tag{98}$$

$$G_5 = \beta_{44}U_5 + \beta_{43}U_2 \tag{99}$$

$$G_6 = 2(\beta_{44}U_6 + \beta_{43}U_3) \tag{100}$$

$$G_7 = (A_3 - \beta_{33}U_1 - \beta_{34}U_4)G_6 - G_3G_2 + \left[\frac{G_4G_6}{2} - G_3G_5\right]C_3 \tag{101}$$

$$G_8 = G_4G_6 - 2G_3G_5 \tag{102}$$

Appendix 2: Notations of the second model

The notations employed to solve the second model which is developed under the second scenario are as follows:

$$K_1 = \frac{2A_1(\beta_3 + L_1) + L_1A_3}{4(\beta_1 + L_1)(\beta_3 + L_1) - L_1^2} \tag{103}$$

Table 10 The sensitivity analysis for the second models profit functions between products 1 and 3 under MS-Stackelberg policy

Parameters	% Changes	Optimal values				% Changes in			
		π'_{m1}	π'_{m3}	π'_{r1}	π'_{r2}	π'_{m1}	π'_{m3}	π'_{r1}	π'_{r2}
A_1	−50	13,621.34	26,129.09	10,872.51	19,801.64	−54.87	−25.83	−55.22	−25.65
	−25	21,089.82	30,508.54	16,911.46	23,091.18	−30.13	−13.39	−30.35	−13.30
	+25	40,904.68	40,284.89	32,975.31	30,428.27	76.42	29.68	77.11	29.45
	+50	53,251.06	45,681.78	43,000.22	34,475.82	76.42	29.68	77.11	29.45
A_3	−50	20,013.16	13,961.28	16,137.56	10,485.95	−33.70	−60.37	−33.53	−60.63
	−25	24,838.40	23,385.57	20,001.17	17,635.55	−17.71	−33.61	−17.62	−33.78
	+25	36,050.76	49,486.00	28,971.21	37,479.43	19.44	40.48	19.33	40.72
	+50	42,437.88	66,162.14	34,077.64	50,173.70	40.60	87.82	40.36	88.39
$\beta_1 = \beta_3$	−50	67,355.56	72,928.53	38,846.42	38,170.52	123.15	107.02	60.00	43.32
	−25	42,543.71	48,121.73	29,292.42	30,943.81	40.95	36.60	20.65	16.18
	+25	22,945.30	27,461.18	21,230.29	23,803.16	−23.98	−22.05	−12.56	−10.63
	+50	18,257.67	22,325.24	19,199.80	21,817.21	−39.51	−36.62	−20.92	−18.08
$\beta_{13} = \beta_{31}$	−50	26,737.51	33,270.70	23,404.89	27,217.93	−11.42	−5.55	−3.60	2.19
	−25	28,759.93	34,464.48	23,976.00	27,002.80	−4.72	−2.16	−1.25	1.39
	+25	31,196.96	35,672.14	24,422.05	26,190.14	3.36	1.26	0.59	−1.66
	+50	31,913.39	35,882.98	24,463.99	25,716.92	5.73	1.86	0.76	−3.44
C_1	−50	31,186.24	35,016.58	25,173.85	26,461.11	3.32	−0.60	3.69	−0.65
	−25	30,683.21	35,121.78	24,724.41	26,547.17	1.65	−0.30	1.83	−0.32
	+25	29,689.42	35,332.66	23,837.81	26,719.73	−1.64	0.30	−1.82	0.32
	+50	29,198.66	35,438.33	23,400.65	26,806.23	−3.27	0.60	−3.62	0.65
C_3	−50	30,033.62	36,092.90	24,147.48	27,358.45	−0.50	2.46	−0.54	2.72
	−25	30,108.90	35,658.71	24,213.22	26,994.69	−0.25	1.23	−0.27	1.36
	+25	30,259.74	34,798.20	24,344.98	26,274.54	0.25	−1.22	0.27	−1.35
	+50	30,335.30	34,371.89	24,411.00	25,918.16	0.50	−2.43	0.54	−2.69

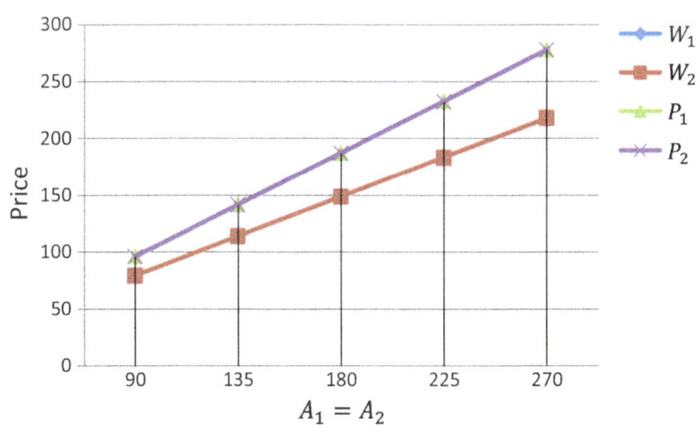

Fig. 4 Changes of optimal prices with respect to $A_1 = A_2$ for the first model under MS-Bertrand policy

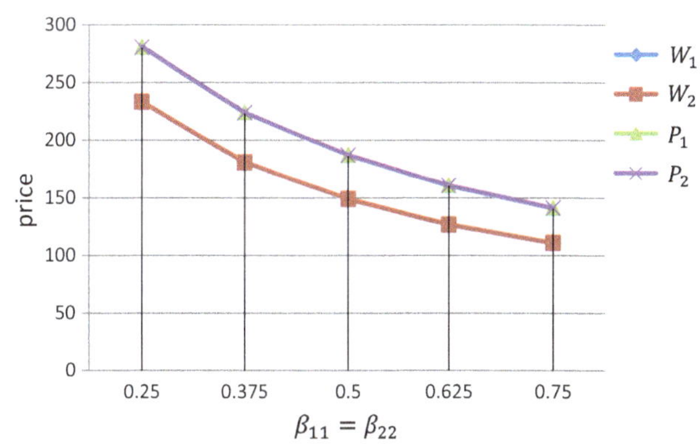

Fig. 5 Changes of optimal prices with respect to $\beta_{11} = \beta_{22}$ for the first model under MS-Bertrand policy

$$K_2 = \frac{L_1(\beta_3 + L_1)}{4(\beta_1 + L_1)(\beta_3 + L_1) - L_1^2} \quad (104)$$

$$K_3 = \frac{2(\beta_1 + L_1)(\beta_3 + L_1)}{4(\beta_1 + L_1)(\beta_3 + L_1) - L_1^2} \quad (105)$$

$$K_4 = \frac{2A_2(\beta_4 + L_2) + L_2 A_4}{4(\beta_2 + L_2)(\beta_4 + L_2) - L_2^2} \quad (106)$$

$$K_5 = \frac{L_2(\beta_4 + L_2)}{4(\beta_2 + L_2)(\beta_4 + L_2) - L_2^2} \quad (107)$$

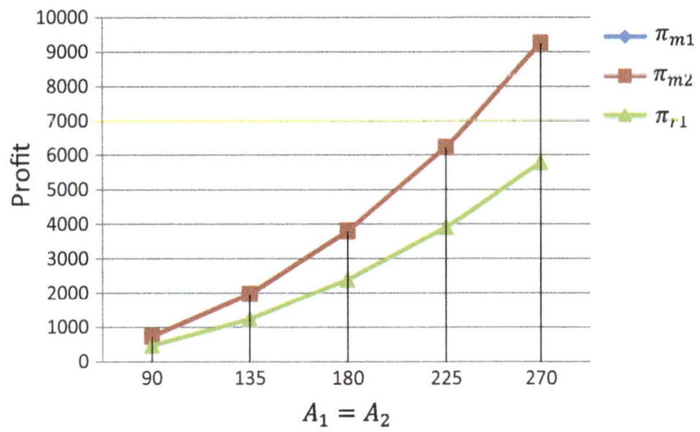

Fig. 6 Changes of maximum profits with respect to $A_1 = A_2$ for the first model under MS-Bertrand policy

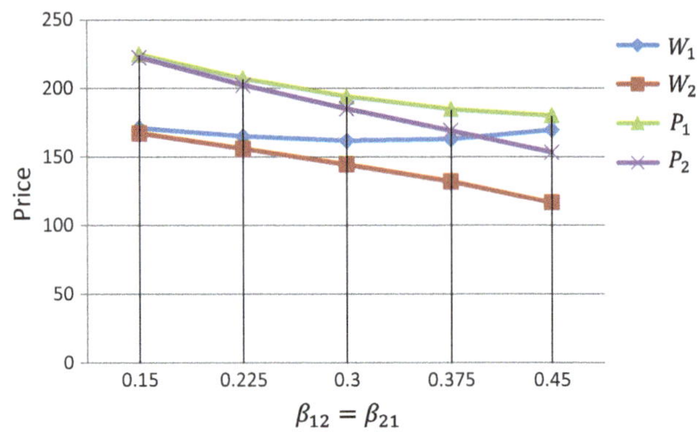

Fig. 7 Changes of optimal prices with respect to $\beta_{12} = \beta_{21}$ for the first model under MS-Stackelberg policy

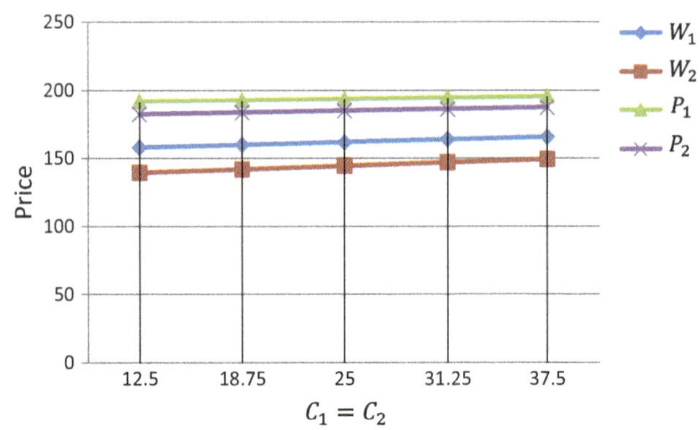

Fig. 8 Changes of optimal prices with respect to $C_1 = C_2$ for the first model under MS-Stackelberg policy

$$K_6 = \frac{2(\beta_2 + L_2)(\beta_4 + L_2)}{4(\beta_2 + L_2)(\beta_4 + L_2) - L_2^2} \tag{108}$$

$$K_7 = \frac{2A_3(\beta_1 + L_1) + L_1 A_1}{4(\beta_1 + L_1)(\beta_3 + L_1) - L_1^2} \tag{109}$$

$$K_8 = \frac{L_1(\beta_1 + L_1)}{4(\beta_1 + L_1)(\beta_3 + L_1) - L_1^2} \tag{110}$$

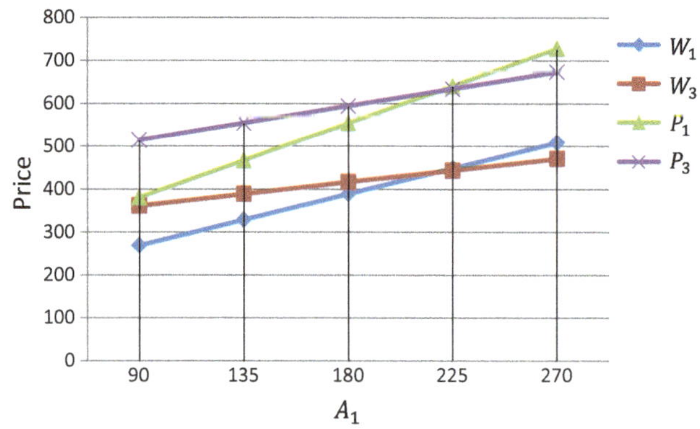

Fig. 9 Changes of optimal prices with respect to A_1 for the second model under MS-Bertrand policy

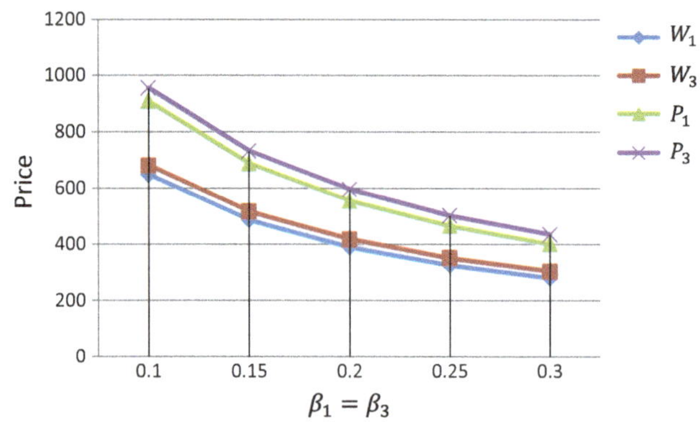

Fig. 10 Changes of optimal prices with respect to $\beta_1 = \beta_3$ for the second model under MS-Bertrand policy

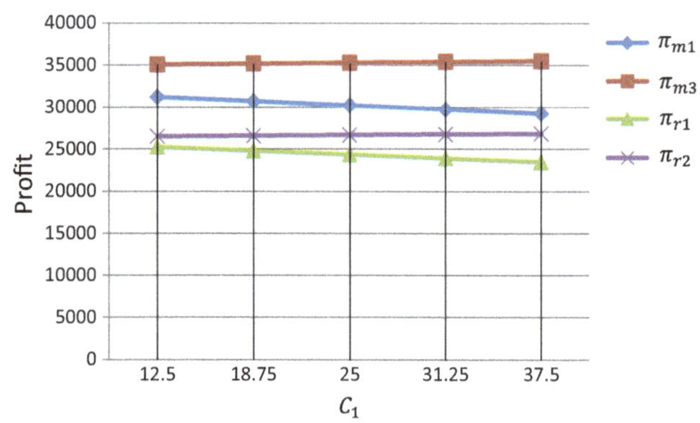

Fig. 11 Changes of maximum profits with respect to C_1 for the second model under MS-Stackelberg policy

$$K_9 = \frac{2A_4(\beta_2 + L_2) + L_2 A_2}{4(\beta_2 + L_2)(\beta_4 + L_2) - L_2^2} \tag{111}$$

$$K_{10} = \frac{L_2(\beta_2 + L_2)}{4(\beta_2 + L_2)(\beta_4 + L_2) - L_2^2} \tag{112}$$

$$N_1 = A_1 - \beta_1 K_1 - L_1(K_1 - K_7) + C_1(\beta_1 K_3 + L_1 K_3 - L_1 K_8) \tag{113}$$

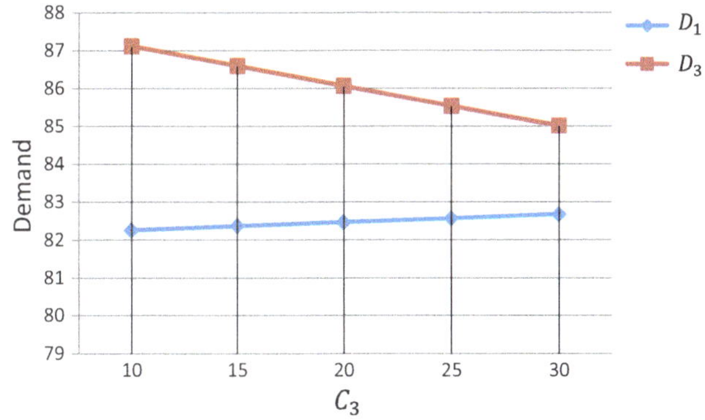

Fig. 12 Changes of maximum demands with respect to C_3 for the second model under MS-Stackelberg policy

$$N_2 = A_3 - \beta_3 K_7 - L_1(K_7 - K_1) + C_3(\beta_3 K_3 + L_1 K_3 - L_1 K_2) \quad (114)$$

$$N_3 = \beta_1 K_2 + L_1 K_2 - L_1 K_3 \quad (115)$$

$$N_4 = 2(\beta_1 K_3 + L_1 K_3 - L_1 K_8) \quad (116)$$

$$N_5 = \beta_3 K_8 + L_1 K_8 - L_1 K_3 \quad (117)$$

$$N_6 = 2(\beta_3 K_3 + L_1 K_3 - L_1 K_2) \quad (118)$$

$$N_7 = A_2 - \beta_2 K_4 - L_2(K_4 - K_9) + C_2(\beta_2 K_6 + L_2 K_6 - L_2 K_{10}) \quad (119)$$

$$N_8 = A_4 - \beta_4 K_9 - L_2(K_9 - K_4) + C_4(\beta_4 K_6 + L_2 K_6 - L_2 K_5) \quad (120)$$

$$N_9 = \beta_2 K_5 + L_2 K_5 - L_2 K_6 \quad (121)$$

$$N_{10} = 2(\beta_2 K_6 + L_2 K_6 - L_2 K_{10}) \quad (122)$$

$$N_{11} = \beta_4 K_{10} + L_2 K_{10} - L_2 K_6 \quad (123)$$

$$N_{12} = 2(\beta_4 K_6 + L_2 K_6 - L_2 K_5) \quad (124)$$

$$N_{13} = (A_3 - \beta_3 K_7 - L_1(K_7 - K_1))N_4 - N_1 N_5 + \left[\frac{N_4 N_6}{2} - N_3 N_5\right] C_3 \quad (125)$$

$$N_{14} = N_4 N_6 - 2N_3 N_5 \quad (126)$$

$$N_{15} = (A_4 - \beta_4 K_9 - L_2(K_9 - K_4))N_{10} - N_7 N_{11} + \left[\frac{N_{10} N_{12}}{2} - N_9 N_{11}\right] C_4 \quad (127)$$

$$N_{16} = N_{10} N_{12} - 2N_9 N_{11} \quad (128)$$

References

Boyacı T, Gallego G (2002) Coordinating pricing and inventory replenishment policies for one wholesaler and one or more geographically dispersed retailers. Int J Prod Econ 77(2):95–111

Chen X, Simchi-Levi D (2004a) Coordinating inventory control and pricing strategies with random demand and fixed ordering cost: the infinite horizon case. Math Oper Res 29:698–723

Chen X, Simchi-Levi D (2004b) Coordinating inventory control and pricing strategies with random demand and fixed ordering cost: the finite horizon case. Oper Res 52:887–896

Chen X, Simchi-Levi D (2006) Coordinating inventory control and pricing strategies: the continuous review model. Oper Res Lett 34(3):323–332

Chen Y, Ray S, Song Y (2006) Optimal pricing and inventory control policy in periodic-review systems with fixed ordering cost and lost sales. Naval Res Logist 53:117–136

Chen YC, Fang SC, Wen UP (2013) Pricing policies for substitutable products in a supply chain with Internet and traditional channels. Eur J Oper Res 224(3):542–551

Dada M, Srikanth K (1987) Pricing policies for quantity discounts. Manag Sci 33(1):1247–1252

Elmaghraby W, Keskinocak P (2003) Dynamic pricing in the presence of inventory considerations: research overview, current practices, and future directions. Manag Sci 49(10):1287–1309

Fei W, Du M, Luo G (2015) Optimal prices and associated factors of product with substitution for one supplier and multiple retailers supply chain. Proc Comput Sci 60:1271–1280

Giri BC, Sharma S (2014) Manufacturer's pricing strategy in a two-level supply chain with competing retailers and advertising cost dependent demand. Econ Model 38(2):102–111

Gurler U, Yilmaz A (2010) Inventory and coordination issues with two substitutable products. Appl Math Model 34(3):539–551

Hsieh CC, Wu CH (2009) Coordinated decisions for substitutable products in a common retailer supply chain. Eur J Oper Res 196(1):273–288

Hsieh CC, Chang YL, Wu CH (2014) Competitive pricing and ordering decisions in a multiple-channel supply chain. Int J Prod Econ 154:156–165

Kim DS, Lee WJ (1998) Optimal joint pricing and lot sizing with fixed and variable capacity. Eur J Oper Res 109(1):212–227

Mahmoodi A, Eshghi K (2014) Price competition in duopoly supply chains with stochastic demand. J Manuf Syst **(in press)**

Maihami R, Karimi B (2014) Optimizing the pricing and replenishment policy for non-instantaneous deteriorating items with stochastic demand and promotional efforts. Comput Oper Res 51:302–312

Mukhopadhyay SK, Yue X, Zhu X (2011) A Stackelberg model of pricing of complementary goods under information asymmetry. Int J Prod Econ 134(2):424–433

Panda S, Modak NM, Sana SS, Basu M (2015) Pricing and replenishment policies in dual-channel supply chain under continuous unit cost decrease. Appl Math Comput 256:913–929

Shavandi H, Mahlooji H, Nosratian NE (2012) A constrained multi-product pricing and inventory control problem. Appl Soft Comput 12(8):2454–2461

Sinha S, Sarmah S (2010) Coordination and price competition in a duopoly common retailer supply chain. Comput Ind Eng 59(2):280–295

Soon W (2011) A review of multi-product pricing models. Appl Math Comput 217(21):8149–8165

Starr M, Rubinson J (1978) A loyalty group segmentation model for brand purchasing simulation. J Mark Res 15(3):378–383

Taleizadeh AA, Noori-daryan M (2014) Pricing, manufacturing and inventory policies for raw material in a three-level supply chain. Int J Syst Sci 47(4):919–931

Taleizadeh AA, Noori-daryan M, Cárdenas-Barrón LE (2014) Joint optimization of price, replenishment frequency, replenishment cycle and production rate in vendor managed inventory system with deteriorating items. Int J Prod Econ 159:285–295

Tang CS, Yin R (2007) Joint ordering and pricing strategies for managing substitutable products. Prod Oper Manag 16(1):138–153

Wang L, Zhao J, Wei J (2014) Pricing decisions of two complementary products in a fuzzy environment. Math Probl Eng 2014:1–8, Art ID 729287. doi:10.1155/2014/729287

Wei J, Zhao J (2011) Pricing decisions with retail competition in a fuzzy closed-loop supply chain. Expert Syst Appl 38(9): 11209–11216

Wei J, Zhao J, Li Y (2013) Pricing decisions for complementary products with firms' different market powers. Eur J Oper Res 224(3):507–519

Wei J, Zhao J, Li Y (2015) Price and warranty period decisions for complementary products with horizontal firms' cooperation/noncooperation strategies. J Clean Prod 105:86–102. doi:10.1016/j.jclepro.2014.09.059

Xiao T, Jin J, Chen G, Shi J, Xie M (2010) Ordering, wholesale pricing and lead-time decisions in a three-stage supply chain under demand uncertainty. Comput Ind Eng 59(4):840–852

Yan R, Bandyopadhyay S (2011) The profit benefits of bundle pricing of complementary products. J Retail Consum Serv 18(4): 355–361

Yu J, Ma S (2013) Impact of decision sequence of pricing and quality investment in decentralized assembly system. J Manuf Syst 32(4):664–679

Yue X, Mukhopadhyay SK, Zhu X (2006) A Bertrand model of pricing of complementary goods under information asymmetry. J Bus Res 59(10):1182–1192

Zhang J, Chiang WYK, Liang L (2014) Strategic pricing with reference effects in a competitive supply chain. Omega 44:126–135

Zhao J, Tang W, Wei J (2012a) Pricing decision for substitutable products with retail competition in a fuzzy environment. Int J Prod Econ 135(1):144–153

Zhao J, Tang W, Zhao R, Wei J (2012b) Pricing decisions for substitutable products with a common retailer in fuzzy environments. Eur J Oper Res 216(2):409–419

Zhao J, Wei J, Li Y (2014) Pricing decisions for substitutable products in a two-echelon supply chain with firm different channel powers. Int J Prod Econ 153(4):243–252

15

A petrol station replenishment problem: new variant and formulation

Abdelaziz Benantar[1] · Rachid Ouafi[1] · Jaouad Boukachour[2]

Abstract One of the most important problems in the petroleum industry is the well-known petrol station replenishment problem with time windows, which calls for the determination of optimal routes by using a fleet of tank trucks to serve a set of petrol stations over a given planning horizon. In this paper, we introduce a model and solve a specific problem that originates from a real-life application arising in the fuel distribution where specific attention is paid to tank trucks with compartments and customers with different types of products and time windows. Literally, we call the resulting problem the multi-compartment vehicle routing problem with time windows (MCVRPTW). To solve the MCVRPTW, we begin by describing the problem, providing its mathematical formulation and discussing the sense of its constraints. As the problem is NP-hard, we propose an efficient tabu search algorithm for its solution. We introduce the Kolmogorov–Smirnov statistic into the framework of the tabu search to manage the neighbourhood size. We evaluate the performance of the algorithm on a set of vehicle routing problems with time windows instances as well as other realistic instances. Our results are compared to CPLEX, to the heuristics reported in the literature and also to those extracted from the company plans.

✉ Abdelaziz Benantar
a_benantar@yahoo.fr

Rachid Ouafi
rouafi@usthb.dz

Jaouad Boukachour
jaouad.boukachour@univ-lehavre.fr

[1] Department of Operational Research, USTHB University, P.O. Box 32, 16111 Bab-Ezzouar, Algeria

[2] LMAH, Normandie University, ULH, 76600, Le Havre, France

Keywords Petrol station replenishment · Vehicle routing · Compartments · Time windows · Tabu search

1 Introduction

The specific problem, which will be discussed in this paper, is a variant of the capacitated vehicle routing problem and occurs in the context of fuel distribution. More precisely, the paper deals with the daily replenishment-planning problem that the biggest Algerian petroleum company is facing. The company is faced with a particular problem which is demonstrated by the following procedures in operations. (a) The company has to deliver different fuel types ordered by a set of petrol stations during the next working day. (b) These stations order one or more fuel types each time and specify the quantity to be delivered for each product ordered. (c) The products are incompatible and must be transported in independent vehicle compartments. In addition, petrol stations specify time windows during which they must be served. The company delivers products from one or more depots. (d) Each depot is responsible for the demand of stations located in the same district as the depot. The company assigns a fleet of vehicles to each depot. (e) These vehicles are compartmentalized and do not have flow metres. This implies that the contents of a compartment cannot be used to replenish more than one underground reservoir. Consequently, each compartment of the delivery vehicle must be filled with one of the products to be delivered on its route. In this context, the company prepares a replenishment plan for their petrol stations for the next day. This plan requires a number of simultaneous and interrelated decisions to be made. To prepare such a plan, the company must determine the routes for the delivery of all the demanded products, assign these routes to vehicles, determine the quantities to be delivered by each route, and load these quantities to the compartments. On this point, it should be noted that, in 2013, an Agreement was concluded between the company and the petrol station managers. According to this Agreement, the quantities loaded in the compartments can be adjusted up to a given threshold. This particular situation occurs quite frequently when the demands of petrol stations are high in winter. The objective of the replenishment plan is to determine a set of routes satisfying all customer orders at a minimal routing and service cost.

Our real-life application can be viewed as a combination of two variants of the vehicle routing problem (VRP): a VRP with multiple compartments (MCVRP) and VRP with time windows (VRPTW). For clarity of exposition, this problem will be called the multi-compartment vehicle routing problem with time windows (MCVRPTW). However, it is different from the problems previously discussed in the literature with respect to the following properties:

1. Besides the common constraints of heterogeneous fleets, multiple compartments and time windows, we incur a penalty cost in connection with the use of private vehicles. This constraint originates from the fact that the company managers would like to promote the use of their own vehicles.
2. Since restrictions on the accessibility of vehicles (see constraints 2 in the problem formulation) are imposed, some vehicles cannot visit some petrol stations, e.g., the petrol stations managed by the army cannot be delivered to by private vehicles.
3. The quantities loaded in the compartments can be adjusted according to an Agreement concluded between the company and the petrol station managers.

Consequently, not only do the vehicle routes have to be determined, but how to maximize the quantities to be delivered also has to be decided for each day.

The problem is NP-hard since it is a combination of the MCVRP and VRPTW, which are known to be NP-hard problems [34]. Furthermore, the MCVRPTW is more complicated than the MCVRP and VRPTW considering that it needs to tackle compartments and time windows constraints simultaneously. Because the practical large-scale MCVRPTW instances are difficult, if not impossible, to tackle efficiently within a reasonable amount of computing time, even by using the most powerful MIP solvers such as CPLEX (see Sect. 5), the purpose of this paper is to propose an effective metaheuristic for the MCVRPTW. To make the implementation simpler, we employ a tabu search as the algorithm framework. Different structural neighbourhood methods are used in the improving phase of the tabu search to broaden the exploration of the search space. Meanwhile, a Kolmogorov–Smirnov statistic (KSS) is incorporated into the framework of the tabu search to manage the neighbourhood size. The main idea of the KSS is that a neighbourhood size is determined according to a probability model that minimizes the distance criterion and decides whether two customers are neighbours or not.

2 Literature review

Several versions of the petrol station replenishment problem have attracted interest over the last three decades. In this section, we present a brief review of the published literature dealing with these versions of the petrol station replenishment problem in a chronological order.

One of the first articles was published by Brown and Graves [6], who developed an automated real-time dispatch system for the distribution of petroleum products for a major US oil company. Each order includes several gasoline products, jointly constituting a full truckload. Brown and Graves proposed a model to assign orders to trucks. The objective was to minimize the sum of travel costs and establish a penalty for trucks that exceed the allowable working hours per day, as well as those that had less than the required minimum working hours.

Brown et al. [7] developed a computerized assisted dispatch system for Mobil Oil Corporation in the United States. The dispatching procedure used by the system was an extension of the one presented by Brown and Graves [6] and allowed visiting more than one customer per trip.

Franz and Woodmanse [17] developed a rule-based semi-automated decision support system for a regional oil company to determine the daily schedule of the drivers and the dispatching of the tank trucks.

Ronen [27] studied the dispatching problem. The main concern was to set up a timely and economic delivery of petroleum products and/or liquid chemicals by a fleet of vehicles. He presented three models, the set partitioning model, the elastic set partitioning model and the set packing model, and showed how they could be used in a petrol product distribution system. Bausch et al. [2] proposed an elastic set partitioning technique to solve the same problem. The candidate schedules are obtained by generating trips using the sweep algorithm. Also, in 1995, Van der Bruggen et al. [36] solved the single period version of the problem as part of a broader study aimed at optimizing the distribution network of a large oil company operating in the Netherlands. They suggested some simple models to assign customers to depots, to determine the fleet size and composition and to restructure the depot network.

Nussbaum and Sepulveda [24] addressed the distribution problem for the biggest fuel company in Chile. The delivery plans are obtained using a heuristic approach and a planning, execution and control system is developed employing a knowledge-based approach that utilizes a graphical user interface, which mimics the mental model of the user.

Several greedy heuristics followed by simple improvement procedures for the multi-period problem were proposed by Taqa allah et al. [33]. They proposed two heuristics for constructing petrol station replenishment plans for the case in which there is only one depot, an unlimited homogeneous tank truck fleet, and no time windows. In 2002, Ben Abdelaziz et al. [3] presented a real-life routing problem in which a variable neighbourhood search heuristic was applied to solve a single period petroleum products delivery problem using a heterogeneous fleet of compartmented tank trucks. Malépart et al. [21] generalized this problem by allowing delivery to more than one station in the same trip. The authors proposed four heuristics for constructing replenishment plans over a horizon of several working days. Their heuristics were tested using some real-life problems obtained from a transport company in Eastern Canada.

Avella et al. [1] studied a petrol replenishment network involving one depot, a heterogeneous tank truck fleet and no time windows. They proposed a heuristic and an exact algorithm based on a route generation scheme and a branch-and-price algorithm. To test the performance of their approach, they used real-world data consisting of 25 customers and six tank trucks of three different types.

Ng et al. [23] studied two small petrol distribution networks in Hong Kong: the Hong Kong Island network and the network for the Kowloon Peninsula and the New Territories. They proposed a model for simultaneously assigning trips to trucks and stations. For this case, station' inventories were managed by the vendor who decided when to replenish each station and what quantities to deliver.

Cornillier et al. [11] tackled the petrol station replenishment problem (PSRP) with one depot. They developed an exact algorithm to solve the single period case with unlimited heterogeneous truck fleet but without time windows. Cornillier et al. [12] extended the PSRP to a multi-period setting and developed a multi-phase heuristic with look-back and look-ahead procedures. Basically, the heuristic first determines the stations to be serviced in each period. Then, the problem reduces to solving multiple PSRPs where the exact algorithm of Cornillier et al. [11] is utilized. Cornillier et al. [13] addressed the PSRP with time windows. They developed two heuristics based on the mixed integer linear programming formulation of the problem. In a different setting, Day et al. [15] implemented a three-phase heuristic for the cyclical inventory routing problem encountered at a carbon dioxide gas distributer in Indiana. Their heuristic determines regular routes for each of three available delivery vehicles over a 12-day delivery horizon while improving delivery labour cost, stockouts, delivery regularity and driver-customer familiarity.

Cornillier et al. [14] published a paper, which was different from the previous ones as it dealt with the multi-depot version of the problem. In addition to the proposed formulation, the authors developed a heuristic, which requires generating trips, not routes as in the previous papers, and trips are generated in a different way. In fact, they suggested a restricted set of promising feasible trips to solve the trip selection model. They used a two-phase procedure in which they first constructed an initial set of trips and then selected a subset of this set to obtain the required set.

Popović et al. [25] developed a variable neighbourhood search (VNS) heuristic model for solving a multi-period multi-product IRP (Inventory Routing Problem) in fuel delivery with multi-compartment homogenous vehicles and deterministic consumption that varies with each petrol station and each fuel type. The stochastic VNS heuristic is compared to a mixed integer linear programming (MILP) model and the deterministic "compartment transfer" (CT) heuristic. For the same problem, Vidović et al. [38] presented two solution approaches: the MIP model and the heuristics approach. The MIP model is formulated as the problem of petrol station assignment to individual routes with consideration of the daily inventory costs. The proposed heuristic includes a relaxed MIP model for obtaining the initial solution, ideas for transferring deliveries over one or more time periods earlier, assignment of petrol stations to a vehicle in the same route (represented through the utilities calculation) and a variable neighbourhood descent (VND) search.

Our aim is to improve the fuel distribution operations of Naftal company using OR techniques. The problem seems similar to that of Cornillier et al. [13] since both attack a petrol station replenishment problem with time windows using multi-compartment vehicles. However, there are some key differences. Firstly, Cornillier et al. [13] consider a case where trucks can visit up to four stations per route, which is justified by the fact that most trucks have from four to six compartments, while stations generally require two or three products, one of which frequently requires two compartments. Moreover, the time windows for servicing stations are very wide. In addition, their aim is to maximize the total profit equal to the sales revenue, minus the sum of routing costs and of regular and overtime costs, whereas, in our case, since restrictions on the accessibility of vehicles (see constraints 2 in the problem formulation) are imposed, a vehicle cannot visit some petrol stations, e.g., the petrol stations managed by the army cannot be delivered to by the private vehicles.

The remainder of this paper is organized as follows. Section 3 describes and formulates the problem. Section 4 describes the details of the proposed approach. The instances used and the results obtained are discussed in Sect. 5. Section 6 presents the conclusion and some suggestions for future work.

3 Problem description and formulation

The MCVRPTW is defined on a complete undirected network graph $G = (V, E)$ where $V = \{0, 1, 2, \ldots, n\}$ is a vertex set, and $E = \{(i,j) \in V \times V; 0 \leq i,j \leq n, i \neq j\}$ is an edge set. Vertex 0 represents the depot while the remaining vertices $N = \{1, 2, \ldots, n\}$ correspond to the customers. The depot stores p types of products. There are two types of vehicles. The number of vehicles for each type is limited. Let P, K_1 and K_2 represent the sets of the p types of products and the two types of vehicles (National company and Private companies vehicles), respectively. Each vehicle k has capacity C_k with the set of Q_1 or Q_2 compartments. Each compartment q is dedicated to product p and has a known capacity c_k^q. Furthermore, a penalty cost f_2 is incurred for each use of vehicle $k \in K_2$ in the routes. The travelling cost of each edge $(i, j) \in E$ is c_{ij}. For each customer $i \in N$, we have a positive demand d_{ip} and a time window $[a_i, b_i]$. Each demand has to be delivered in total. However, this demand may be adjusted according to the Agreement that was concluded between the company and the petrol station managers. Hence, the resulting new demand may be up to λ_{ip} % less than the ordered demand d_{ip}, i.e., $d'_{\mathrm{ip}} = (1 - \lambda_{\mathrm{ip}}) \times d_{ip}$, where $\lambda_{\mathrm{ip}} \in [0, 0.10]$. This particular problem, called the MCVRPTW-AD (MCVRPTW with adjustment), occurs quite frequently when the demands of petrol stations are high in winter.

The time window at the depot $[a_0, b_0]$ corresponds to the feasible scheduling horizon for each vehicle route. Associated with each arc $(i, j) \in E$, t_{ij} represents the travel time along that arc. Note that the service time at customer i must start within the associated time window and the vehicle must stop for a t_i time. To take into account the restrictions imposed on the accessibility of vehicles (see constraint 2 in the problem formulation), we define $\{0, 1\}$ matrix $A = (a_k^i)$, which equals 1 if and only if customer i can be served by vehicle k and which equals 0 otherwise.

The MCVRPTW requires the following three types of variables:

- $x_{ijk} = \begin{cases} 1 & \text{if } i \text{ precedes } j \text{ in the route of vehicle } k. \\ 0 & \text{otherwise.} \end{cases}$
- $y_{ipk} = \begin{cases} 1 & \text{if customer } i \text{ receives product } p \text{ from vehicle } k. \\ 0 & \text{otherwise.} \end{cases}$
- s_{ik} specifies the arrival time at i when serviced by vehicle k. In case of vehicle k does not service i, s_{ik} has no meaning and consequently its value is considered irrelevant.

Given all the parameters and variables defined above, the MCVRPTW can be formulated as follows:

$$Minimize \sum_{k \in K_1} \sum_{i \in V} \sum_{j \in V} c_{ij} x_{ijk} + \sum_{k \in K_2} \sum_{i \in V} \sum_{j \in V} (c_{ij} + f_2) x_{ijk} \quad (1)$$

Subject to

$$y_{ipk} \leq a_k^i, \quad i \in N, \quad k \in \{K_1 \cup K_2\}, \quad p \in P \quad (2)$$

$$\sum_{i \in V} x_{ihk} - \sum_{j \in V} x_{hjk} = 0, \quad h \in N, k \in \{K_1 \cup K_2\} \quad (3)$$

$$x_{ijk} + x_{jik} \leq 1, \quad i,j \in N, i \neq j, k \in \{K_1 \cup K_2\} \quad (4)$$

$$\sum_{j \in N} x_{0jk} \leq 1, \quad k \in \{K_1 \cup K_2\} \quad (5)$$

$$\sum_{i \in N} x_{i0k} \leq 1, \quad k \in \{K_1 \cup K_2\} \quad (6)$$

$$\sum_{i \in N} d_{\mathrm{ip}} y_{ipk} \leq c_k^q, \; k \in \{K_1 \cup K_2\}, \quad q \in \{Q_1 \cup Q_2\}, \; p \in P \quad (7)$$

$$\sum_{k \in \{K_1 \cup K_2\}} y_{ipk} = 1, \quad i \in N, \quad p \in P \quad (8)$$

$$y_{ipk} \leq \sum_{j \in V} x_{jik}, \quad i \in N, \quad k \in \{K_1 \cup K_2\}, \quad p \in P \quad (9)$$

$$s_{ik} + t_i + t_{ij} - M(1 - x_{ijk}) \leq s_{jk}, \quad i,j \in V, \quad k \in \{K_1 \cup K_2\} \quad (10)$$

$$a_i \leq s_{ik} \leq b_i, \quad i \in V, \quad k \in \{K_1 \cup K_2\} \quad (11)$$

$$x_{ijk} \in \{0, 1\}, \quad i,j \in V, \quad k \in \{K_1 \cup K_2\} \quad (12)$$

$$y_{ipk} \in \{0, 1\}, \quad i \in N, \quad k \in \{K_1 \cup K_2\}, \quad p \in P \quad (13)$$

$$s_{ik} \geq 0, \quad i \in N, \quad k \in \{K_1 \cup K_2\} \quad (14)$$

Objective (1) is to minimize the total cost, which consists of the travelling costs and penalty costs of private company vehicles used for service. Constraint (2) considers the accessibility restrictions, where some vehicles cannot serve some customers. Constraints (3)–(4) guarantee that one vehicle arrives at each customer, leaves it and does not return to the previous customer. Constraints (5)–(6) impose that the start and end of the route for each vehicle must be the depot. Constraint (7) ensures that the compartment capacity is not exceeded. In constraint (8), each product ordered by the customer is brought by one single vehicle. Constraint (9) sets y_{ikp} to zero for each product p if customer i is not visited by vehicle k. Constraint (10) represents the relationship between the starting time of the service to one customer and the departure time from its predecessor. In constraint (10), parameter M is an arbitrarily large value. Constraint (11) ensures that the service takes place at each customer with respect to the time window. Constraints (12)–(14) define the decision variables, which are all binary except for variable s_{ik}.

4 Solution approach

To solve the problem modelled above, we describe our approach from its general structure to its main components.

4.1 General structure

The detailed steps of this approach are completely described as follows:

- *Step 1. Initialization*
  1. Generate one initial feasible solution by combining the loading aspect with the modified Solomon heuristic [28] as described in Sect. 4.2.
  2. Initialize parameters.
- *Step 2. Local search*
  2-opt*, relocate and swap operators are applied to explore the search space. In fact, the search process is restricted to a set of elite neighbouring solutions, and the criterion used to select them is based on the Kolmogorov–Smirnov statistic. At each iteration, one of these three operators is randomly chosen and applied to the current solution S.
- *Step 3. Penalty component*
  In order to enlarge the search space by visiting infeasible solutions, the capacity and time constraints are relaxed, i.e., the capacity and time violations are penalized by two coefficients, and the penalized term is added to the objective function.
- *Step 4. Tabu list*
  The tabu list is implemented as an upper triangular matrix where each element contains a set of attributes able to characterize the solution and records the iteration in which an arc is moved from one route to another. The size of the tabu list is updated according to the quality of the solutions obtained during the recent moves.
- *Step 5. Diversification and intensification strategies*
  Four mechanisms are used for diversifying and intensifying the search (see Sect. 4.7 for more details).
- *Step 6. Termination*
  The process can be stopped if the termination conditions are completed; otherwise go to step 2 and continue the search.

4.2 Route construction heuristic

Usually, the methods used to generate an initial solution are simple and fast to compute because it is assumed that the task of producing a good solution is mainly handled by the tabu search algorithm [28]. In our work, the search process for the initial solution includes two main phases. In the first phase, called loading, an initial loading solution is obtained. This solution is then used to generate the initial routes in the routing phase. The details of these two phases are explained as follows:

1. In the loading phase, our aim is to determine the subsets of orders that will be loaded into the same vehicle. More precisely, one must determine which of the products is assigned to a certain compartment. The compartments of each vehicle are loaded so that none of the routing constraints is considered. This can be done by solving the following compartment-loading problem (CLP):

$$Minimize \sum_{k\in\{K_1\cup K_2\}} \sum_{q\in\{Q_1\cup Q_2\}} \sum_{p\in P} z_{pqk} \tag{15}$$

subject to

$$d_{ip} \leq \sum_{q\in\{Q_1\cup Q_2\}} c_k^q z_{pqk}, \quad k \in \{K_1 \cup K_2\}, \quad i \in N, \quad p \in P \tag{16}$$

$$\sum_{p\in P} z_{pqk} \leq 1, \quad k \in \{K_1 \cup K_2\}, \quad q \in \{Q_1 \cup Q_2\} \tag{17}$$

Binary variable z_{pqk} indicates whether product p is assigned to compartment q in vehicle k. The objective function in (15) expresses the fact that we wish to minimize the number of loaded compartments. Constraint (16) states that the quantities to load for one product must not exceed the sum of the capacity of compartments assigned to that product. Constraint (17) imposes that each compartment is dedicated at most to one product per route.
2. In the above initial loading solution, each vehicle has a list of customers to visit. For each list, we apply the modified nearest neighbour heuristic proposed by Solomon [28] to generate our initial routes. As graphically described in Fig. 1, our heuristic constructs the routes by first visiting the customer closest to the depot, where only temporal closeness is taken into account. At each iteration, a vehicle with its subset of customers is selected. The route starts with the customer who has the earliest time a_i. The next customer to be visited in the route will be the one that is (1) not yet visited and (2) closest to the last customer in the current route. This process is repeated until there is no customer to visit in the current route. When this happens, the whole process is repeated until all the customers are visited.

4.3 Neighbourhood structure

In the tabu search algorithm, it will take a long time to compute the values of all moves that allow one to pass

(a)

(b)

Node	a_i	b_i	Visiting Order
0	0	--	--
1	3	6	2
2	5	15	3
3	2	10	1
4	10	17	8
5	14	18	9
6	15	19	10
7	9	13	7
8	8	12	5
9	8	16	6
10	7	11	4

(c) (d)

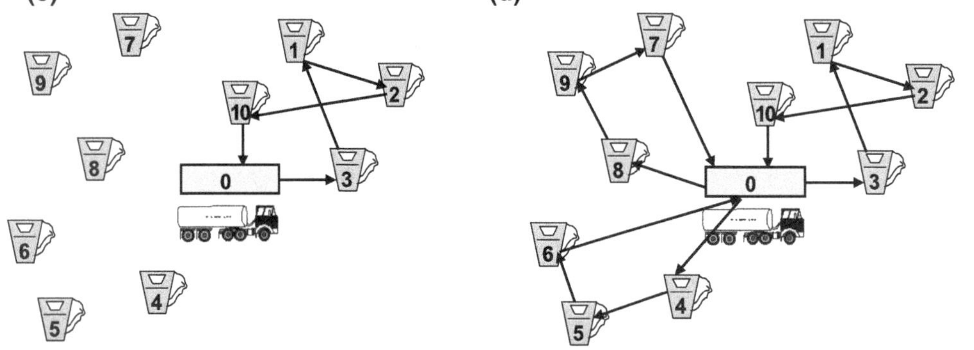

Fig. 1 Route construction heuristic: **a** Spatial location of the depot and customers. **b** Customers' time windows and visiting order. **c** Customers visited by one vehicle. **d** Initial solution

from one solution to a neighbouring one. One of the reasons for these large computing-time requirements is that the mechanisms of generating candidate solutions normally need to perform several thousand iterations to obtain high-quality solutions [34]. Therefore, the strategy of selecting the nearest neighbours was adopted to improve the convergence speed in this paper. Within the general structure of the neighbourhood, this strategy may be seen as an implementation of a candidate list. In fact, the search is restricted to a set of elite neighbouring solutions, and the criterion used to select them is fixed in advance. The size of the candidate list will be determined in the following sections by applying the Kolmogorov–Smirnov statistic. The introduction of this statistic is an advance over existing work in which most of the relevant algorithms tend to use classical moves. To improve the clarity of exposition, in Sect. 4.4, we describe in detail the Kolmogorov–Smirnov statistic by applying it to a given instance of Solomon. As is shown in Fig. 4, with 100 customers, the size of the candidate list is restricted to a value of 21. Our approach employs three move types; each explores a restricted search space by embedding these lists in the search process. At each iteration, the algorithm randomly chooses one operator from these three operators and applies it to the current solution S. The details of these operators are listed and described below.

a. *2-opt**: Two customers, $i \in R_k$ and $j \in R_{k'}$, are chosen. Then, the edges connecting i to $i+1$ and j to $j+1$ are removed. Two new edges are added adjoining i with $j+1$ and j with $i+1$. See Fig. 2a.
b. *Relocate*: Two customers, $i \in R_k$ and $j \in R_{k'}$, are selected and i is removed from its original route R_k and inserted following j in the second route $R_{k'}$. The relocate operator may reduce the number of vehicle routes. See Fig. 2b.
c. *Swap*: Two customers, $i \in R_k$ and $j \in R_{k'}$, are chosen and exchanged between two routes. See Fig. 2c.

4.4 Neighbourhood size

The tabu search can be very time consuming due to the large size to the neighbourhood N(S) and also to the cost functions that must be constantly reevaluated. Thus, we propose a neighbour reduction strategy designed to reduce the computing time. Within the general settings of tabu

Fig. 2 Neighbourhood structure

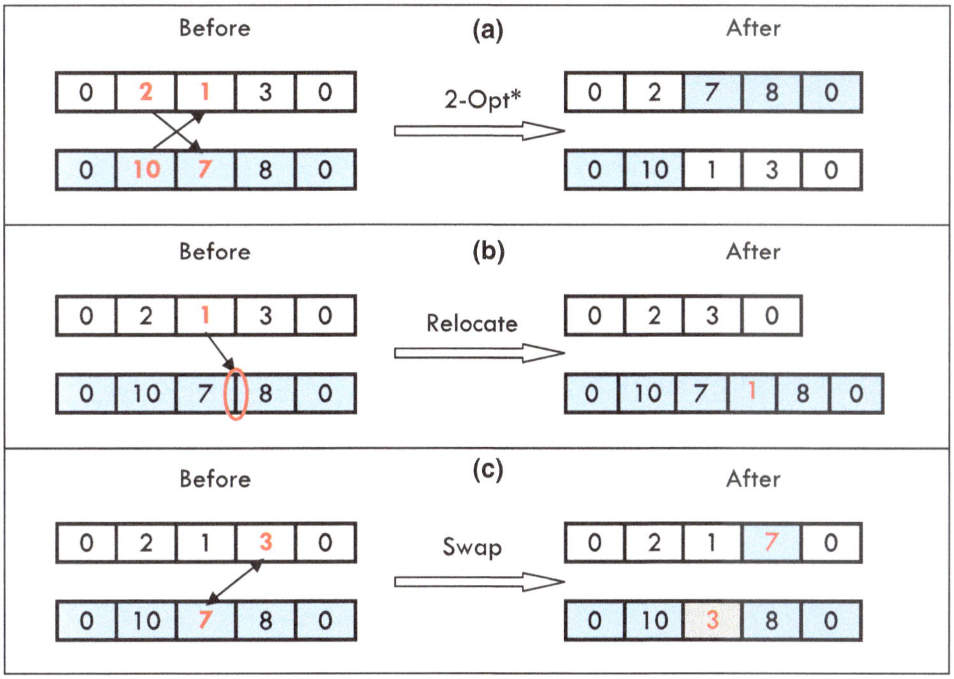

search, this strategy may be seen as an implementation of candidate list [19]. In fact, the search is restricted to a set of elite neighbouring solutions, and the criterion used to select them is fixed in advance. Similar ideas have been used by various authors to speed up local-search algorithms [35]. The most widely used strategy, called Random, involves randomly selecting some neighbours from N(S) for evaluation, thus allowing only a predefined proportion of the total neighbourhood to be considered. By forcing a decrease in the number of neighbours, the strategy is able to decrease the time needed to find a solution. However, because the strategy does not necessarily select the best neighbour, it could lead to a different, and maybe worse, solution. Another frequently used strategy, called Distance, relies on taking the problem configuration into account [26]. The idea is simple: a given neighbour is evaluated only if the customer under consideration is within a fixed distance from the nearest customer on the route into which it will be inserted. For example, in [26], the fixed distance is predefined as being equal to a fraction of the longest distance on the geographical map. Apart from the mentioned strategies, the candidate list is also restricted to a fraction of the total number of customers [8].

By analysing these strategies, some drawbacks can be underlined. In fact, their way of fixing the number of candidates does not take into account the instance's density and consequently no guarantee is provided regarding the compromise between solution quality and the time to find it. This means that if the customer under consideration is located in a remote location, the algorithm can evaluate some unpromising moves; or if it is in a cluster, it might still result in too many moves to be evaluated. In order to overcome these drawbacks, our paper attends to additional aspects which, according to the best of our knowledge, have not been considered in the literature before. These aspects include (1) the analysis of the instance configuration, (2) the introduction of the Kolmogorov–Smirnov distance criterion, and (3) the implementation of candidate list for each customer. The second aspect defines a new feature with which we suggest a heuristic (Algorithm 1) for the neighbourhood size.

The purpose of introducing the Kolmogorov–Smirnov statistic (KSS) is to find the preferable customers to visit (when moving from one solution to a neighbouring one), taking into to account their geographical positions in the instance. The idea is simple: (1) select the model that minimizes the distance criterion, (2) deduce its parameters and (3) decide whether two customers are neighbours or not. The customers are considered neighbours of a given customer i, if and only if the distance between them is less or equal to a given parameter A_i. This parameter is calculated by using the pth quantile formula, which depends on the probability distribution of its distance sample L_i. The use of such a probability aspect in computing A_i is justified by the aim of checking whether one of the known distributions can be suited to each distance sample. In fact, the KSS fits the distance sample by selecting a list of candidate distributions, estimating their parameters and giving their ranking. Once the best statistical model is identified, the parameter A_i is defined as follows: $F(A_i) = \Pr(L_i \leq A_i) = p \Rightarrow A_i = F^{-1}(p)$.

The whole procedure of fit is briefly explained in Algorithm 1.

Algorithm 1 Procedure of fit.

Step 1.
for each customer $i = 1, ..., n$ **do**
(a) Calculate $l_{ij} = \sqrt{(x_i - x_j)^2 + (y_i - y_j)^2}, j \in N$
(b) Build the set $L_i = \{l_{ij}, j \in N\}$
end for
Step 2.
Over the candidate distributions do:

(a) Fit the distributions to the sample (L_i).
(b) Calculate the Kolmogorov-Smirnov statistics, give their ranking and parameters.
(c) Choose the best distribution that minimizes the distance criterion.
(d) Use the parameters of the best distribution to determine $A_i = F^{-1}(p)$.

Step 3.
Repeat steps 2 for each sample L_i.

To improve the clarity of exposition, we give an illustrative application in which we show how to apply the Kolmogorov–Smirnov statistic on a given instance. Let $R101$, $i = 26$ and L_{26} represent the instance of Solomon (Fig. 4), the selected customer and its distance sample, respectively. With this sample design, we test the following set of distributions: Normal, Exponential, Weibull, GEV, Gamma, Fréchet and Pareto. Table 1 summarizes the goodness of fit for this example and shows that the GEV model is the best. This result is confirmed when the GEV quantile-quantile plot of the L_{26} variable is displayed as in Fig. 3. As can be seen in this figure, our distribution is very closely related to the empirical cumulative distribution function. Hence, the L_{26} variable seems to follow a GEV distribution which is defined as follows:

$$\begin{cases} P(L_i \leq A_i) = F(A_i, \mu_i, \sigma_i, \xi_i) \\ P(L_i \leq A_i) = \exp\left\{-\left[1 + \xi_i\left(\dfrac{A_i - \mu_i}{\sigma_i}\right)\right]^{-1/\xi_i}\right\} \end{cases} \quad (18)$$

where F is the Cumulative Distribution Function (CDF), $\mu_i \in \mathbb{R}$ is the location parameter, $\sigma_i > 0$ is the scale parameter and $\xi_i \in \mathbb{R}^*$ is the shape parameter. By inverting (18), we compute the parameter GEV A_i as follows:

Table 1 Goodness of fit-synthesis

Distribution	Kolmogorov–Smirnov		Corresponding parameters
	Statistic	Rank	
GEV	0.0372	1	$\mu = 22.697$, $\sigma = 10.923$, $\xi = -0.1901$
Weibull	0.0530	2	$\alpha = 2.6739, \beta = 30.999$
Gamma	0.0567	3	$\alpha = 5.4814, \beta = 5.0205$
Normal	0.0602	4	$\mu = 27.244, \sigma = 11.471$
Fréchet	0.1327	5	$\alpha = 1.9407, \beta = 19.648$
Exponential	0.2949	6	$\lambda = 0.0367$
Pareto II	0.3141	7	$\alpha = 197.26, \beta = 5048.2$
Pareto I	No fit	8	No fit

$$F(A_i, \mu_i, \sigma_i, \xi_i) = p \Rightarrow A_i = \left[\left(-\frac{1}{\ln p}\right)^{\xi_i} - 1\right]\frac{\sigma_i}{\xi_i} + \mu_i. \quad (19)$$

With $(\mu_{26}, \sigma_{26}, \xi_{26}) = (22.697, 10.923, -0.1901)$ and $p = 25\ \%$ for example, we get $A_{26} = 19.016$. Hence, the set of nearest neighbours of the 26th customer will be noted by $V(26)$ and Fig. 4 offers its representation. $V(26) = \{j \in N | l(26,j) \leq 19.016\}$. This set, considered as the restricted neighbourhood of the current solution, will be used by the three operators to evaluate the neighbourhood $N(26)$.

4.5 Penalty component

To develop the approach for the MCVRPTW, possible violations of both capacity and time window constraints need to be addressed. Let S be an infeasible solution that violates capacity constraints and/or time window constraints. The penalized cost function $F'(S)$ of solution S is defined in Eq. (20). It consists of the total travel distance $F(S)$ and the penalty terms $C(S)$ and $T(S)$ for the violations of the capacity and time window constraints multiplied by the penalty coefficients α and β respectively. Initially set equal to 1, these coefficients are periodically divided by $1 + \rho$ ($\rho \in]0, 1[$) after each block of ϕ feasible solutions and multiplied by $1 + \rho$ after each block of ϕ infeasible solutions. This way of proceeding produces a mix of feasible and infeasible solutions, which acts as a diversification strategy.

$$F'(S) = F(S) + \alpha C(S) + \beta T(S). \quad (20)$$

$C(S)$ is straightforwardly defined in Eq. (21) as the sum of the total demand excess in all routes [35].

$$C(S) = \sum_{k \in \{K_1 \cup K_2\}} \max\left\{\sum_{p \in P}\sum_{i \in N} d_{ip} y_{ipk} - C_k, 0\right\} \quad (21)$$

As for the T(S) penalty term, variants of the time window penalty for the VRP with soft time windows are

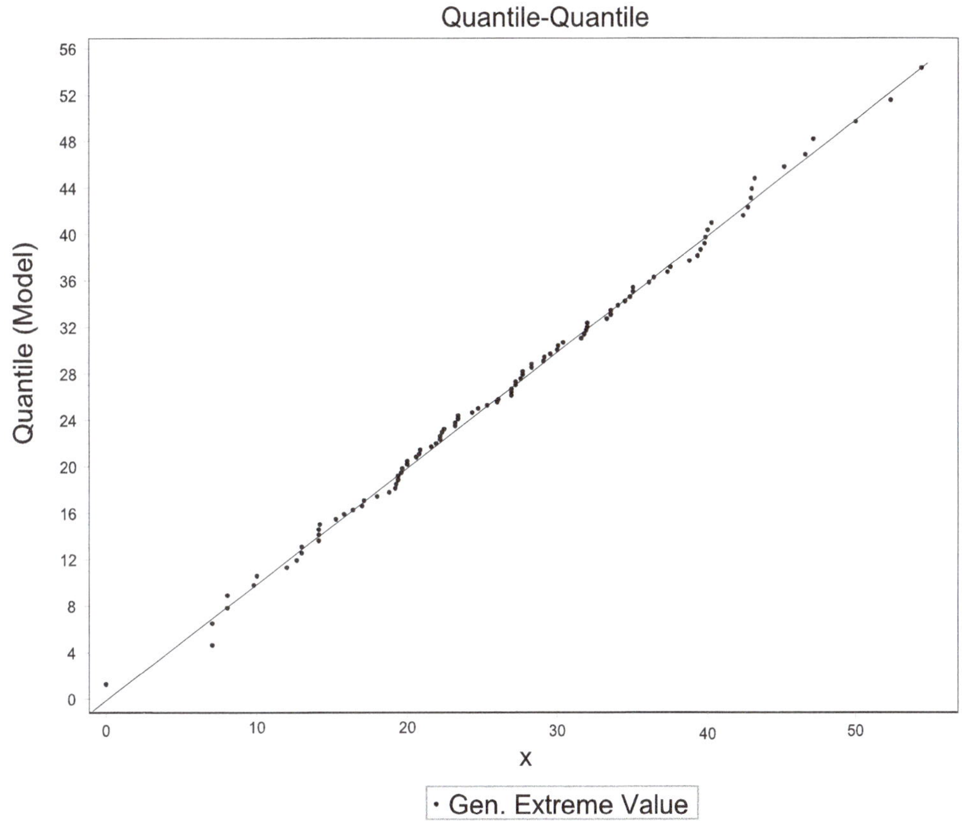

Fig. 3 GEV Quantile–Quantile plot

Fig. 4 An example of candidate list

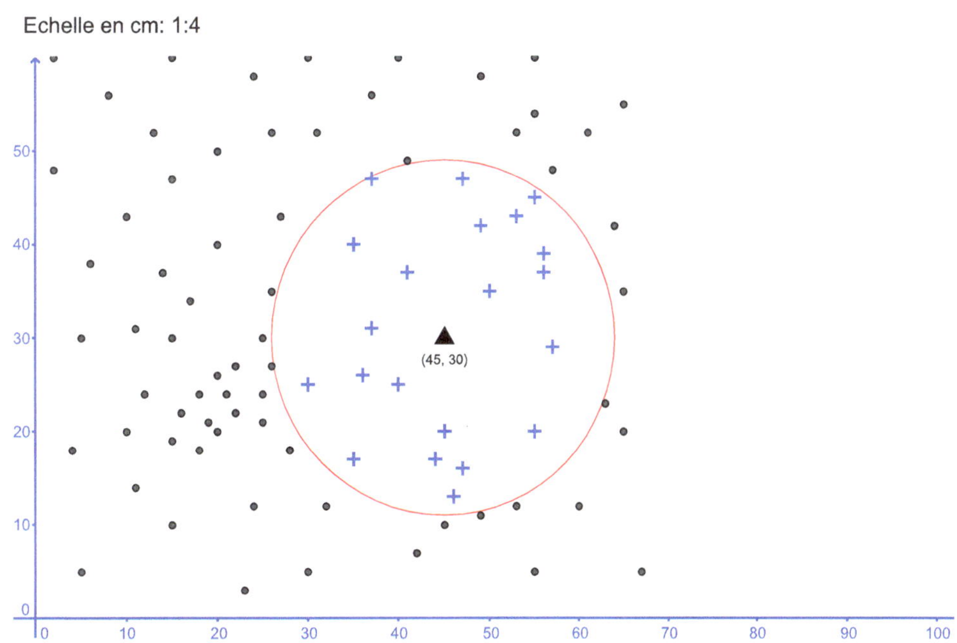

employed in Berger and Barkaoui [4]. We suggest the definition of Nagata et al. [22] for the time window penalty structure defined in Eq. (23) whose change can be computed in O(1) time for most traditional neighbourhood operators. Moreover, this penalty measures the amount of the time window violation more appropriately as described below.

Given a solution S, the extended earliest departure time at a depot $\widetilde{s}_{0k}$, the extended earliest start time of service at a customer i, $\widetilde{s}_{ik}$ are defined recursively in Eq. (22) and the suggested time window penalty of the solution, denoted as T(S), is defined in Eq. (23).

$$\begin{cases} \widetilde{s'}_{0k} = \widetilde{s}_{0k} = a_0 \\ \widetilde{s'}_{ik} = \widetilde{s}_{i-1,k} + t_{ik} + t_{i-1,i} \\ \widetilde{s}_{ik} = \max(\widetilde{s'}_{ik}, a_i) & \text{if} \quad \widetilde{s'}_{ik} \leq b_i \\ \widetilde{s}_{ik} = b_i & \text{if} \quad \widetilde{s'}_{ik} > b_i \end{cases} \tag{22}$$

$$T(S) = \sum_{k \in K} \sum_{i \in V} \max(\widetilde{s'}_{ik} - b_i, 0\} \tag{23}$$

Note that $\widetilde{s}_{ik}$ is equal to s_{ik} if the route of vehicle k is feasible with respect to the time window constraint. $\widetilde{s'}_{ik}$ refers to the extended earliest arrival time of vehicle k at customer i, and the time window constraint is violated at customer i if $b_i < \widetilde{s'}_{ik}$. In this case, we assume that the vehicle can travel back in time to b_i to start the service of customer i without delay, but at the expense of paying a penalty $(\widetilde{s'}_{ik} - b_i)$. Therefore, in this case, $\widetilde{s}_{ik}$ is set to b_i. The total time window violation in solution S, $T(S)$, is defined by the sum of the penalties that the vehicles must pay in all the routes to service all customers and to arrive at the depot without delay.

4.6 Tabu list

The tabu list is implemented as an upper triangular matrix L of $K \times K$ dimensions where each element $L(k, k'), (k, k' = 1, ..., K, k < k')$ is associated with the pair of routes R_k and $R_{k'}$. Each element $L(k, k')$ contains a set of attributes able to characterize the solution and also records the iteration in which the arc (i, j) has been moved from route R_k to route $R_{k'}$. The element $L(k, k')$ has the structure:

$$(R_k,\ R_{k'},\ i,\ position\, i,\ j,\ position\, j,\ F(S),\ t).$$

where R_k and $R_{k'}$ are the two routes under operation, i is a customer from R_k and *position i* is the service order of i in R_k. The case is likewise for j and *position j*. $F(S)$ is the solution value, and t is the iteration in which the arc (i, j) has been moved from route R_k to route $R_{k'}$. An arc moved at iteration t cannot be reinserted in the solution before iteration $t + \theta$.

This notion provides a guideline to avoid making similar moves in the near future. Such representation does not uniquely describe a move, because a full description is very complicated and its use increases the computation tremendously. Therefore, when an exchange between routes R_k and $R_{k'}$ is accepted, we just update the information corresponding to line k and column k'. Thus, we avoid calculating information above the other pairs of routes, which do not contain either R_k or $R_{k'}$. The size of tabu list θ takes its values in $[\theta_{\min}, \theta_{\max}]$ starting from θ_{init}. Parameter θ is updated according to the quality of the solutions obtained during the recent moves. After each improvement of the current objective function, θ is updated as $\theta = \max(\theta - 1, \theta_{\min})$ with the aim of intensifying the search around this solution. Otherwise, after ϕ_{LT} consecutive moves deteriorating the value of the visited solutions, the size of the tabu list is updated as $\theta = \min(\theta + 1, \theta_{\max})$.

4.7 Diversification and intensification

There are two diversification strategies for the proposed algorithm to guide the search into less explored regions. (1) A neighbourhood is randomly selected from the moves described in Sect. 4.3. (2) The idea of passing through infeasible regions is introduced and defined in Sect. 4.5.

The intensification is mainly achieved by using the following two strategies. (1) The search is accentuated around the best-known solution by increasing or decreasing the size of the tabu list as explained in Sect. 4.6. (2) The full search proceeds starting from the most promising solution. In fact, the tabu search restarts the exploration of the solution space from the best feasible solution evaluated, but not visited $\overline{S}$. To update $\overline{S}$ at each iteration, the algorithm generates two solutions, S' and S'', from the neighbourhood $N(S)$ of the incumbent solution S. S' represents the best non-tabu solution in $N(S)$, and it is used to continue the search process, while S'' represents the best non-tabu feasible solution obtained in $N(S) \backslash S'$. Note that S' can be infeasible, since the solutions visited may violate capacity or time constraints. After $\gamma_{\max}$ iterations without improving the best feasible solution found so far or after $v_{\max}$ iterations since the last restart, the search process "jumps" to $\overline{S}$ and restarts with an empty tabu list. The maximal number of restarts is fixed as $\eta_{\max}$, but this process can be stopped if after $\overline{\eta_{\max}}$ restarts the best solution is not improved.

4.8 Algorithm overview

Algorithm 2 gives the skeleton of the proposed tabu search. Before describing its general structure, we first give some notations as follows:

- η: Number of restarts.
- $\overline{\eta}$: Number of restarts without improvement.
- γ: Number of iterations without improvement.
- γ_S: Number of iterations without improvement the incumbent solution.
- v: Total number of iterations.
- χ: Number of consecutive feasible solutions.
- $\overline{\chi}$: Number of consecutive infeasible solutions. $\overline{\chi} = \overline{\chi_c} + \overline{\chi_t}$.
- $\overline{\chi_c}$: Number of consecutive infeasible solutions that violate capacity constraint.
- $\overline{\chi_t}$: Number of consecutive infeasible solutions that violate time window constraint.

Algorithm 2 Procedure Search

(1) Generate one initial solution S_0 by combining the loading aspect with the modified nearest neighbor heuristic as described in Section 4.2.
(2) Set $\overline{S} := S_0, S^* := S_0$
(3) initialize the parameters: $\theta = \theta_{init}, \eta = \overline{\eta} = \gamma = \nu = \chi = \overline{\chi} = \gamma_S = 0$.

```
repeat
  η = η + 1
  while ν ≤ ν_max and γ ≤ γ_max do
    Randomly choose an operator from the set of three operators and apply it to the
    current solution S. Two solutions S' and S'' are generated from N(S):
    S' : the best non-tabu solution in N(S).
    S'' : the best feasible non-tabu solution in N(S) \ S'.
    Set S̄ = argmin(F(S'), F(S''))
    if F(S') < F(S) then
      Update the size of the tabu list. Set θ := max(θ − 1, θ_min)
      Set γ_S := 0
    else
      Update γ_S := γ_S + 1
    end if
    if γ_S = φ_LT then
      Update the size of the tabu list. Set θ := min(θ + 1, θ_max)
      Set γ_S := 0
    end if
    if F(S') < F(S*) and S' is feasible then
      Set S* := S', γ := 0, χ := χ + 1.
    else
      γ := γ + 1
    end if
    S := S'
    Update the tabu list
    if S is infeasible then
      χ̄ := χ̄ + 1. Update χ̄_c, χ̄_t.
    else
      χ̄ := 0. Update χ̄_c, χ̄_t.
    end if
    check the variables corresponding to the number of consecutive feasible and non
    feasible solutions and update the penalization parameters.
    if χ = φ then
      Set α := α/(1 + ρ), β := β/(1 + ρ).
    end if
    if χ̄ = φ then
      Set ρ_c := χ̄_c/φ, ρ_t := χ̄_t/φ.
      Update α := α * (1 + ρ_c), β := β * (1 + ρ_t).
    end if
  end while
  if S* is updated then
    η̄ := 0
  else
    η̄ := η̄ + 1
  end if
  S = S̄
until (η = η_max) or (η̄ = η̄_max)
Return S*.
```

5 Computational results

5.1 Overview

Because the problem under study is a new problem in the literature, there is still a lack of benchmark instances to test the behaviour of the algorithm. As a result and with the aim of avoiding indirect or limited comparisons, we performed our computational experiments on the well-known data sets given by Solomon [28]. These instances (http://www.neo.lcc.uma.es/vrp/vrp-instances/), developed for the classical VRPTW, are transformed to fit our problem. There are, in total, 56 different instances, which can be classified into six categories. In sets R1 and R2, the customer locations are randomly generated. Problem sets C1 and C2 have clustered distributions of customers. Sets RC1 and RC2 are semi-clustered with a mix of randomly distributed and clustered customers. Sets R1, C1 and RC1 have a shorter route horizon compared with those of sets R2, C2 and RC2, which have longer scheduling horizons. In order to accommodate these instances to the MCVRPTW with two products and two compartments, we adopted the idea proposed by El Fallahi et al. [16] to derive the MCVRP instances from the VRP data sets. More precisely, we split each customer demand into two parts. The demands of customer i for the two types of products are $d_{i1} = [d_i/\omega]$ and $d_{i2} = d_i - d_{i1}$ respectively where ω is a random integer in [2, 3], $[x]$ stands for the integer part of x and d_i indicates the demand of customer i in Solomon instances. The capacity of compartments is set as $c_{q1} = (C \times \overline{d_1})/(\overline{d_1} + \overline{d_2}); c_{q2} = (C \times \overline{d_2})/(\overline{d_1} + \overline{d_2})$ where $\overline{d_1}$, $\overline{d_2}$ and C are the average demand for the first product, the average demand for the second product and the vehicle capacity in the original VRPTW, respectively.

Although our implementation handles more compartments, we took the case of two compartments for testing because a) the problem is already difficult enough with two compartments, and b) the demands obtained by splitting the original demands into three or more demands would be too small.

To further assess the performance of the algorithm, we solve a real case where the company managers provided us with daily data for a period of 15 days.

All instances can be downloaded from the VRP-REP website (http://www.vrp-rep.org/).

The algorithm proposed in our solution procedure is coded in C++, and the mathematical model is solved by using IBM ILOG CPLEX 12.6 for a 7200s CPU. All experiments are conducted on a laptop computer with an Intel Core i7 2.9 GHz processor with 8 GB RAM and operating with Windows® 7 Professional edition.

5.2 Parameters

After some preliminary experiments, the parameter configurations for the tabu search algorithm have been set to the values reported in Table 2. α, β and ρ are three penalty factors. ϕ and ϕ_{LT} have been set to the same value and, respectively, count the number of consecutive moves updating or deteriorating the value of the visited solutions. The size of the tabu list is updated in the interval $[\theta_{\min}, \theta_{max}]$ starting from θ_{init}. The maximum number of moves without improving the best feasible solution in each restart is $\gamma_{\max}$. $v_{\max}$ is the number of iterations of each restart. $\eta_{\max}$ fixes the maximal number of restarts while $\overline{\eta_{max}}$ is the maximal number of restarts without improvement. λ_{ip} is the threshold used as an adjustment parameter in Eq. (24). p is the pth GEV quantile used to find the neighbourhood size.

Table 2 Parameters used in tabu search algorithm

Parameter	Value
α	1
β	1
ρ	0.3
$\phi = \phi_{LT}$	15
θ_{init}	10
$\theta_{\min}$	5
$\theta_{\max}$	15
$\gamma_{\max}$	700
$v_{\max}$	2500
$\eta_{\max}$	100
$\overline{\eta_{\max}}$	5
λ_{ip}	0.04
p	0.25

5.3 Results

5.3.1 Results for the MCVRPTW instances

We started by comparing the MCVRPTW with what we call the MCVRPTW-AD (MCVRPTW with adjustment) in which the quantities loaded in the compartments can be adjusted up to a given threshold (see Sect. 3). To handle this MCVRPTW-AD, we simply modified the tabu search algorithm to get two versions in which constraint (7) is replaced by constraint (24):

$$\sum_{i \in N} (d_{\text{ip}} - \lambda_{\text{ip}}) y_{ipk} \leq c_k^q, \quad k \in \{K_1 \cup K_2\}, \quad q \in \{Q_1 \cup Q_2\}, \quad p \in P \tag{24}$$

Table 3 provides the solutions obtained by the tabu search algorithm for each instance. In Table 3, we report the total distance (TD), the number of vehicles used for the service (NV), the CPU times in seconds (CPU) and the final gain in percentage (*Gain %*) when the adjustment option is permitted. Solutions in this table indicate that, in all problem instances, the adjustment option proves to be favourable, both in terms of the total distance (7 %) and the number of vehicles used (10 %). Another interesting thing we have found is that the tabu search converges much faster for the MCVRPTW than for the MCVRPTW-AD. This may be explained by the fact that, in the MCVRPTW-AD, we have the possibility of moving customers among the routes without violating any capacity constraint. Conversely, when excluding the adjustment option, the set of feasible solutions for the MCVRPTW becomes very restricted, resulting in a rapid search.

We then tried to solve the formulation (1)–(14) directly within CPLEX 12.6. CPLEX failed to solve any of the MCVRPTW instances to optimality and produced an out of

Table 3 Best solutions on the MCVRPTW instances

Data	MCVRPTW			MCVRPTW-AD			Gain (%)	
	NV	TD	CPU (s)	NV	TD	CPU (s)	NV	TD
C1	13	1131.39	432	12	1030.74	541	9	10
C2	7	1056.80	359	6	986.31	491	13	7
R1	17	1496.71	543	16	1422.73	731	7	5
R2	6	1163.03	459	5	1102.62	601	11	5
RC1	16	1728.05	364	15	1588.31	541	9	9
RC2	7	1356.80	295	7	1302.41	441	9	4
Avg.	11	1322.67	825	10	1240.97	1069	10	7

memory error after about 16,000s of computation time. Because of this, we compared the solutions obtained for the small-sized instances by our algorithm to those obtained by CPLEX 12.6. We conducted a set of experiments by randomly selecting 15 customers from each MCVRPTW instance. These instances are denoted as in the following example. C101-15 corresponds to the instance of class "C1" where only 15 customers are considered. In our solution, each of the generated instances is resolved in the same way, as for the instances with 100 customers considered above. For CPLEX, a maximum CPU of 7200s is imposed on the solution time. Table 4 shows the solutions obtained by our algorithm and the optimal solutions obtained by CPLEX. For completeness, the final optimality gap in percentage (Gap_f (%)) is also provided.

Results in this table indicate that our algorithm provides the optimal solutions to C102-15, R101-15 and RC102-15 instances with a substantially lower computation time. Furthermore, CPLEX cannot provide a feasible solution (within the time limit) to six problem instances (C104-15, R103-15, R104-15, R107-15, R108-15 and RC107-15). These results confirm that the solutions obtained by our algorithm are better than those obtained by CPLEX. Moreover, the average computation time required by our solution procedure to solve these instances is much less than that required by CPLEX.

Table 4 Comparison between our solutions and optimal solutions obtained by CPLEX

Data	Our solutions			CPLEX solutions			Gap_f(%)
	NV	TD	CPU	NV	TD	CPU	
C101-15	3	222.61	13	3	220.92	7200	70.34
C102-15	**2**	**207.84**	**24**	**2**	**207.84**	**2593**	
C103-15	3	227.68	32	3	233.68	7200	84.04
C104-15	3	220.62	48		***NO SOL***		
C105-15	3	229.88	15	3	229.88	7200	51.32
C106-15	3	239.02	19	3	239.02	7200	45.35
C107-15	2	198.08	23	2	194.95	7200	86.79
C108-15	3	270.96	22	3	273.78	7200	84.16
C109-15	3	233.44	31	3	243.21	7200	87.33
R101-15	**4**	**359.66**	**39**	**4**	**359.66**	**6025**	
R102-15	4	338.06	47	4	353.11	7200	78.12
R103-15	3	278.88	90		***NO SOL***		
R104-15	3	280.54	43		***NO SOL***		
R105-15	3	298.92	32	4	295.77	7200	72.86
R106-15	3	305.41	37	3	322.19	7200	78.37
R107-15	3	297.50	57		***NO SOL***		
R108-15	3	284.22	45		***NO SOL***		
R109-15	3	299.71	36	4	318.44	7200	69.95
R110-15	3	271.52	41	3	311.28	7200	88.15
R111-15	3	300.46	42	4	329.57	7200	75.43
R112-15	3	270.25	42	3	428.92	7200	82.03
RC101-15	4	370.26	26	4	370.26	7200	80.10
RC102-15	**3**	**367.34**	**26**	**3**	**367.34**	**791**	
RC103-15	3	319.78	34	3	324.55	7200	89.76
RC104-15	3	326.87	53	3	398.20	7200	79.92
RC105-15	3	393.84	41	3	400.41	7200	88.17
RC106-15	3	319.56	33	3	320.80	7200	90.12
RC107-15	3	339.49	26		***NO SOL***		
RC108-15	3	322.23	38	3	323.92	7200	89.77
Avg.	3	289	36	3	307	6670	78.60

Bold indicates optimal solutions

5.3.2 Results for the VRPTW instances

For a more meaningful comparison of results, we interpreted the classical VRPTW instances as MCVRPTW instances with one product and one compartment. To this end, we compared our results with nine other meta-heuristic approaches proposed by Taillard et al. [29] (TBGGP), Chiang and Russell [9] (CR), Gambardella et al. [18] (GTA), Tan et al. [30, 31] (TLO), Cordeau et al. [10] (CLM), Braysy and Gendreau [5] (BG), Lau et al. [20] (LST), Tan et al. [32] (TCL) and Vidal et al. [37] (VCGP). The comparison of the results of each approach is shown in Table 5. The first row gives the name of the authors of the study. Rows C1, C2, R1, R2, RC1 and RC2 present the average number of vehicles (NV) and average total distance (TD) with respect to the six groups of problem instances, respectively. The performance of our algorithm

Table 5 Comparison among different heuristics

Data set		TBGGP [29]	CR [9]	GTA [18]	TLO [30, 31]	CLM [10]
C1	NV	10	10	10	10	10
	TD	828.45	828.38	828.38	828.74	828.38
C2	NV	3	3	3	3	3
	TD	590.30	591.42	589.86	590.69	589.86
R1	NV	12.25	12.17	12	12.92	12.08
	TD	1216.70	1204.19	1217.73	1187.35	1210.14
R2	NV	3	2.73	2.73	3.51	2.73
	TD	995.38	986.32	967.75	960.83	969.58
RC1	NV	11.88	11.88	11.63	12.74	11.50
	TD	1367.51	1397.44	1382.42	1355.37	1389.78
RC2	NV	3.38	3.25	3.25	4.25	3.25
	TD	1165.62	1229.54	1129.19	1068.26	1134.51

		BG [5]	LST [20]	TCL [32]	VCGP [37]	Our solution
C1	NV	10	10	10	10	10
	TD	828.38	832.13	828.38	828.38	828.38
C2	NV	3	3	3	3	3
	TD	589.86	589.86	589.86	589.86	589.86
R1	NV	11.92	12.16	12	11.92	12.08
	TD	1222.12	1211.55	1217.73	1210.69	1197.97
R2	NV	2.73	3	2.73	2.73	2.73
	TD	975.12	1001.12	967.75	951.51	952.17
RC1	NV	11.50	12.25	11.63	11.50	11.50
	TD	1389.58	1418.77	1382.42	1384.17	1355.97
RC2	NV	3.25	3.37	3.25	3.25	3.25
	TD	1128.38	1170.93	1129.19	1119.24	1116.38

appears satisfactory, considering that all approaches in the literature were tailored for the VRPTW, most of them actually aiming first to reduce the travelled distance.

5.3.3 Results for the real-life instances

To further prove the feasibility and effectiveness of the presented algorithm under real situations, we collected and investigated the real data of the NAFTAL petroleum company. This company is responsible for delivering various kinds of fuel (gasoline, premium gasoline, aviation gasoline, kerosene and diesel fuel) from 70 depots, serving more than 3500 petrol stations and using around 1500 tank trucks. In this experiment, we used instances that we obtained from the regional depot of petroleum products, Caroubier depot, in the city of Algiers in Algeria. For this depot, the company provided us with daily data for a period of 15 days.

The data consist of the cities where the customers are located and the associated distance matrix, the order quantities with their time windows and tank truck related information, such as the number of tanks and their capacities. The fleet dedicated to the replenishment of petrol stations consists of 20 tank trucks, five of which are owned by private companies. The travel time in minutes between each pair of petrol stations is calculated by multiplying the travel distance in kilometres by a definite constant 1.200 based on the average travel speed, 50 km per hour or 0.833 km per minute. The travel cost between each pair of petrol stations is calculated by multiplying the travel distance by a definite constant 0.685 given by the company. As mentioned earlier, the private trucks are hired with a penalty cost. This cost is calculated by multiplying the travel distance between each pair of petrol stations by a definite constant $f_2 = 0.497$ based on the average rental price of private vehicles. In Table 6, the instances are denoted by a name that allows one to identify their customers per day. In particular, the names have the form $D - n$ where D is the working day and n is the number of customers. For example the instance "1–41" denotes the 41 customers to be delivered to on the first working day.

In the first place, we tried to solve these 15 instances using CPLEX 12.6. Unfortunately, it failed to find the optimal solution for most of these trials and sometimes produced an out of memory error. Because of this, we completed our assessment by comparing our solutions with those provided by the company. For CPLEX, a maximum of a 7200s CPU is imposed on the solution time, and the final optimality gap in percentage (Gap_f (%)) is provided.

Table 6 shows the solutions obtained by our algorithm, the optimal solutions obtained by CPLEX and the solutions extracted from the company plans over a testing period of 15 days. In this table, we report the number of vehicles used for the service (NV), the total distance (TD), the cost (u), the CPU times in seconds (CPU), and the final optimality gap in percentage (Gap_f (%)) when CPLEX 12.6 is used. Results in this table indicate that our algorithm provides the optimal solutions to instances 3–25, 6–32 and 8–27 with a substantially lower computation time. Furthermore, CPLEX cannot provide a feasible solution (within the time limit of 7200 seconds) to seven instances (4–62, 5–83, 7–89, 10–79, 12–71, 14–61 and 15–83).

As a comparison with the solutions extracted by the company, we may conclude that our algorithm has a better performance on every measure. The main reason that the company system does not calculate the CPU times is because it develops the daily delivery plans manually using MS Excel.

5.4 Sensitivity analysis of the adjustment option on a real case

We have noted earlier that threshold parameter λ_{ip} is an important and integral component affecting the

Table 6 Daily results using the real data of Caroubier depot

Day	Method	NV	DT	*Cost* ($\times 10^2$)	CPU (s)	Gap_f (%)
1–41	Cplex	4	536	47,570	7200	82.17
	Company	6	610	56,390	/	
	Our solution	4	527	47,292	315.43	
2–46	Cplex	6	881	121,766	7200	86.24
	Company	8	991	151,494	/	
	Our solution	6	881	121,766	243.12	
3–25	Cplex	3	379	23,694	6578	
	Company	4	443	36,512	/	
	Our solution	**3**	**379**	**23,694**	**317.82**	
4–62	Cplex	***NO SOL***				
	Company	9	1014	107,938	/	
	Our solution	7	918	90,747	403.82	
5–83	Cplex	***NO SOL***				
	Company	11	1475	11,382		
	Our solution	9	1323	101,745	515.39	
6–32	Cplex	3	488	29,178	4915	
	Company	4	510	40,129	/	
	Our solution	**3**	**488**	**29,178**	**182.25**	
7–89	Cplex	***NO SOL***				
	Company	15	1577	151,436	/	
	Our solution	12	1479	129,305	398.76	
8–27	Cplex	3	283	25,886	3755	
	Company	5	356	31,500	/	
	Our solution	**3**	**283**	**25,886**	**143.29**	
9–40	Cplex	6	853	55,350	7200	79.12
	Company	8	1030	77,892	/	
	Our solution	6	841	52,773	204.74	
10–79	Cplex	***NO SOL***				
	Company	11	1692	213,915	/	
	Our solution	9	1537	174,676	479.38	
11–28	Cplex	6	911	79,000	7200	89.12
	Company	7	1112	91,503	/	
	Our solution	6	913	79,023	207.19	
12–71	Cplex	***NO SOL***				
	Company	9	1233	147,419	/	
	Our solution	8	1129	130,771	210.16	
13–33	Cplex	7	1014	95,310	7200	69.23
	Company	9	1084	99,459	/	
	Our solution	7	1003	95,163	309.80	
14–61	Cplex	***NO SOL***				
	Company	10	1305	90,417	/	
	Our solution	8	1091	71,460	92.01	
15–83	Cplex	***NO SOL***				
	Company	13	1567	128,979	/	
	Our solution	10	1387	110,388	429.52	

Bold indicates optimal solutions

Table 7 Impact of the adjustment option on the daily costs

λ_{ip}	Total	Quantity (m^3)			NV		
(%)	*Cost* ($\times 10^2$)	Requested	Delivered	Gap (%)	Company	Private	Total
0	129,305	975	975	0.00	9	3	12
1	129,191	975	972.25	−0.28	9	3	12
2	128,609	975	969.75	−0.54	9	2	11
3	128,005	975	962.5	−1.28	9	2	11
4	127,319	975	955	−2.05	8	2	10
5	126,903	975	951.25	−2.44	8	2	10
6	126,180	975	944.75	−3.10	8	2	10
7	125,813	975	939.25	−3.67	8	2	10
8	124,298	975	918	−5.85	8	1	9
9	123,677	975	898.25	−7.87	8	1	9
10	121,963	975	890.5	−8.67	8	0	8

performance of the adjustment option in the MCVRPTW. To observe its role in the solution's quality, we perform a sensitivity analysis by solving the problem on the daily data for varying values of λ_{ip} between 0 and 10 %. Note that we did not perform the algorithm to better observe the effect of λ_{ip} values. With these values, we implemented the plan of the *seventh* day only and resolved the problem when excluding and including the adjustment option. In Table 7, we report the daily results obtained and we show the effect of threshold parameter λ_{ip} on (a) the total cost figures, (b) the quantities delivered and (c) the vehicles used. The results show that the solution quality is very sensitive to the threshold parameter. This is indeed an expected result since the adjustment option attempts to assign the demands taking more account of the capacity constraints. From this table, the following three points can be observed. (1) All the total costs decrease when customer quantities are adjusted, and this decrease is inversely proportional to the increase of the adjustment rate. (2) The total delivered quantity and the vehicles used decrease when λ_{ip} values become increasingly large. (3) Overall, the solution quality varies in different ratios, indicating that the solution quality is very sensitive to travelled distance, quantities delivered and even to the type of vehicle used.

6 Conclusion

The specific problem we tackled in this paper, called the multi-compartment vehicle routing problem with time windows (MCVRPTW), originated from a real-life application concerning the distribution of fuel. In this problem, we focused on the daily replenishment-planning problem that the biggest Algerian petroleum company is facing. In particular, the vehicles have multiple compartments and customers require to be served during a given time window. Because of the loading aspect, an additional question arises concerning the assignment of product types to vehicles.

The main contributions of this paper include (1) a description and formulation of the problem inspired by a real-life application, and (2) the development of an effective heuristic solution procedure, which combines the loading and the routing problems. This is an advance over existing work, in which most researchers take a two-stage framework where the routing problem acts as the main problem and iteratively calls for specific procedures to deal with the loading sub-problem. (3) The introduction of a Kolmogorov–Smirnov statistic in order to explore the solutions space is used, unlike most of the relevant approaches, which tend to use the classical moves. (4) The conduct of a series of numerical experiments on benchmark instances and an analysis of a real case in fuel distribution to demonstrate the effectiveness of the proposed approach are adopted.

Concerning the experiments on benchmark instances, Solomon's 56 VRPTW 100-customer instances have been modified in a way that reflects real-life situations. For this purpose, a comparison is made between the real MCVRPTW and what we call the MCVRPTW-AD (MCVRPTW with adjustment) in which the quantities loaded in the compartments can be adjusted up to a given threshold. This particular problem occurs quite frequently when the demands of petrol stations are high in winter. Under this scenario, we conducted experiments on how the algorithm performs when excluding and including the adjustment option. We demonstrated how the number of vehicles and the total distance can be reduced when the adjustment option is allowed and how this reduction depends on the fixed threshold.

Our solutions are also compared to CPLEX and to the heuristics reported in the literature. The obtained results show that our approach is competitive for the VRPTW and highly effective for the MCVRPTW instances. As for the realistic instances, we solved a real case where petrol stations need to be replenished over a planning horizon of 15 days. The comparative analysis shows that our results are better than those produced by the dispatcher on every measure in terms of total travel distance and number of vehicles.

As for prospects, we envision producing more efficient solutions by adapting known metaheuristics, such as Particle Swarm Optimization (PSO), a Genetic Algorithm (GA) or Simulated Annealing (SA), to the problem and by adjusting the parameters of the algorithm because, often, in a metaheuristic, the selection of good parameter values significantly affects the quality of solutions. A hybridization of clever heuristics with complex search methods and an examination of penalty functions are also on the agenda.

Acknowledgments We gratefully acknowledge the anonymous referees for carefully reading the paper and for their helpful comments and suggestions.

References

1. Avella P, Boccia M, Sforza A (2004) Solving a fuel delivery problem by heuristic and exact approaches. Eur J Oper Res 152(1):170–179. doi:10.1016/S0377-2217(02)00676-8
2. Bausch DO, Brown G, Ronen D (1995) Consolidating and dispatching truck shipments of Mobil heavy petroleum products. Interfaces 25:117. doi:10.1287/inte.25.2.1
3. Ben Abdelaziz F, Roucairol C, Bacha C (2002) Deliveries of liquid fuels to SNDP gas stations using vehicles with multiple compartments. In: International conference on systems man and cybernetics 2002 IEEE, Hammamet, Tunisia. doi:10.1109/ICSMC.2002.1168021
4. Berger J, Barkaoui M (2004) A parallel hybrid genetic algorithm for the vehicle routing problem with time windows. Comput Oper Res 31(12):2037–2053. doi:10.1016/S0305-0548(03)00163-1
5. Braysy O, Gendreau M (2002) Tabu search heuristics for the vehicle routing problem with time windows. TOP 10(2):211–237. doi:10.1007/BF02579017
6. Brown GG, Graves WG (1981) Real-time dispatch of petroleum tank trucks. Manag Sci 27(1):19–31. doi:10.1287/mnsc.27.1.19
7. Brown GG, Ellis CJ, Graves WG, Ronen D (1987) Real-time, wide area dispatch of mobil tank trucks. Interfaces 17(1):107–120. doi:10.1287/inte.17.1.107
8. Bullnheimer B, Hartl RF, Strauss C (1999) An improved ant system algorithm for the vehicle routing problem. Ann Oper Res 89:319–328
9. Chiang WC, Russell RA (1997) A reactive tabu search metaheuristics for the vehicle routing problem with time windows. INFORMS J Comput 9(4):417–430. doi:10.1287/ijoc.9.4.417
10. Cordeau JF, Laporte G, Mercier A (2001) A unified tabu search heuristic for vehicle routing problems with time windows. J Oper Res Soc 52(8):928–936. doi:10.1057/palgrave.jors.2601163
11. Cornillier F, Boctor FF, Laporte G, Renaud J (2008a) An exact algorithm for the petrol station replenishment problem. J Oper Res Soc 59(5):607–615. doi:10.1057/palgrave.jors.2602374
12. Cornillier F, Boctor FF, Laporte G, Renaud J (2008b) A heuristic for the multi-period petrol station replenishment problem. Eur J Oper Res 191(2):295–305. doi:10.1016/j.ejor.2007.08.016
13. Cornillier F, Boctor FF, Laporte G, Renaud J (2009) The petrol station replenishment problem with time windows. Comput Oper Res 36(3):919–935. doi:10.1016/j.cor.2007.11.007
14. Cornillier F, Boctor FF, Renaud J (2012) Heuristics for the multi-depot petrol station replenishment problem with time windows. Eur J Oper Res 220:361–369. doi:10.1016/j.ejor.2012.02.007
15. Day JM, Wright PD, Schoenherr T, Venkataramanan M, Gaudette K (2009) Improving routing and scheduling decisions at a distributor of industrial gasses. Omega 37(1):227–237. doi:10.1016/j.omega.2006.11.007
16. El Fallahi A, Prins C, Wolfler Calvo R (2008) A memetic algorithm and a tabu search for the multi-compartment vehicle routing problem. Comput Oper Res 35(5):1725–1741. doi:10.1016/j.cor.2006.10.006
17. Franz LS, Woodmanse J (1990) Computer-aided truck dispatching under conditions of product price variance with limited supply. J Bus Logist 11(1):127–139
18. Gambardella LM, Taillard E, Agazzi G (1999) MACS-VRPTW: a multiple ant colony system for vehicle routing problems with time windows. In: New ideas in optimization, pp 63-76
19. Glover F, Laguna M (1997) Tabu search. Kluwer, Boston
20. Lau HC, Sim M, Teo KM (2003) Vehicle routing problem with time windows and a limited number of vehicles. Eur J Oper Res 148(3):559–569. doi:10.1016/S0377-2217(02)00363-6
21. Malépart V, Boctor FF, Renaud J, Labilois S (2003) Nouvelles approches pour l'approvisionnement des stations d'essence. Revue Franaise de Gestion Industrielle 22:15–31
22. Nagata Y, Braysy O, Dullaert W (2010) A penalty-based edge assembly memetic algorithm for the vehicle routing problem with time windows. Comput Oper Res 37(4):724–737. doi:10.1016/j.cor.2009.06.022
23. Ng WL, Leung S, Lam J, Pan S (2008) Petrol delivery tanker assignment and routing: a case study in Hong Kong. J Oper Res Soc 59:1191–1200. doi:10.1057/palgrave.jors.2602464
24. Nussbaum M, Sepulveda M (1997) A fuel distribution knowledge-based decision support system. Omega 25(2):225–234. doi:10.1016/S0305-0483(96)00059-X
25. Popović D, Vidović M, Radivojević G (2012) Variable neighbourhood search heuristic for the inventory routing problem in fuel delivery. Expert Syst Appl 39(18):13390–13398. doi:10.1016/j.eswa.2012.05.064
26. Renaud J, Bolduc MC, Laporte G, Boctor F (2010) A tabu search heuristic for the split delivery vehicle routing problem with production and demand calendars. Eur J Oper Res 202:122–130
27. Ronen D (1995) Dispatching petroleum products. Oper Res 43(3):379–387. doi:10.1287/opre.43.3.379
28. Solomon M (1987) Algorithms for the vehicle routing and scheduling problems with time window constraints. Oper Res 35(2):254–265. doi:10.1287/opre.35.2.254
29. Taillard E, Badeau P, Gendreau M, Geurtin F, Potvin JY (1997) A tabu search heuristic for the vehicle routing problem with time windows. Transp Sci 31(2):170–186. doi:10.1287/trsc.31.2.170
30. Tan KC, Lee LH, Ou K (2001a) Artificial intelligence heuristics in solving vehicle routing problems with time window constraints. Eng Appl Artif Intell 14(6):825–837. doi:10.1016/S0952-1976(02)00011-8

31. Tan KC, Lee LH, Ou K (2001) A messy genetic algorithm for the vehicle routing problem with time window constraints. In: IEEE congress on evolutionary computation, vol. 1, pp 679–686. doi:10.1109/CEC.2001.934457
32. Tan KC, Chew YH, Lee LH (2006) A hybrid multiobjective evolutionary algorithm for solving vehicle routing problem with time windows. Comput Optim Appl 34(1):115–151. doi:10.1007/s10589-005-3070-3
33. Taqa allah D, Renaud J, Boctor FF (2000) Le problme d'approvisionnement des stations d'essence. J Eur des Syst Autom 34:11–33
34. Toth P, Vigo D (2014) Vehicle routing: problems, methods, and application, 2nd edn. Society for Industrial and Applied Mathematics, Philadelphia
35. Toth P, Vigo D (2003) The granular tabu search and its application to the vehicle routing problem. INFORMS J Comput 15(4):333–348. doi:10.1287/ijoc.15.4.333.24890
36. Van der Bruggen L, Gruson R, Salomon M (1995) Reconsidering the distribution of gasoline products for a large oil company. Eur J Oper Res 81(3):460–473. doi:10.1016/0377-2217(94)00189-J
37. Vidal T, Crainic TG, Gendreau M, Prins C (2013) A hybrid genetic algorithm with adaptive diversity management for a large class of vehicle routing problems with time windows. Comput Oper Res 40(1):475–489. doi:10.1016/j.cor.2012.07.018
38. Vidović M, Popović D, Ratković B (2014) Mixed integer and heuristics model for the inventory routing problem in fuel delivery. Int J Prod Econ 147:593–604. doi:10.1016/j.ijpe.2013.04.034

Permissions

All chapters in this book were first published by Springer; hereby published with permission under the Creative Commons Attribution License or equivalent. Every chapter published in this book has been scrutinized by our experts. Their significance has been extensively debated. The topics covered herein carry significant findings which will fuel the growth of the discipline. They may even be implemented as practical applications or may be referred to as a beginning point for another development.

The contributors of this book come from diverse backgrounds, making this book a truly international effort. This book will bring forth new frontiers with its revolutionizing research information and detailed analysis of the nascent developments around the world.

We would like to thank all the contributing authors for lending their expertise to make the book truly unique. They have played a crucial role in the development of this book. Without their invaluable contributions this book wouldn't have been possible. They have made vital efforts to compile up to date information on the varied aspects of this subject to make this book a valuable addition to the collection of many professionals and students.

This book was conceptualized with the vision of imparting up-to-date information and advanced data in this field. To ensure the same, a matchless editorial board was set up. Every individual on the board went through rigorous rounds of assessment to prove their worth. After which they invested a large part of their time researching and compiling the most relevant data for our readers.

The editorial board has been involved in producing this book since its inception. They have spent rigorous hours researching and exploring the diverse topics which have resulted in the successful publishing of this book. They have passed on their knowledge of decades through this book. To expedite this challenging task, the publisher supported the team at every step. A small team of assistant editors was also appointed to further simplify the editing procedure and attain best results for the readers.

Apart from the editorial board, the designing team has also invested a significant amount of their time in understanding the subject and creating the most relevant covers. They scrutinized every image to scout for the most suitable representation of the subject and create an appropriate cover for the book.

The publishing team has been an ardent support to the editorial, designing and production team. Their endless efforts to recruit the best for this project, has resulted in the accomplishment of this book. They are a veteran in the field of academics and their pool of knowledge is as vast as their experience in printing. Their expertise and guidance has proved useful at every step. Their uncompromising quality standards have made this book an exceptional effort. Their encouragement from time to time has been an inspiration for everyone.

The publisher and the editorial board hope that this book will prove to be a valuable piece of knowledge for researchers, students, practitioners and scholars across the globe.

List of Contributors

D. Mourtzis, M. Doukas and E. Vlachou
Laboratory for Manufacturing Systems and Automation, Department of Mechanical Engineering and Aeronautics, University of Patras, 26500 Rio, Greece

Arian Hafezalkotob
Department of Mechanical Engineering, Islamic Azad University, South Tehran Branch, Ahang Ave, Mahalati Blv, Tehran, Iran

Ashkan Hafezalkotob
Department of Industrial Engineering, Islamic Azad University, South Tehran Branch, Entezari Aly, Oskoui St, Choobi Brg, Tehran, 1151863411, Iran

Mehdi Seifbarghy
Department of Industrial Engineering, Alzahra University, Tehran, Iran

Masoud Mirzaei Kalani and Mojtaba Hemmati
Faculty of Industrial and Mechanical Engineering, Qazvin Branch, Islamic Azad University, Qazvin, Iran

Carl Marcus Wallenburg and Thorsten Schäffler
Kühne-Foundation Chair of Logistics and Services Management, WHU - Otto Beisheim School of Management, Erkrather Straße 224a, Düsseldorf, Germany

Michael Kanisuru Adeyeri and Khumbulani Mpofu
Department of Industrial Engineering, Tshwane University of Technology, Pretoria, South Africa

Buliaminu Kareem
Department of Mechanical Engineering, The Federal University of Technology, Akure, Nigeria

Meysam Ghaffari and Ashkan Hafezalkotob
Department of Industrial Engineering, Islamic Azad University, South Tehran Branch, Entezari Aly, Oskoui St, Choobi Brg, 1151863411 Tehran, Iran

Ahmad Makui
Department of Industrial Engineering, Iran University of Science and Technology, Hengam St, Resalat Sq, 1684613114 Tehran, Iran

Madhu Jain
Department of Mathematics, IIT Roorkee, Roorkee 247667, India

Ragini Mittal
Department of Mathematics, St. John's College, Agra 282002, India

Moghadaseh Vafaeinezhad, Reza Kia and Parisa Shahnazari-Shahrezaei
Department of Industrial Engineering, Firoozkooh Branch, Islamic Azad University, Firoozkooh, Iran

S. A. Oke
Department of Mechanical Engineering, Faculty of Engineering, University of Lagos, Room 10, Mezzanine Complex, Akoka-Yaba, Lagos, Nigeria

D. E. Ighravwe
Department of Mechanical Engineering, Faculty of Engineering, University of Lagos, Room 10, Mezzanine Complex, Akoka-Yaba, Lagos, Nigeria
Department of Mechanical Engineering, Faculty of Engineering, Ladoke Akintola University of Technology, Ogbomosho, Nigeria

K. A. Adebiyi
Department of Mechanical Engineering, Faculty of Engineering, Ladoke Akintola University of Technology, Ogbomosho, Nigeria

Morteza Shafiee and Mehdi Ghaderi
Department of Industrial Management, Shiraz Branch, Islamic Azad University, Shiraz, Iran

Farhad Hosseinzadeh Lotfi
Department of Mathematics, Science and Research Branch, Islamic Azad University, Tehran, Iran

Hilda Saleh
Department of Mathematics, Central Tehran Branch, Islamic Azad University, Tehran, Iran

Saurabh Agrawal, Rajesh K. Singh and Qasim Murtaza
Mechanical, Production and Industrial Engineering Department, Delhi Technological University, Delhi 110042, India

S. Ziari
Department of Mathematics, Firoozkooh Branch, Islamic Azad University, Firoozkooh, Iran

Madhu Jain
Department of Mathematics, Indian Institute of Technology Roorkee, Roorkee 247 667, Uttarakhand, India

Chandra Shekhar
Department of Mathematics, Birla Institute of Technology and Science, Pilani 333 031, Rajasthan, India

Shalini Shukla
Department of Mathematics, D.A.V. (PG) College, Dehra Dun 248 001, Uttarakhand, India

Afshin Esmaeilzadeh and Ata Allah Taleizadeh
School of Industrial Engineering, College of Engineering, University of Tehran, Tehran, Iran

Abdelaziz Benantar and Rachid Ouafi
Department of Operational Research, USTHB University, P.O. Box 32, 16111 Bab-Ezzouar, Algeria

Jaouad Boukachour
LMAH, Normandie University, ULH, 76600, Le Havre, France

Index